MADE TO
MEASURE

MADE TO MEASURE

NEW MATERIALS FOR THE 21ST CENTURY

Philip Ball

PRINCETON UNIVERSITY PRESS · PRINCETON, NEW JERSEY

Library of Congress Cataloging-in-Publication Data

Ball, Philip
Made to measure : new materials for the 21st century / Philip Ball
p. cm.
Includes bibliographical references and index.
ISBN 0-691-02733-1 (cloth : alk. paper)
1. Materials—Technological innovations.   I. Title.
TA403.B2247   1997
620.1′1—dc21   97-4027

This book has been composed in Times Roman

Princeton University Press books are printed on acid-free paper and meet the guidelines for
permanence and durability of the Committee on Production Guidelines for Book
Longevity of the Council on Library Resources

http://pup.princeton.edu

Printed in the United States of America

1 2 3 4 5 6 7 8 9 10

CONTENTS

ACKNOWLEDGMENTS   vii

INTRODUCTION
The Art of Making   3

CHAPTER ONE
Light Talk: Photonic Materials   15

CHAPTER TWO
Total Recall: Materials for Information Storage   63

CHAPTER THREE
Clever Stuff: Smart Materials   103

CHAPTER FOUR
Only Natural: Biomaterials   143

CHAPTER FIVE
Spare Parts: Biomedical Materials   209

CHAPTER SIX
Full Power: Materials for Clean Energy   244

CHAPTER SEVEN
Tunnel Vision: Porous Materials   282

CHAPTER EIGHT
Hard Work: Diamond and Hard Materials   313

CHAPTER NINE
Chain Reactions: The New Polymers   344

CHAPTER TEN
Face Value: Surfaces and Interfaces   384

BIBLIOGRAPHY   429

FIGURE CREDITS   445

INDEX   447

ACKNOWLEDGMENTS

THAT THE writing of this book was a far less fraught experience than was my last is due to the generosity of my colleages at *Nature* to permit me to work a three-day week. I am deeply grateful for comments on the text from John Angus, Robert Birge, C. Jeffrey Brinker, Paul Calvert, Leigh Canham, Don Eigler, Jean Fréchet, Paul Gourley, Larry Hench, Allan Hoffman, Mark Kuzyk, Robert Langer, Stephen Mann, Jeffrey Moore, Stanford Ovshinsky, Geoffrey Ozin, Wim Sinke, and Kenji Uchino. I am also indebted to Frank Bates, R. Malcolm Brown, James Coleman, Norbert Hampp, Martin Pollack, Keith Prater, Deanna Richards, and David Tirrell for supplying useful reference material, and to the many others who have generously provided figures.

Emily Wilkinson, Kevin Downing, and their colleagues at Princeton University Press have been consistenly supportive and tolerant. For ensuring that my writing hours remain within sensible bounds I have again to thank Julia.

MADE TO
MEASURE

The Art of Making

As to the inventions of printing and of paper, we generally consider
these in the wrong order, attributing too much importance
to printing and too little to paper.

—Norbert Wiener, *Invention*

THIS BOOK is made possible by the leaking of one of the best-kept industrial secrets of all time. It happened twelve hundred years ago in Samarkand, and it was not a pleasant affair. Chinese prisoners captured during an attack on the Arab city were coerced, by means that we can only guess at but which were clearly persuasive, into revealing how to make a coveted material. Using local flax and hemp, the prisoners showed their captors the art of papermaking—an art developed in China during the first two centuries of that millennium, and jealously guarded ever since. When the Moors invaded Spain in the eighth century, they brought with them a culture that had many things to teach the Europeans, and papermaking was not the least of them. Around 1150, the first paper mill in Europe was built in Valencia, and after that the word was out.

I can think of no better illustration of the power of materials technology as a force for social change. The invention of the printing press is widely and rightly held to have heralded the beginning of a revolution in information that today is accelerating as never before; but like all conceptual advances in technology, its realization required the right fabric. Paper was surely to medieval information technology what silicon is to today's, and what optical fibers and so-called photonic materials will be to tomorrow's.

But we have been taught to revere ideas more than fabrics. That's a habit acquired from ancient Greece, where the artisans and craftsmen were at best humble members of society, and at worst slaves. Because the Chinese were not infected by this attitude, their materials technology was far richer than that of the West for centuries, so that we would go begging, or more often battling, for silks, for ceramics, for explosives. Today, I suspect that as a result we take materials for granted—we appreciate their benefits, perhaps, but how often do we wonder where they come from? Sold on the idea of science as discovery and revelation, we have relegated mere invention—mere creation—to the realm of engineering, a grubby business for sure. "Invention," says Norbert Wiener, the founder of cybernetics theory and a mathematician of a rare practical persuasion, "as contrasted with the more general process of discovery, is not complete until it reaches the craftsman." By that stage, it no longer seems heroic, and the rest of the world has generally lost interest.

This is a book about invention, and I think also about a craft: the craft of making new materials, of designing new fabrics for our world. I find these fabrics astonishing. We can make synthetic skin, blood, and bone. We can make an information superhighway from glass. We can make materials that repair themselves, that swell and flex like muscles, that repel any ink or paint, that capture the energy of the Sun. I'd like to tell you how.

ADVANCED MATERIALS

It has been said that, while historical periods may define their own, unique style, a living culture never reflects just the most contemporary of these. Life in the 1990s differs from that in previous decades largely by the addition of a few new artifacts and ideas to the vast collection of cultural baggage that has been accumulated over centuries. Visitors to Britain can fly supersonically to see a twelfth-century church, yet the church is still here—it has not (one hopes) been replaced by a hypermarket. Since materials are as much a part of this cultural baggage as are music, architecture, and philosophies, they too reveal a mix of the old and the new. The houses that have appeared across the road as this book has been written have wooden timbers, cement foundations, steel joists. There are no fancy new materials that threaten to replace these trusty items. And yet I suspect that the floors are carpeted with synthetic textiles, the bathrooms contain a rich selection of plastics and plastic coatings, and the central heating system may house a silicon microchip or two.

The encroachment of new materials into the marketplace is generally slow and subtle, and never complete. I don't think that we shall ever see wood replaced as a building material, nor stone blocks, bricks, and mortar. They are simply too cheap to be threatened—the supply is abundant, the processing is minimal. For a while in the 1950s and 1960s it might have seemed as though plastics would one day replace everything, but that is clearly not going to happen. On the other hand, I think it is safe to say that this century has seen a shift in the use of materials that is like nothing that has gone before. Not only do we have a far, far greater range of materials from which to choose in fabricating any artifact, but the whole decision-making process is radically different. For the first time in history, materials are *designed* for particular applications. Often the application, the requirements, come first—"I want a material that does this and that"—and the material will then be concocted, invented if you will, to meet those demands.

This is true even for materials that we might imagine are off-the-shelf items. You want to make steel suspension springs? It is no good telling your production manager to go out and order a hundred tons of steel—that is like an interior decorator requesting a dozen cans of paint. Will that be mild steel, stainless steel, medium- or high-carbon steel, nitrided steel, steel with nickel, chromium, manganese, titanium . . .? Steels today are designed materials, a delicate blend of ele-

ments whose strengths span a factor-of-ten range—and whose cost varies likewise. While in one sense we might imagine that making steel boats is a traditional
use of an old material, you can bet that the stuff of today's metal vessels is a far
more carefully selected and more skillfully engineered material than that which
Isambard Kingdom Brunel, the first iron-boat builder, had at his disposal.

But the development of new steels is nothing compared with the way that some
of today's new materials are put together. They are literally designed on the drawing board in the same way that a house or an electronic circuit is designed. The
difference is that the designers are working not with skylights and alcoves, not
with transistors and capacitors, but with atoms. The properties of some new materials are planned from and built into their atomic structure. This means, of course,
that we have to be able to understand how the characteristics of a particular molecular constitution translate into the bulk properties that we wish to obtain. In
practice, it means that materials scientists must enlist the help of physicists,
chemists, and, ever increasingly, biologists to be able to plan successfully. Frequently the strategy is a modular one—in this regard it is not really so different
to the circuit designer who knows what combinations of components will give her
an oscillator or a memory unit. You want a flexible molecule? Then let's insert
some flexible molecular units here. You want it to absorb green light? Then we'll
graft on these light-absorbing units here, equipped with atomic constituents that
tune the absorption properties to the green part of the spectrum. Alternatively, the
design process might involve a careful adjustment of a material's crystal structure—for example, to place the atoms in a crystal a certain distance apart, or to
ensure that the crystal contains gaps or channels of specified dimensions.

In this book I will talk largely about materials whose properties are designed in
this way—whose composition and structure are specified at the smallest scales,
right down to the atomic, so as to convey properties that are useful. On the whole,
this control requires clever chemistry (to arrange the molecular components how
we want them), physics (to understand which arrangements will lead to which
properties), and fabrication methods (for example, to pattern materials at microscopic scales). What all of this means is that such materials are generally expensive to make. Most are not materials for building bridges with—their applications
will be highly specialized, and will require only small amounts of the material.
The high cost, it is usually hoped, will be bearable because the materials will do
things that no others can. In other words, they will find new niches on the market,
rather than replacing older, cheaper materials. These new materials will augment
our technological palette, not replace the old primary colors with new, subtler
shades. Many will scarcely be noticed by the user, at least in a tangible sense.
While you will appreciate it when your bicycle frame is made of a lightweight
fiber composite rather than steel, you will be less likely to recognize that your
desktop computer contains photonic semiconductors, which process light signals, rather than silicon chips. But you *will* notice the change in speed and data-
handling capacity that this will bring.

These new, sophisticated, designed materials are often called *advanced materials*. That is an ambiguous term, and I don't suppose that it tells one anything

much more than does the label "modern art." Will today's art still be "modern," and our latest materials still be "advanced," in a hundred years' time? But it might help to draw the distinction that advanced materials are generally costly, created by rather complex processing methods (at least in comparison to cutting down trees) and aimed at highly specialized applications. They are, in the parlance of economics, "high-value-added"—their uniqueness and the consequent high commercial cost of the products that use them offset the high cost of their production. In contrast, older materials like brick, wood, and cast iron are "low-value-added," available in large quantities at a low cost for a broad range of applications in which there is usually a considerable tolerance to variability of properties and performance.

A word of caution is needed. I have attempted here to skim across the top of the breaking wave of the new materials science, and to pick off some morsels that I hope will be appealing. But inevitably, when the current wave breaks, not all of these will surface. At the forefront of any science are ideas and enthusiasms that have not yet been exposed to the exacting test of time. A road that looks exciting today may turn out to end in a cul-de-sac next week, or next year. In short, I can be certain that not all (perhaps not even many) of the new materials that I discuss will ever find their way into the commercial world (although some have already). But that is not the point. What I hope to show is the way that materials science works at the frontier: how a problem, a need, is identified, and how researchers might then go about developing a material that will solve that problem, meet that need. I hope to capture emerging strategies and trends rather than to alight on specific materials that will become marketable items in the next few years. It might be as well, then, to say something very briefly about that long and rocky road from the laboratory to the corner store.

MAKING IT WORK

All Part of the Process

Materials scientists are pretty good at figuring out how to make things, and that is a skill worth having. But most are not industrialists, and this can be something of a hindrance. Let us say that a materials scientist has just figured out how to make a plastic that will turn blue when warmed past water's freezing point, and realizes that this is just what the Plaxpax company wants for packaging its frozen foods; you can see at a glance when it has become too warm, he tells them. So the Plaxpax chemists come to see how the stuff is made, and the scientist explains that you dissolve this organic material in that solvent, heat it to 500 degrees Celsius under pressure, and an amorphous sticky substance will separate out on cooling—at least it will usually, but sometimes not (on those occasions the whole mixture just turns to a black goo).

The Plaxpax people love the product, but the synthetic method is useless. The solvent is toxic, the high pressures are hazardous, and success is variable. So the Plaxpax industrial chemists face a challenge every bit as daunting as the original synthesis: to turn it into a process that can be conducted safely and economically on an industrial scale.

The processing route used to turn a material into a commercial product is generally as important for its success in the marketplace as the properties of the material itself. Scientists can conduct syntheses in the lab that no one would dream of doing in an industrial plant, because they are too costly, too dangerous, or simply impossible to scale up. A material can switch from being a lab curiosity to a crucial company asset merely through the identification of a processing method that is industrially viable. The choice of material for a particular application can depend as much on the availability of a suitable processing technique for forging that material into the required form as on the properties of the candidate materials. Alternatively, even when a given material has been selected for an application, the engineer may be faced with a further choice of processing method best suited to that situation.

Nowhere is the importance of processing more clear than in metallurgy. In recent years, new methods of processing metals have substantially improved the performance that can be extracted from metal parts, and this in turn has presented subtle economic questions in metals manufacturing. To the old-style fabrication methods of casting and forging have been added new methods whose application requires a balancing of cost against performance of the products. A technique called powder-metallurgy forging (also known as hot isostatic pressing) makes components from a metal powder (usually an alloy), which is loaded into a mold and subjected to high temperatures and pressures. Because the shape of the cast product can be made very close to that of the final metal part, less subsequent machining of the cast object is needed, reducing both labor and materials wastage. Moreover, by using different powders to fill different parts of the mold, a single component can be fabricated from two different metal alloys. But a disadvantage that must be weighed into the balance is the high cost of the molds.

If cost is less critical than performance (durability and strength, say), a new processing method called directional solidification is often used. Here the metal part is formed by pouring the molten metal into a mold that is subjected to a highly controlled heating and cooling regime to influence the way that the metal crystallizes, so as to remove the microscopic flaws that limit the strength of conventional cast components. This process is expensive but is used to make turbine blades for jet engines, where long life and strength at high temperatures are critically important.

The importance of manufacturing methods extends not only to a material's consumer (insofar as the processing method plays a part in determining the material's cost and properties) but to everyone affected by an industrialized society—and today no one is any longer excluded from that category. For manufacturing has an environmental cost as well as a financial one. There can be no denying that in the past these two costs were frequently traded against one another to the detriment of the former. Making materials can be a messy business, and manufacturing companies have often been none too careful with their wastes. Toxic organic solvents have made their way into water supplies. Thousands of tons of toxic heavy metals, including lead, cadmium, thallium, and mercury, are emitted every year into the atmosphere from smelting, refining, and manufacturing processes. The CFCs used as refrigerants, foam-blowing agents, and solvents have proved to be far from the inert, harmless compounds originally envisioned: when they reach the stratosphere, they fall apart into ozone-destroying chemicals.

So materials manufacturing has a bad name, and not without cause. In the United States alone, something like eleven billion tons of nonhazardous waste and three quarters of a billion tons of hazardous (inflammable, corrosive, reactive, or toxic) waste are generated each year. Around 70 percent of the hazardous waste is produced by the chemical industry, and most of it is dealt with by physical, chemical, or biological treatment of the water streams that contain it. But there are signs that these dirty habits are changing. Some engineers are beginning to talk about "industrial ecology," which is concerned with developing industrial systems that make optimal use of energy, minimize or ideally eliminate (or make beneficial use of) their wastes, and are ultimately sustainable rather than simply consuming available resources. Industrial ecologists recognize the futility (indeed, the danger) of looking at a manufacturing plant in isolation, in terms of bare economic indices such as productivity and overheads—just as it makes no sense to look at one niche of a natural ecosystem, or one trophic level of a food web, as if it were independent of the rest of the system. They recognize that there are human and social facets to manufacturing systems, and that here, as in the economic sphere, there are costs, benefits, and risks to be evaluated.

This is not an exercise in altruistic idealism. It is becoming increasingly apparent that an industrial ecosystem view makes commercial sense too. By reducing their waste emissions by nearly 500,000 tons in 1988, the 3M company actually saved $420 million.

Increasingly, legislation punishes polluters with taxes, levies, and fines. (And as demonstrated by the recent boycotting of Shell gasoline stations in Europe over the threatened dumping of the Brent Spar oil rig in the North Sea, the public is prepared to punish them too—regardless, perhaps, of the scientific pros and cons.) But in addition, profligate use of raw materials and energy, and disregard for products labeled as waste, can be economically foolish. Many so-called waste products contain potentially valuable materials. Depending on the value of the material, its concentration in the waste, and its ease of extraction, there will be some threshold at which waste becomes a viable materials resource. Thousands of tons of heavy metals such as mercury, copper, cadmium, and silver are discharged as hazardous industrial waste each year when analyses suggest that they could be *profitably* recovered and recycled.

Within the paradigm of industrial ecology, the ideal is to move beyond waste reduction and recycling to its eventual elimination. This implies a shift in the whole concept of manufacturing. At present, most attempts to deal with manufacturing pollution are "end-of-pipe" methods, which look at the noxious substance dribbling from the waste pipe and worry over what to do about it. But we would like to have no need for that pipe at all. Commonly this requires the development of entirely new processing methods. A major source of hazardous waste is organic solvents such as hexane, benzene, and toluene, which are used in all manner of processes ranging from the manufacture of electronic printed circuit boards to paints. There is now much interest in developing "dry" processes for circuit-board manufacture, which involve no solvents at all. One of the most striking advances in this arena in recent years is the appearance of nontoxic solvents called super-critical fluids: these are commonly benign fluids such as water or carbon dioxide which, when heated and pressurized to a "supercritical" state (described on page 308), are able to reproduce many of the characteristics of the toxic organic sol-vents. Union Carbide has introduced a paint-making process that reduces the use of volatile organic solvents by 70 percent by thinning the paint with supercritical carbon dioxide.

But the environmental cost of materials in fact extends far beyond the effects of their manufacture. The raw materials have to be mined, refined and transported, and the final products might ultimately have to be disposed of. All of this has an environmental price, and it is frequently met not by the supplier, manufacturer, or consumer, but by the world—all too often by disadvantaged parts of it. Within the viewpoint of industrial ecology, these "hidden costs" are no longer ignored but are weighed into the balance in the choices that are made.

Spoiled for Choice

You want to make an engine part? A vacuum cleaner? A coat hanger? Then take your pick—at a very rough count, you have between 40,000 and 80,000 materials to chose from. How do you cope with that?

Well, I don't propose to answer this question. It's simply too big. Primarily I want to demonstrate only that it is into a crowded marketplace that the new mate-rials described in this book are entering. That is why it pays to specialize, to be able to do something that no other material can, or at least to find one of the less-congested corners of the market square. It is seldom a good idea, however, to focus single-mindedly on refining just one aspect of a material's behavior until it outperforms all others in that respect. The chances are that you'd find you have done so only at the expense of sacrificing some other aspect (commonly cost) that will prevent the wonderful material from becoming commercially viable. For while in the laboratory there may be a certain amount of academic pride and kudos to be gained by creating, say, the material with the highest ever refractive index, in practice the engineer will be making all manner of compromises in selecting a material for a particular application. He might want a strong material, say—but the strongest (diamond) is clearly going to be too expensive for the large components he wants to make. And he doesn't want a material that will be too heavy, for it is to be used in a vehicle and so he wants to keep the weight down. And the material has to be reasonably stiff too—strength against fracture will be no asset if the material deforms too easily. Then he has to think about whether corrosion will be a problem . . . and how about ease of finding a reliable supplier? Will the cost stay stable in years to come? How easy is the material to shape on a lathe?

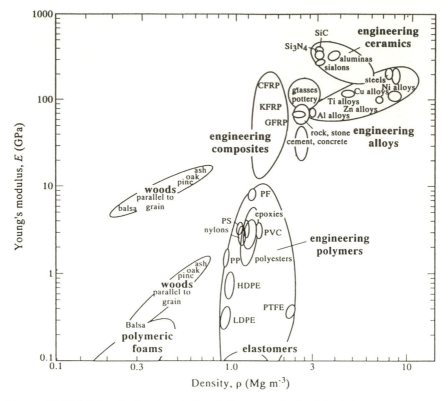

FIGURE I.1 Materials selection charts help a designer to make the right choice from the bewildering array of engineering materials now on offer. The best choice generally represents a compromise between different factors, such as weight (density), strength, stiffness, or cost. A single chart displays the ranges of two such factors spanned by different materials, and so allows the designer to determine the permissible options. Here I show a chart depicting the relationships between density and stiffness, as quantified by a parameter called the Young's modulus. You can see that stiffer materials (toward the top) are generally also denser (toward the right). (CFRP, KFRP, and GFRP are carbon-, Kevlar-, and glass-fiber-reinforced plastics.) (Figure courtesy of Michael Ashby, University of Cambridge.)

To guide the engineer through this jungle of choices, Michael Ashby at Cambridge University in England has championed the use of materials selection charts, which attempt to render on a single graph those properties, for a range of materials, that are most salient to a particular application. The engineer can then circumscribe his design parameters on the chart and see which choices that leaves him. The selection charts plot two relevant materials properties—say, density and strength—along the two axes, and the ranges of these two properties for all manner of materials are depicted by closed curves (fig. I.1). Assume, for example, that we are seeking to choose a material for making table legs. The prime considerations, at least initially, may be stiffness (which is quantified by a parameter called the Young's modulus) and density (the legs should be lightweight). So we would take a look at the chart shown in figure I.1. The stiffer the material, the thinner we can afford to make the legs, and so the more we can sacrifice in terms of density. So we can draw a diagonal line across the plot, above which the materials are stiff enough, for their respective density, to do the job. This shows us which materials to focus on; typically, we can then do a similar exercise for other design constraints (such as cost) until we have narrowed the choice down to a small short list.

Promises, Promises

It is a common complaint that, of the scientific "breakthroughs" proclaimed in tomorrow's headline news, most will have vanished from sight a year hence. This is true, and largely inevitable. "Breakthrough" is not a very helpful word for scientists, although it has an unfortunate tenacity for science journalists. While it conveys the impression of revolutionary new technologies just around the corner, the reality is that almost all scientific "breakthroughs" are beginnings. They are seldom the final, critical step that will allow some fantastic new product to impinge on our lives, but more often the first firm step in a new direction. Breakthroughs usually come suddenly, from some unexpected direction; the hard work comes after, not before. It can take years, decades, for an exciting new discovery to lead to a useful application—if it ever gets there at all. For a breakthrough is usually a result pregnant with possibilities, but there is never any telling whether some very mundane hitch will subsequently make itself manifest and spoil the fun.

It is in this spirit that I suggest you read this book. For I will often be talking about research that is at the so-called breakthrough stage, at the breaking edge where scientists are still excited and have not yet gotten down to the graft of figuring out how to convert the possibilities of their findings into reality.

I'd like to illustrate this with an example. One of the most prominent advanced materials that I have not discussed elsewhere in the book is the class of solids called high-temperature superconductors. This omission is partly because they are one of the very few new materials to have received wide attention elsewhere, but also partly because they have reached the "graft" stage; after intense excitement in the mid-1980s, researchers are now laboring at the difficult business of turning them into useful materials. This example is instructive because, to have heard the story at the peak of the excitement, one would have thought that this was a new material that just couldn't fail.

Superconductors carry electrical currents without resistance. As a consequence, they do not dissipate electrical energy as heat—a superconducting power line would not lose power over large distances, as conventional power lines do. It is a dramatic property: a current circulating around a superconducting ring will, in theory, circulate forever without dissipating its energy. Surely there must be valuable uses for a material that conducts without resistance? And what is more, a superconductor expels a magnetic field and so repels magnets: a magnet will hover above a superconductor, levitated by this repulsive force. This effect has conjured up visions of magnetically levitated trains, running almost friction-free on superconducting rails.

Superconductivity is not a new discovery; it was first seen by the Dutch physicist Heike Kamerlingh Onnes in 1911. But the excitement of the 1980s came from the discovery of a class of materials that become superconductors at much higher temperatures than those known previously. Kamerlingh Onnes had to cool mercury to just 4 degrees Celsius above absolute zero before it became superconducting, and until the 1980s no material was known that would superconduct at a temperature greater than about 23 degrees above absolute zero. The need for expensive cooling systems restricted superconductors to rather specialized applications, for example in the coils of electromagnets that produce very strong magnetic fields, or in devices called superconducting quantum interference devices (SQUIDs) that detect very small magnetic-field fluctuations such as those that occur in the brain.

In 1986 Georg Bednorz and Alex Müller of IBM's research laboratories in Zurich, Switzerland, found a ceramic oxide material that became superconducting at 35 degrees Celsius above absolute zero. So dramatic was this jump above the previous record that laboratories all around the world immediately began experimenting with other, related oxide ceramics. By 1987 the record had shot up to 93 degrees above zero, and a year later it rose a further 32 degrees. These latter temperatures were well above the boiling point of liquid nitrogen (77 degrees above absolute zero), which meant that this could be used as a coolant rather than the liquid helium necessary for the old superconductors, making the refrigeration technology cheaper.

The field looked set to produce levitating trains, ultrafast superconducting circuits, loss-free power lines, and who knew what else. A decade later, none of these things have materialized; so far, the only significant application of the high-temperature superconductors is in a new generations of SQUIDs, used for geological prospecting and for magnetic scanning of brain activity.

What happened? It turns out that the "hotness" of the superconducting transition is not the only, or even the most crucial, factor that determines the materials' usefulness. In most prospective applications, including transmission lines and levitation devices, superconducting wires are needed that carry large current densities. But as the current through the high-temperature ceramic superconductors is increased, there comes a threshold (a critical current) above which the superconductivity breaks down. For most applications, the critical current of available superconducting wires is too low.

It appears that this problem is mainly one of materials processing. Being ceramics rather than (like the older superconductors) metals, the new materials are brittle and not easily formed into wires. They are usually fashioned instead into tapes, made from powders pressed into hollow tubes of silver, and pressed and rolled flat. These tapes have some flexibility, but their superconducting core is a composite of tiny crystalline grains. Measurements on individual single crystals suggest that the high-temperature superconductors can in principle carry appreciably higher critical currents than the tapes, and it seems that the boundaries

SIZE AND STRUCTURE

Throughout this book I will use conventional metric units of length when talking about the microscopic structure of materials, since the alternative of defining lengths in, say, millionths of a millimeter is not only cumbersome but no more enlightening.

A **micrometer** is a thousandth of a millimeter. A single transistor on a microchip is typically about two to ten micrometers across, similar in size to a red blood cell. Most bacteria are one or two micrometers long.

A **nanometer** is a thousandth of a micrometer, or a millionth of a millimeter. It is about the size of a small protein molecule, such as insulin. Current microelectronic technology allows us to fabricate circuit elements no smaller than about 200 nanometers across.

An **angstrom** is a tenth of a nanometer, and is about the length of a typical chemical bond, such as that between a carbon and a hydrogen atom. A carbon atom itself is about one and a half angstroms in diameter.

I shall depict the molecular structure of many of the materials that I discuss by using spheres to represent the constituent atoms. I shall use a scheme in which white spheres represent carbon atoms, light gray spheres nitrogen, dark gray spheres oxygen, and small black spheres hydrogen. Atoms of other elements will be labeled with the appropriate chemical symbol—for instance, S for sulfur and Si for silicon. On the whole I will be aiming to place clarity of presentation foremost, and to indicate only approximately the real shapes of molecules. Spheres in contact will represent atoms connected by chemical bonds; but I shall occasionally depict these bonds more explicitly as sticks between spheres, when this helps the clarity of the presentation (for example, when I wish to distinguish between single, double, and triple bonds).

between crystal grains in the tapes degrade their performance. While researchers labor to find a way to ameliorate the problem of grain boundaries by new processing methods, it remains unclear whether or not these practical problems will undermine a materials "breakthrough" that, at the outset, looked so enticing.

SNAPSHOTS

I have tried to assemble a collection of snapshots of materials science, and not to paint the full picture. Each chapter is intended to be self-contained, although I have also tried to choose an order that minimizes any need for forward referencing. The choices and omissions will not please everyone—in particular, I must make excuses to those who work on advanced structural engineering materials such as alloys and ceramics. I have attempted, where it seemed appropriate, to identify trends in the development of new materials; the most prominent of these, which bear repeating at the outset, seem to me to be the tendency toward functional materials and the increasing use of composites. Functional materials are more than structural fabrics—they are devices of a sort, substances that do not simply hold things together or perform some passive role such as insulation but which carry out a task. Maybe they emit light, or change shape. They *respond*. Composites are old news, in one sense—the archers of middle and western Asia had developed composite bows by the third millennium B.C., gluing animal sinew to wood. But composite advanced materials are often fabricated from materials whose interfaces and interrelations are engineered at the molecular level so as to combine the favorable characteristics of several materials. This is something that until recently only Nature herself knew how to do.

Light Talk

PHOTONIC MATERIALS

Every day you play with the light of the universe.

—Pablo Neruda

The next revolution in information technology will dispense with the transistor and use light, not electricity, to carry information. This change will rely on the development of photonic materials, which produce, guide, detect, and process light.

BY A few years into the twenty-first century, the whole world will be "online." Just about every nation on Earth will be linked up to a communications network in which information can flow in the blink of an eye between computer terminals in Denver and Beijing, Mombasa and Copenhagen. This is the information superhighway, a web of information channels that knows no territorial, cultural, or political barriers. That it will coexist with the most appalling poverty in some parts of the world, with wars and ethnic conflicts, is a stark reminder that information alone solves no human problems. Yet however you regard it, a communications system of this sort will be like nothing we have seen before, and it will change our lives.

The flow of data that this system will have to support is immense. Many millions of individual messages will be routed along the superhighway's arteries, simultaneously and without interfering with one another. Their transmission must take place over long distances without deterioration of the signal. Computer networks like the Internet create an ever-expanding demand for efficient communications systems, and already threaten to strain existing systems to overload. The nascent digital video technology will add to the pressure; sending digital video data "down the line" so that distant viewers can receive live pictures from a video camera requires around five hundred times more data-transmission capacity than a telephone call.

All this is simply the latest development in a succession that has led from the telegraph of the early nineteenth century to the telephone, the television, the communications satellite, and the fax machine. Until the early 1970s, the demand on long-distance communications could be met by the electronics industry. But it has become ever more clear that electronic transmission of information will be unable to accommodate the growth in data flow that the future promises to bring. A new technology is needed.

That technology is with us already, but only in a form comparable to that of the early days of electronic communications. It is called *photonics*, and it replaces electrical currents with light: instead of being conveyed by electrons in a copper wire, information is borne by photons, the particles of light. The first long-distance photonic transmission cable was laid down in 1988; today such cables are replacing copper telecommunications cables in just about all long-distance and most short-distance applications. These cables, made from glass optical fibers, can carry many thousands of times more information than electrical wires, and at lower power consumption.

At present, just about all of the data handling at each end of a fiber-optic transmission cable is still done by electronics. So it has been necessary to devise ways of converting an electronic signal into a series of light pulses that are fed into the optical cable, and to turn those pulses back into electricity at the other end. This integration of optical and electronic data processing is called *opto-electronics*.

Optoelectronic circuits are now an essential part of information technology. The practical difficulties of making optoelectronic devices that can be integrated with silicon-based circuits on a single microchip are far from trivial, however, and are still being tackled. Quite aside from this integration problem, the use of electronics will ultimately set a speed limit on the rate with which data can be handled—photonics alone could do it much faster. So engineers are now asking whether this cumbersome method of converting a signal first to one form and then to another is really the best way of going about the problem. Why not, they suggest, do it *all* with light? That is to say, why not dispense with electronics altogether and make chips that perform data processing purely by photonic means?

The scientific underpinnings of an all-photonic technology are already in place: we know how to make miniaturized components that guide beams of light and use them to perform logical operations—the central steps of computation. A photonic transistor, a device that is still in the early stages of development, would be switchable much more quickly than the electronic varieties, and this might allow a photonic computer to run a thousand times more speedily than modern electronic computers. Moreover, photonic devices permit engineers to contemplate entirely new types of circuit design and architecture. Optical circuit components should in principle contain fewer constituent parts than their electronic counterparts, making them cheaper and easier to package onto chips. All in all, photonics should be a cleaner, faster, more compact, and more versatile form of information processing.

None of these developments can happen without the right materials. For optical communications, the optical properties of glass fibers have been honed to an astonishing degree. Optoelectronics has been wholly dependent on the identification of suitable materials for making the solid-state lasers that act as light sources and photodetectors for converting light back to electricity. Performing information processing with light requires materials whose response to light is highly unusual and very different from that of our everyday experience. When the photonic era arrives, it will be materials scientists who will act as the midwife.

A REVOLUTION WRITTEN IN SILICON

Telecommunications—literally, long-distance discourse—became an instant affair only with the advent of the electronic age. First came the telegraph, tapped out in code in the manner beloved of movies of the Old West; then in the 1870s Alexander Graham Bell's telephone, regarded in its early days with almost superstitious awe; and in the 1890s Guglielmo Marconi's "wireless telegraph," which showed that words could be sent through the air rather than through copper wire. Electronic communications, then and now, use modulated electrical signals—a current that varies in time—to carry information down copper wires. By the 1970s, the U.S. telecommunications industry was consuming around 200,000 tons of copper per year in cabling.

As the traffic of information grew, the task of processing it—modulating the signal at the transmitting end, routing the data correctly, and interpreting it at the receiving end—became ever more challenging. The turning point in electronic data processing came in 1947 with the invention of the transistor by John Bardeen and Walter Brattain at Bell Telephone Laboratories. Previously, the modulation and amplification of electrical signals were performed by vacuum tubes, which were fragile, cumbersome, and consumed a lot of power. Transistors did away with all of these problems in a single swoop—they are compact and robust and consume a minuscule fraction of the power of vacuum tubes (even the very first transistor ran on a millionth of the power of a tube). What is more, they are much faster and more reliable switches. It is no coincidence that the invention of the transistor was soon followed by a rapid growth in the power and commercialization of computers—automated devices for handling and processing electronic information. The earliest computers, such as the ENIAC device developed by engineers at the University of Pennsylvania in the 1940s, were tube-driven analog machines that occupied entire rooms and were of questionable reliability. Today many thousands of transistors and other electronic devices can be carved into semiconducting materials on a single chip no more than a millimeter square (fig. 1.1), and computers can fit into a briefcase.

The transistor's central place in modern electronics has been gained only through diligent research on the materials from which it is made, of which the most important is silicon. It is hard to think of any other industry that has become more intimately associated with the material on which it depends. We hear talk of the silicon revolution and of silicon chips pouring out of America's heartland of

FIGURE 1.1 A silicon microchip manufactured by Digital Equipment Corporation. This chip, the Alpha 21164, is the world's fastest single-chip microprocessor, able to execute over one billion instructions per second. (Photograph courtesy of Digital Equipment Corporation.)

information technology, Silicon Valley in California. So closely has silicon become linked with "thinking" machines that it is the staple of science-fiction writers searching for plausible life forms not based on carbon.

The key to silicon's central role in microelectronics is the fact that it is a *semiconductor*—a material whose electrical properties can be influenced in a variety of subtle ways. A material's electrical conductivity is determined by its electronic structure, by which I mean the disposition of its electrons. The chemical bonds that hold materials together are formed by overlap of the veils of electrons (called orbitals) that surround atoms; these are called covalent bonds.[1] In solids these overlapping electron orbitals give rise to extended networks of "electron density" throughout the material; in general, different networks can be ascribed to the overlap of different sets of atomic orbitals. The energies of electrons in these extended states, or "bands," are restricted by quantum mechanics to a certain range of values, and so the electronic structure of solids can be depicted as electronic bands separated by gaps of forbidden energies, called *band gaps* (fig. 1.2a).

An electrical current corresponds to the flow of electrons (or sometimes of other charged particles). Although electronic bands are notionally extended throughout a solid, the mobility of the electrons that each contains depends on the extent to which the band is filled. Each band has only a limited electron capacity; once a band is filled, additional electrons in the material have to go into the band of next highest energy. Electrons in filled bands are relatively immobile, being constrained to stay more or less in the vicinity of individual atoms. Electrons in bands that are only partially filled, on the other hand, can move throughout the solid when a voltage is applied across it. So solids with only fully filled electronic bands cannot conduct—they are insulators—whereas those with partly filled bands (a category that includes most metals) are electrical conductors.

In all solids, the fully filled electronic band that has the highest energy is called the *valence band*. (Valence electrons are those that are available for forming chemical bonds; this naming of the uppermost filled band reflects the fact that it is these higher-energy electrons that are primarily responsible for the bonds between neighboring atoms.) The next band above the valence band is called the *conduction band*; in insulators this is empty, in metals it is partly filled (fig. 1.2a). A voltage applied across a material makes the electrons' energies vary in space; they lower in energy close to the positive terminal and higher close to the negative terminal. So a voltage introduces a tilt to the band structure (fig. 1.2b), and electrons that are free to move (that is, those that are in a partially filled band) flow down the slope.

Semiconductors typically have an electrical conductivity somewhere between metallic conductors such as copper and insulators such as diamond. This suggests that they have some mobile charge carriers, but far fewer than metals. The electronic band structure of pure semiconductors like silicon is of the same type as that of insulators: the uppermost electronic band (the valence band) is completely filled, and a band gap separates this from a completely empty conduction band. But the crucial distinction between a semiconductor like silicon and an insulator like diamond is the size of this gap: in silicon it is small enough that a few electrons can pick up enough thermal energy to hop up into the conduction band, where they are free to move (fig. 1.2a). This hopping leaves behind an electron

---

[1] There are other types of chemical bond too. Many solids are held together by ionic bonds, in which charged atoms of opposite charge attract one another. The distinction between ionic and covalent bonds is not absolute—in fact, most bonds between different atoms have some ionic (electrostatic) and some covalent (electron-cloud overlap) character.

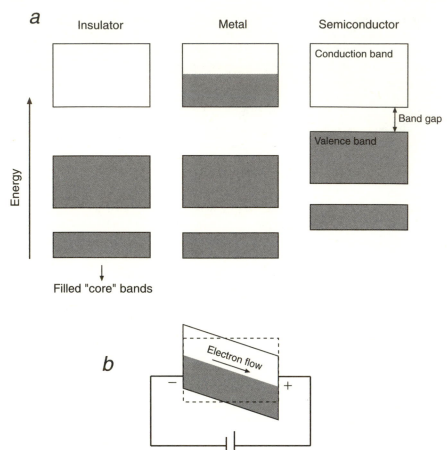

FIGURE 1.2 (*a*), The overlap of electron clouds around atoms in solids gives rise to "electronic bands" in which the electrons' energies lie between well-defined values. Each band has a certain capacity for electrons, and so the electrons in the solid fill bands of successively higher energies. Electrons in fully filled bands are not mobile and so cannot carry an electrical current. The fully filled band of highest energy is called the *valence band*, and the next highest band is the *conduction band*. If the conduction band is partly filled with electrons, these can move through the solid and the material is a metal. If the conduction band is empty, the material is an insulator—unless the valence band below is close enough in energy for a few electrons to be thermally excited into the conduction band, in which case it is a semiconductor. The difference in energy between the top of the valence band and the bottom of the conduction band is the band gap. (*b*), When an electric field is applied across a material, mobile charge carriers will move in the direction of the field. The energies of electrons closest to the positive terminal are lowered and those closest to the negative terminal are increased, so the overall effect of the field is to tilt the band structure. Crudely speaking, the electrons can then be considered to "flow downhill."

vacancy—a hole—in the valence band, which can be conveniently regarded as a kind of virtual particle with an electrical charge opposite to that of an electron. So in a semiconductor like silicon, electrical current is carried by a few energetic electrons in the conduction band moving in one direction, and by positively charged holes in the valence band moving in the other.

In truth, the characteristic that defines a semiconductor more formally is not its absolute conductivity but the fact that this increases as the temperature rises. This is because the charged particles that give rise to the electric current in a semiconductor are thermally excited. The hotter the material, the more charge carriers there are. This situation contrasts with that in metals, where heat degrades the conductivity by causing the atoms of the material to vibrate more vigorously, making them larger obstacles to the motion of charge carriers through the solid. (This thermal jostling occurs in semiconductors too, but there it is more than compensated by the increase in charge carriers.)

The conductivity of silicon can be enhanced by adding to it certain foreign atoms that provide additional charge carriers. These atoms are called dopants, and it is this ability to fine-tune the electronic properties of silicon by doping that makes it of such value to the microelectronics industry. If we insert into the silicon crystal lattice an atom of arsenic in place of silicon, the lattice acquires a surplus electron. Each atom of silicon has four valence electrons, which together fill up the valence band. But arsenic has five valence electrons, so there is not room in the valence band for the extra electron. It therefore sits in an energy state of its own within the band gap; physically, we can consider that the electron remains close to the arsenic dopant atom. This electron has to acquire even less energy to reach the conduction band than those in the valence band, and so it readily becomes a thermally excited charge carrier. Because this kind of doping introduces negative charge carriers, it is called n-type (fig. 1.3).

A similar situation is created if we use as the dopant atoms those that have one fewer valence electrons than silicon—that is, atoms from group III of the periodic table, such as boron. Then, the electron deficiency creates a hole in the valence band, which acts as a positive charge carrier. This is p-type doping.

Manipulating the electronic properties of silicon by doping provides the basis of silicon microelectronics. The central leitmotif of the silicon industry is the p–n junction, in which slabs of p-doped and n-doped silicon are placed back to back. In this configuration, thermally excited, mobile conduction electrons (in the n-doped material) and holes (in the p-doped material) can meet at the interface, whereupon they annihilate one another—the electrons fall into the holes (fig. 1.4). This means that there is net flow of charge—a net current—across the junction, because holes moving in one direction are equivalent to electrons moving in the other. To fall into a hole, an electron must lose an amount of energy more or less equivalent to the energy of the band gap. This can happen in several ways: the energy can be dissipated as heat, or can be radiated as light.

But the recombination of charge carriers at the interface cannot be sustained, because their migration across the interface sets up an electric field which pre-

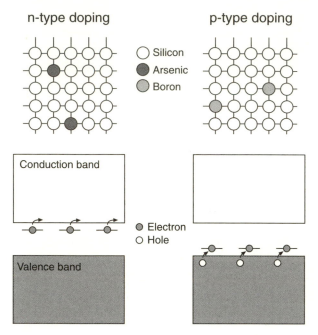

FIGURE 1.3 The conductivity of semiconducting materials can be enhanced by adding dopant atoms that inject more electrons into the conduction band. The dopant atoms have extra electrons, relative to the atoms of the bulk material. These sit in energy levels in the band gap, close to the bottom of the conduction band, and can be readily excited thermally into this band. This is called n-type doping. Alternatively, dopant atoms with a deficit of electrons provide energy levels into which electrons from the conduction band can jump, leaving behind mobile holes (which can be considered as positively charged pseudo-particles) in the conduction band. This is p-type doping.

vents further transfer of electrons into the valence band of the p-type region. Processes of this kind in semiconductor devices are often easier to envision in terms of a diagram of energies of the charge carriers. As charge migration across the interface sets up an electric field, the electronic energy bands across a p–n junction are tilted, effectively pulling the bands of the p-type and n-type regions out of alignment (fig. 1.4). This means that, in order to continue recombining with holes, free electrons have to first mount the slope up to the conduction band of the p-doped region, something for which they have insufficient energy. Recombination is therefore stopped.

But it can be switched on again by applying a voltage across the junction with the negative terminal attached to the n-doped side and the positive terminal to the p-doped side. This counteracts the field at the interface and pulls the bands back into alignment. Migration and recombination of electrons and holes across the interface can then take place. But if the direction of the voltage is reversed, the two charge carriers are both drawn away from the interface, so no current passes.

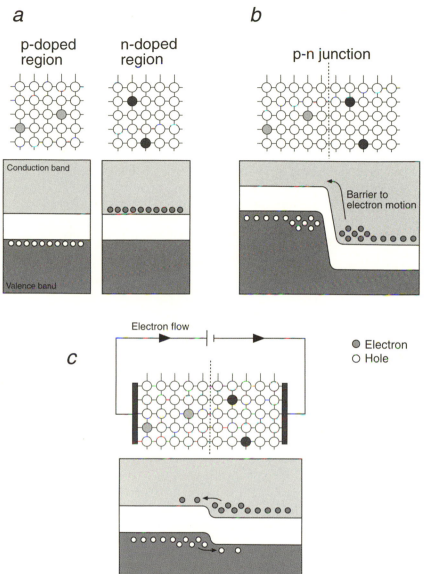

FIGURE 1.4 At a p–n junction, a p-type and n-type semiconductor (*a*) are placed back to back. Electrons in the n-type material and holes in the p-type material can meet at the interface and annihilate each other—the electrons fall into the holes, a process called recombination. The electrons lose energy in doing so, and this can be carried off as heat or light. Because there is a passage of electrons from the n-type side to the p-type side, there is a net current flow, in one direction only, across the junction until an internal electric field is set up that opposes this flow (*b*). By applying a voltage across the junction to counteract the internal field, the flow of charge is resumed (*c*). This allows a p–n junction to behave as a diode.

The p–n junction is therefore a kind of gate which lets current through in one direction but not in the other. This kind of behavior is called rectification, and is characteristic of a device called a diode.

WIRED FOR LIGHT

The transmission of information via pulses of light is a technology far older than electronic communication: it was used by the ancient Greeks, whose winking heliographs turned the Sun's rays into a coded photonic signal. Nor did this mode of communication cease at sunset; beacons burning on hilltops would also broadcast a message far and wide. But this approach needed an efficient system of relays to get over the horizon. Today we can channel light signals right around the world by using optical fibers, wires that carry light rather than electricity.

One advantage of transmitting information in this way is that optical fibres can potentially carry much more information than copper wires. Imagine all of the telephone conversations taking place across the United States at any one instant passing between your fingertips. That's one busy wire! If you have in mind a copper telecommunications cable, you can forget it—you'd be unable to get both arms around the cable needed to carry that much information. But in theory, a single optical fiber can do the job with room to spare—it can carry up to twenty-five trillion bits per second, one of those numbers too large to be meaningful (unless we can accommodate the awesome thought of all those chattering voices). In practice, however, the capacity of optical fibers falls considerably short of this theoretical maximum, although it still exceeds that of current-carrying wires. The very first transatlantic optical telephone cable, which was installed by the AT&T Bell Corporation and began operating in 1988, straightaway boosted the number of phone conversations that a single cable could carry by a factor of four, relative to its electronic counterparts. Fibers for carrying optical signals are now rapidly replacing electrical cables for all long-distance communications.

I should say a few words about light itself at this point. It is an electromagnetic wave, in the form of oscillating electric and magnetic fields perpendicular to one another. The frequency of these undulations is related to the wavelength: the higher the frequency, the shorter the wavelength. Within the visible spectrum, red light has the longest wavelength (around 700 nanometers) and violet the shortest (around 400 nanometers). At still longer wavelengths are infrared waves (0.8 nanometers to hundreds of micrometers), then microwaves (millimeters to centimeters) and radio waves (up to many kilometers). And beyond the short-wavelength (violet) end of the visible spectrum are ultraviolet light, X rays and gamma rays. The wavelengths of X rays are typically of the same order as the distance between two atoms in a crystal.

But within the quantum-mechanical description of light that was developed in the early part of this century, it has an alternative character: a collection of "light particles" called *photons*. These are discrete packets of electromagnetic energy, each with a characteristic wavelength and frequency. The energy of a photon increases in proportion to its frequency. In talking about photonic technology, I will sometimes need to use the wave picture of light and sometimes the particle picture. They are simply two ways of describing the same thing.

Optical fibers work by trapping light within a transparent central core (generally of silica glass, in essence the same material as is used for windowpanes) which is surrounded by a cladding of a material with different optical properties—specifically, with a lower refractive index than the core. The refractive index of a material is the ratio of the speed of light in a vacuum to the speed of light in the material. When light passes from one material into another with a different refractive index, the differing speeds cause light rays to bend at the interface: this is the phenomenon of refraction, which causes the distortion of objects seen through water. If the angle at which a light ray impinges on an interface between two substances of different refractive index is oblique enough, it can be completely reflected; this angle of total reflection depends on the difference in refractive index. An optical fiber confines light by total reflection—rays in the core are reflected back completely when they encounter the interface with the cladding, and the rays therefore bounce back and forth along the axis of the fiber (fig. 1.5).

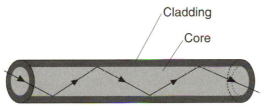

FIGURE 1.5 Optical fibers confine light through total reflection at the walls, a consequence of the difference in refractive index between the fiber core and its cladding.

Silica-glass optical fibers are made by inserting a glass rod into a glass tube with a slightly lower refractive index. This composite, called a preform, is then heated and drawn out into very thin fibers, as little as a tenth of a millimeter in diameter. Preforms are now routinely made from high-purity glass by burning a volatile silicon compound such as silicon tetrachloride to deposit a cladding of silica (silicon dioxide) on the surface of an inner glass core. A difference in refractive index between the core and the cladding is created by doping the former with germanium or phosphorus oxides, or the latter with fluorine. Because of the abrupt change in refractive index between core and cladding, these are called stepped-index fibers.

Although a single optical fiber can replace several copper cables, they are more expensive to produce, and the economics depend rather critically on the performance—the transmission efficiency—of the fibers. One of the key issues is the extent to which the intensity of a light signal becomes reduced over long transmission distances; this is equivalent to electrical losses suffered in electrical cables due to resistive heating. Light is lost from optical fibers primarily by being scattered by impurities and defects within the core material. Some scattering is inevitable in glass fibers, because the disordered atomic structure sets up small, random variations in composition and density which act as scatterers. But because the extent of scattering increases as the wavelength of the light decreases, this is not too great a limitation for optical communications that use infrared signals, which have a longer wavelength than visible light. And although impurities such as metals, water, and air bubbles are common in window glass, the more advanced procedures used to make glass fibers reduce or even eliminate these.

Some light is also lost by absorption by the glass. All materials absorb radiation at frequencies that match those with which the bonds between the constituent atoms vibrate, since light at these frequencies excites the bonds into resonant vibration. For silica glass these resonant absorption frequencies correspond to wavelengths of about 8 to 15 micrometers, in the near-infrared part of the spectrum; but absorption already starts to become significant at rather shorter wavelengths, so silica glass fibers can be used only for infrared signals with wavelengths no longer than about 2.5 micrometers. To make optical fibers that can carry longer-wavelength signals, the only option is to use another material with markedly different bond-vibration frequencies. Such materials are not common, since many interatomic bonds vibrate in about the same frequency range. Some metal fluorides, however, have vibration frequencies at considerably longer infrared wavelengths: zirconium tetrafluoride, for instance, exhibits resonant absorption only at wavelengths of 17 to 25 micrometers. Glasses made from mixtures of metal fluorides can provide optical fibers capable of conveying near-infrared signals, in the wavelength range from 0.3 to 8 micrometers. Because these materials have a lower melting temperature than silica, a different fabrication route becomes possible in which the core and cladding are melted in separate crucibles and drawn out together into strands from concentric nozzles. Metal fluoride glass optical fibers should theoretically be able to attain a transparency around twenty times smaller than that of the best silica glass fibers currently available. You would be able to see through window panes 200 kilometers thick if they were made with this degree of transparency. But at present we are far from this theoretical limit—the best fluoride glass fibers are only half as transparent as the best silica fibers.

Zirconium fluoride has also been used in *crystalline* optical fibers, which, because they have an ordered atomic structure, contain none of the random compositional fluctuations that scatter light in glass fibers. Crystalline fibers have also been fabricated from arsenic triselenide and potassium iodide, which possess large atoms connected by bonds with sluggish, low-frequency vibrations. Arsenic triselenide remains transparent to wavelengths up to about 10 micrometers. But making crystalline fibers that are free from defects is a slow and expensive process; the fibers have to be grown as single crystals, for which the growth rates can be as small as a few centimeters per minute. So this approach faces formidable obstacles before it can become practical.

Another problem that crops up in fiber-optic communications is that light pulses have a tendency to be smeared out as they travel through a stepped-index fiber. This is because in all but the narrowest fibers, the photons in a given light pulse follow many different paths. Some pass straight along the fiber's axis, while others bounce back and forth from the core/cladding interface at various angles. But the steeper this angle, the further the photon has to travel as it makes its way down the fiber, just as a wobbly cyclist who veers from one side of the road to the other ends up travelling farther than one whose path follows the curb. The photons that bounce most steeply from the walls find themselves lagging behind those whose route is more direct, and an initially tight cluster of photons begins to spread out, an effect called dispersion (fig. 1.6*a*). This smearing-out of the signal imposes limitations on the rate and distance that it can be transmitted without becoming indistinct. To avoid this problem, optical fibers have been developed in which the refractive index of the core varies smoothly from the center outward along its radius; these are called graded-index fibers. Light in the outer part of the core then travels faster than that near the core's axis. This compensates for the slowing down of rays that spend more of their time in the outer part (that is, those reflected at higher angles), and keeps the photons bunched together (fig. 1.6*b*). To achieve a smooth modulation of the refractive index in silica glass fibers, impurities such as boron oxide or germanium dioxide are introduced in quantities that increase with radial distance from the core's axis, for example by gradually mixing more of these compounds with the volatile silicon compound from which the fiber is deposited.

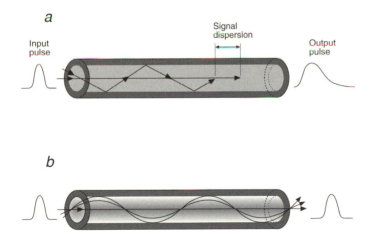

FIGURE 1.6 Because the distance that a light beam travels along a stepped-index optical fiber in a given time depends on the angle with which it is incident on the reflective interface, a pulse of light (containing beams traveling at many different angles) gradually gets smeared out as it passes along the fiber (*a*). This is called dispersion. It can be greatly reduced by varying the refractive index of the fiber smoothly from the inside to the outside, for example by gradually adding more of a dopant species as the fiber is deposited around a central core. This ensures that the speed of a light beam decreases the closer it is to the fiber's central axis, compensating for the slowing down of more oblique beams (*b*). These are called graded-index fibers.

Because some signal losses in fibers are inevitable, at present optical signals transmitted across long distances (such as under the Atlantic) need to be amplified every 50 kilometers or so to prevent the signal from petering out before it reaches its destination. Optical amplifiers used for this purpose, called repeaters, can themselves be made from optical fibers, and so the amplifier can be inserted directly into the transmission line. Amplification of an optical signal can be achieved by doping a glass fiber with certain metals, such as the lanthanide elements, which fluoresce (re-emit absorbed light) in the infrared region of the spectrum. The fluorescent emisson is generated by a "pump" light source, generally a laser, and the brightness (intensity) of the emission is modulated by the input signal coming into the fiber amplifier. In this way, a weak input signal can be used to produce a strong output signal with the same pulse pattern, so that it carries the same information. An alternative to this composite system of transmission cables interspersed with amplifiers is a fiber system in which the entire cable provides constant amplification along its length, so that losses in the signal are compensated continually as the light travels down the cable. Fibers of this sort have been prepared by doping a lanthanide metal at very low concentrations all along the core; the amplification is then pumped by light fed into each end of the cable, at the right frequency to excite fluorescence in the lanthanide atoms: this pump signal travels alongside the information-bearing signal. In this way, cables up to 50 kilometers long have been prepared that suffer virtually no losses at all.

While most optical fibers in commercial use are made from silica glass and doped variants thereof, the appearance of new polymer materials with good optical properties promises to bring important changes in fiber-optic technology. Polymer fibers can be very strong yet highly flexible, and, most important, they can be made cheaply, using abundant raw materials and simple technologies that do not require high temperatures or vacuum equipment. On the other hand, it is very difficult to make polymer fibers with the optical purity of glass, so that the former tend to incur greater loss of signal. For this reason, polymer optical fibers seem likely to find application mainly in short-distance communications, for example in distributing optical signals within a town or locality. The standard material for such fibers is currently poly(methyl methacrylate) (PMMA), which is coated with a fluorinated polymer of lower refractive index to provide stepped-index fibers. Researchers at Keio University in Japan have developed graded-index PMMA fibers that reduce signal dispersion.

LET IT GLOW

Just as large-scale electronic processing and transmission had to await the invention of the transistor, so photonic information technology had to await a particular technological breakthrough before it could get off the ground. This was the development of the laser, the first of which was demonstrated in 1960. Lasers provide the very special kind of light that is needed for the rapid encoding, transmission, and processing of photonic information. As was the transistor, the laser was primarily a triumph for physicists; but its practical application, especially in information technology, quickly became a challenge that required the expertise of materials scientists. The earliest lasers were bulky devices of bench-top size, but to integrate lasers into microelectronics they have to be much more compact—small enough, indeed, to fit on a silicon chip.

The light that lasers produce is not the same as that given out by an electric bulb or by the Sun—you could call it "designer" light. The photons in a sunbeam oscillate out of step with one another: this is called *incoherent light*. But in laser light, the oscillations are all synchronized: the light is *coherent*. This synchrony means that the photons do not disperse as they do in a beam of normal light; to put it crudely, they do not tread on each other's toes because they are all in step. The consequence is that a laser beam does not spread out in the same way that normal light does; it remains pencil-thin over long distances.

The physical principles that give rise to the emission of coherent light were elucidated in the first half of this century, by Albert Einstein among others. But not until 1954 were these principles put into practice in a working device that emitted coherent radiation. This device, developed by Charles Townes of the University of California at Berkeley, was not really a laser at all but a *maser*, which generated coherent microwave radiation rather than light. Townes and others strove in the ensuing years to create devices that would emit coherent light at visible wavelengths, and the first genuine laser was demonstrated in 1960 by H. Maiman, who obtained red laser light from a ruby laser.

A laser produces coherent light by *stimulated* emission of photons from excited atoms or molecules. An excited atom or molecule (which has an energy greater than its equilibrium thermal energy) can cast off this excess energy packaged in a photon—this is called radiative decay. The rules of quantum mechanics dictate that the energies of molecules cannot take arbitrary values, but are restricted to a set of discrete *energy levels* between which there are gaps of forbidden energies (these same rules give rise to the discrete energy bands and band gaps in solids). The energy levels define the rungs of an energy ladder (fig. 1.7). Emission from an excited species will inevitably take place spontaneously sooner or later—an excited species cannot hold on to extra energy indefinitely, just as a red-hot poker will gradually lose its heat. This radiative decay is governed by chance, like the decay of a radioactive nucleus. But emission can also be *induced* by exposing the excited species to light of the same wavelength as the photon that will be emitted. This is called stimulated emission (fig. 1.7). Its origin lies in quantum mechanics, which says that the probability of emission is enhanced in the presence of an electromagnetic field oscillating at the emission frequency. The field stimulates the excited species into resonant oscillation, much as a sung note will excite an undamped piano string into vibration. The resonant oscillation is synchronized with the stimulating oscillation—that is, the electromagnetic undulations of a photon of stimulated emission are in step with those that stimulated it. The two photons are, in other words, coherent.

The phenomenon of stimulated emission leads to the possibility of a "feedback" process for accelerating the radiative decay of a whole ensemble of excited

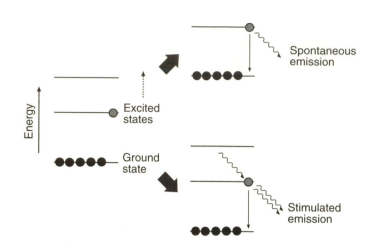

FIGURE 1.7 The energies of atoms and molecules are quantized—they can take only certain discrete values. An atom or molecule that is excited—for example, by absorbing a photon or by collisions—to a high-energy state will ultimately decay back to its original (ground) state by throwing off its excess energy. The energy can be dissipated as heat, carried off by collisions, or radiated in the form of a photon. The latter is called spontaneous emission (top). A photon can stimulate the decay of an excited atom or molecule if the decay generates a second photon of the same energy. This process, called stimulated emission (bottom), is the basis of laser action. The stimulated photon has its electromagnetic oscillations in step with the stimulating photon—they are coherent.

species. If a single excited entity undergoes spontaneous emission, the photon that it emits has the potential to *stimulate* emission from another excited molecule. This then creates two (coherent) photons, which can stimulate emission of two more and so forth. In a laser, the emitting ("active") material, which contains energetically excited atoms or molecules, is enclosed in a cavity between two mirrors that reflect emitted photons back into the cavity. This ensures that, once emission occurs, the light stays within the active material to stimulate more emission rather than escaping out into space. As more and more emission is stimulated, more photons bounce back and forth in the cavity until very rapidly all of the excited species undergo radiative decay in an avalanche that produces a burst of coherent light (fig. 1.8). Of course, if we are to make use of this light it has to be able to get out of the cavity somehow, so one of the mirrors is only partially reflective: enough to stimulate the avalanche, while being transparent enough to let the burst of laser light escape.

In order to create this avalanche process, one needs to start with a large number of excited species. This, however, is not a stable situation. An ensemble of molecules at thermal equilibrium with its surroundings contains a distribution of energies; the number of molecules with energies greater than average falls off very rapidly with increasing energy. So at equilibrium there are ever fewer molecules on each successive rung of the energy ladder. But for the avalanche of stimulated

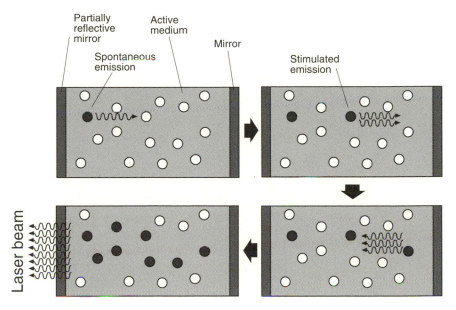

FIGURE 1.8 Stimulated emission is turned into laser action by confining the photons emitted from an excited ("active") material within a cavity. This leads to an avalanche-like amplification of the stimulated emission process—each photon stimulates the emission of others as they bounce back and forth in the cavity, and all of them are coherent. A burst of light is produced, which escapes through a partially mirrored cavity wall.

emission that leads to laser light, we need a disproportionate number of excited species perched on an upper energy level. In other words, rather than the energy ladder being populated in decreasing numbers the higher we go, it must have a region in which the population increases with increasing energy. This highly out-of-equilibrium situation is called a population inversion. It can be brought about by applying some kind of energetic pulse to kick a large number of species up onto a higher rung. In many lasers the pulse is an optical one: a pulse of light is applied through the active medium, and the absorption of this light creates many excited species in a population inversion. Sometimes laser pulses are themselves used to drive other lasers by supplying this optical kick. Population inversions can also be created electronically, while in lasers in which the active medium is a gas, atomic collisions may do the job.

The earliest lasers used a variety of active materials: Maiman's laser used a ruby crystal, which can be excited to emit red light, while other lasers used gases such as carbon dioxide or argon, or liquids such as solutions of dyes. The ruby laser was the forerunner of one of the mainstay lasers of modern technology, the neodymium-YAG laser. In this device, the active material is closely related to ruby, being a garnet mineral containing the elements yttrium and aluminum, to which small quantities of the "dopant" neodymium are added. The neodymium-YAG laser provides high-power emission at infrared wavelengths.

Drawing Light from a Well

For photonic technology, the attractive features of laser light are that it has a very narrow spread of photon frequencies (it is single-colored, or monochromatic), that it is coherent and can be emitted as a parallel-sided beam, and most of all that it can be modulated (switched on and off) extremely rapidly. But the high-power lasers used for bench-top science are of little value for a technology in which one wants devices little bigger than a bacterium. To make miniature lasers for photonic information processing, it was necessary to find a material that could be stimulated by electronic means to emit laser light efficiently from cavities of microscopic dimensions.

Lasing action in any medium is achieved by setting up a population inversion. Semiconducting materials have an electronic structure that allows one to do this electronically, by simultaneously feeding a disproportionate number of electrons into the conduction band and holes into the valence band. Provided that these charge carriers are mobile enough to migrate through the material until they encounter one another, the excess electrons can then fall back into the excess holes, releasing a photon as they do so: this is the process of electron-hole recombination. It happens spontaneously, but can also be stimulated by a photon of an energy equal to that which is lost in recombination.

To make lasers that are compatible with existing microelectronic circuitry, one would ideally like to be able to use silicon as the active material. But it is very hard to make silicon emit light efficiently, for reasons that I shall discuss later. So photonics has relied so far on different kinds of semiconducting materials, and in particular on alloys of elements from columns III and V of the periodic table, called III–V semiconductors. These materials are good light emitters. The most widely used of III–V semiconductors are the binary alloy gallium arsenide (GaAs) and the ternary (three-component) alloy gallium aluminum arsenide (GaAlAs).

One can think of a III–V semiconductor as an extreme form of both n- and p-type doped silicon, in which the dopants have essentially replaced all the silicon. As elements from group III have three valence electrons and those from group V have five, an alloy of an equal amount of the two elements has, on average, four valence electrons per atom, just like silicon. Such an alloy has an electronic structure that is entirely analogous to silicon—a filled valence band separated from an empty conduction band by a small band gap—but with the difference that the material can absorb and emit light efficiently in the near-infrared region of the spectrum (that is, between wavelengths of around 800 to 1,500 nanometers). This is because photons with this wavelength have energies corresponding to the size of the band gap, so an electron can be shunted between the valence and conduction bands by absorption and emission of one such photon. The precise width of the band gap, and thus the color of the emission, can be adjusted by varying the composition of the material, either by changing the relative proportions of the two elements or by adding additional elements from the same groups of the periodic table. Alternatively, composite "designer" materials with a specified band gap can be engineered by sequentially depositing very thin layers of different semiconductor alloys in sandwich structures, an approach called band-gap engineering. The ability to tune the color of emission more or less continuously by altering the nature of the emitting material is one of the great advantages of III–V semiconductors as photonic materials.

To make a laser based on recombination processes within one of these semi-conductors, one has to be able to control the recombination electrically (in other words, to switch the laser emission on and off) and to find a way of triggering stimulated emission (giving coherent laser light) from the spontaneous emission generated by random recombination events. What the first requirement means in practice is that one has to be able to keep the electrons and holes separated until one wishes them to begin recombining. As we saw earlier, a semiconductor can be given excess electrons in the conduction band by n-type doping, and can be enriched in holes in the valence band by p-type doping. Semiconductor lasers generally contain electron and hole reservoirs, consisting of thin films of n-doped and p-doped semiconductor alloys, between which is sandwiched a layer of an-other semiconductor with a smaller band gap. These structures are called quantum wells, because their well-like band structure (fig. 1.9) has its origin in quantum

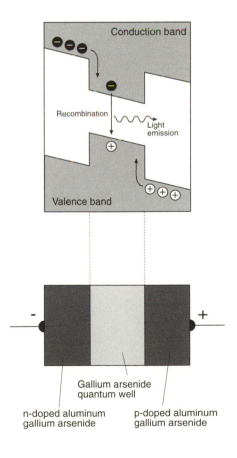

FIGURE 1.9 In semiconductor diode lasers, a "quantum well" of a photonic semiconductor such as gallium arsenide is sandwiched between two semiconductors with a larger band gap. One side of the sandwich is n-doped, and injects electrons into the quantum well when a voltage is applied; the other is p-doped, and injects holes from the other direction. Within the well, these charge carriers are trapped by the step in band gaps, so they can recombine efficiently with consequent light emission.

mechanics. The quantum well is the region in which electron-hole recombination takes place and within which light is emitted. Electrons from the n-doped region and holes from the p-doped region on each side are injected into the well by applying a voltage across the sandwich structure, which causes the charge carriers to migrate in the direction of the electric field. Because they have opposite charges, holes and electrons move in opposite directions, so an applied voltage tips them both into the well region from either side. Once there, the charge carriers cannot get back out again because of the difference in band gaps; the electrons, for instance, would have to climb back up an energy step in the conduction band. Layered structures of this sort are called semiconductor heterostructures (*hetero* being derived from the Greek root meaning "different," since the devices are comprised of layers of different semiconductors, with different band gaps). In many semiconductor lasers the quantum-well layer is comprised of gallium arsenide, and the charge-carrier reservoirs to either side are made of aluminum gallium arsenide, which has a larger band gap.

The recombination events in the quantum well are initially spontaneous, random occurrences. But each photon is capable of stimulating additional recombination events as it passes through the well. As in other types of laser, an avalanche of stimulated emission can be triggered by trapping the spontaneously emitted photons within the active cavity by placing mirrors at either end. In conventional semiconductor lasers, called laser diodes, the mirrors are simply cleaved ends of the layered structure. The refractive-index difference between the active semiconductor medium and air is sufficient to reflect typically 30 percent of the light that impinges on the interface, and this is enough to provide lasing action.

These laser diodes are brick-shaped devices a few micrometers wide and several hundred micrometers long (fig. 1.10). That may sound small, but it is hundreds of times bigger than the electronic devices, such as transistors, that are fabricated on integrated-circuit microchips. The first laser diodes were constructed in the 1960s from III–V semiconductors; by the 1970s they were reliable and powerful enough to find applications, and are now the mainstay light sources for fiber-optic communications systems as well as for compact-disc players and a host of other commercial products. Most GaAs/GaAlAs laser diodes emit laser light in the near-infrared region, but by using different semiconductor alloys as the quantum-well sandwich "filling" (such as aluminum gallium indium phosphide) it is possible to make lasers that emit visible red light.

Photonic engineers would like to extend this range of colors to still shorter wavelengths, toward the blue part of the spectrum. The are not looking simply to bring more color into their lives; the shorter wavelength of blue lasers allows the light to be more tightly focused, which means that they can manipulate higher densities of stored information. The use of blue lasers for data writing and readout in compact-disk and magneto-optic storage devices (discussed in the next chapter) would give these devices much greater storage densities. In the past few years green and blue lasers have been made from a class of alloys that contain elements from groups IIB and VI of the periodic table, in particular zinc selenide (plate 1). And very recently, nitrogen-containing III–V semiconductor alloys such as gal-

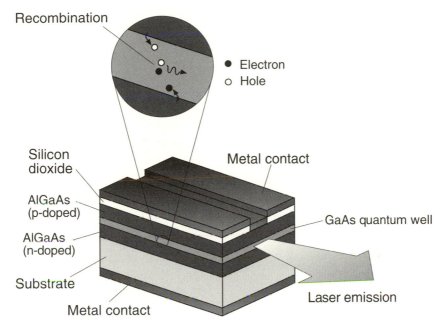

Recombination

● Electron
○ Hole

Silicon dioxide

Metal contact

AlGaAs (p-doped)

GaAs quantum well

AlGaAs (n-doped)

Substrate

Laser emission

Metal contact

FIGURE 1.10 Semiconductor laser diodes are layered, brick-shaped devices several hundred micrometers long. Recombination processes in the quantum-well layer give rise to stimulated emission, as the photons are confined to a lasing cavity by reflection at the ends of the slabs. The color of the laser light depends on the band gap of the material that constitutes the quantum-well layer.

lium nitride have been synthesized that emit blue light without suffering from the lack of long-term stability that has hampered attempts to make practical devices from zinc selenide. The first blue-light laser based on these nitride materials was announced in 1996 by Shuji Nakamura and colleagues at Nichia Chemical Industries in Japan.

Atomic Sandwich-Making

Making these heterostructure devices represents an astonishing feat of engineering. The layers of semiconducting materials are often extremely thin—perhaps just a hundred nanometers (about two hundred atoms) across. But their thickness must be controlled with great accuracy, and the films must be very flat. To make layered structures to these specifications, highly sophisticated processing techniques have been developed. In general these involve the deposition of an atomic gas or a beam of atoms onto a surface, while the composition of the gas or the beam is adjusted very precisely to give the required mixture of elements in the deposited film (recall that this is what determines the doping level). The two most prominent techniques for making these thin films are vapor-phase epitaxy and molecular-beam epitaxy.

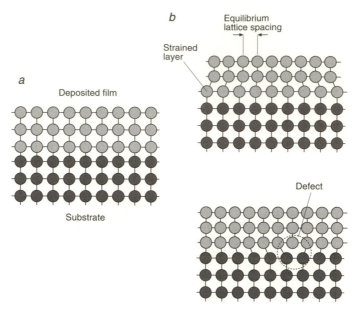

FIGURE 1.11 One crystalline material can be deposited on another without strain or defect proliferation if the spacing between atoms (the lattice spacing) is the same in both materials. This is called epitaxial growth (*a*). If, as is usually the case, the lattice spacings of the two materials do not match exactly, the deposited overlayer is strained—either compressed or expanded (*b*). This strain can be relieved by the formation of defects, where the regular crystal structure breaks down. Defects are detrimental to a semiconductor's electrical conductivity.

"Epitaxy" refers to the relationship between the atomic structure of the deposited film and that of the substrate on which it grows. To obtain good electronic properties in the semiconductor films, they must have highly ordered, crystalline structures. Defects such as misalignments of atoms degrade the conductivity of the material, for example by trapping charge carriers in their vicinity. If the lattice spacing of atoms in the deposited film matches that in the substrate, the film is said to be epitaxial (fig. 1.11*a*). This situation encourages the growth of highly ordered films, because it means that atoms in the deposited film can form bonds with atoms on the surface of the substrate without having to distort the normal crystal structure. But when the film and the substrate are comprised of different materials, their equilibrium lattice spacings will generally differ, and atoms of the two materials at the interface can then bond to one another only at the cost of either altering the lattice spacing in the first few layers of the overlying film (which imposes a strain on the material) or putting up with a few defects every so often to absorb the mismatch (fig. 1.11*b*).

If epitaxial growth is possible, then the substrate can act as a template which ensures that the deposited film grows in an ordered manner. Atoms will not necessarily impinge on the surface in exactly the right place for forming an epitaxial layer, however, and so the substrate is generally kept hot so that the deposited

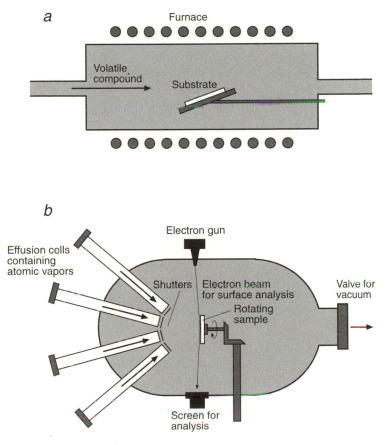

FIGURE 1.12 Microscopic semiconductor devices are generally prepared by depositing thin films from vapors of the relevant elements onto a substrate. In chemical vapor deposition (CVD; *a*) each element is introduced to a heated vacuum chamber in the form of a volatile compound, which is broken down into its constituent atoms or small molecular units by the heat of the furnace. In molecular beam epitaxy (MBE; *b*) the elements are introduced in the form of a pure elemental vapor, generated by heating a lump of the pure element in a separate chamber. The composition of the deposited film is controlled by varying the mix of vapors in the deposition chamber. These techniques allow the thickness of the deposited films to be specified with a precision of virtually a single atomic layer.

atoms have enough thermal energy to shuffle around into the right positions. In vapor-phase epitaxy (VPE, a form of the more general technique called chemical vapor deposition) the substrate is placed in a chamber under high vacuum, to eliminate gaseous contaminants that might otherwise stick to the surface and degrade the purity of a deposited film. Volatile compounds of the elements to be deposited are then injected as vapors into the chamber (fig. 1.12*a*). For instance, to deposit films of an arsenic-containing alloy, the arsenic atoms might be provided in the form of the volatile compound arsine, AsH_3. The deposition chamber is placed inside a furnace, which raises the temperature to a point where the volatile compounds decompose into their constituent parts. If the substrate is kept cooler than the rest of the furnace, the atoms in the vapor will stick to it to form a film. The composition of the film can be controlled by injecting different amounts of the various volatile compounds.

Early forms of vapor-phase epitaxy had the drawback that the volatile compounds used were highly toxic, necessitating careful and expensive safety measures. A more recent variant, called metalorganic chemical vapor deposition (MOCVD), makes use of somewhat less dangerous compounds in which the metallic or "semimetallic" elements from group III of the periodic table are introduced in the form of organometallic compounds, with organic groups bound to the metals. Aluminum, for instance, is supplied in the form of trimethylaluminum, $(CH_3)_3Al$.

Molecular beam epitaxy (MBE) differs from these approaches in that the material to be deposited is supplied in the form of beams of the pure elements, made by heating lumps of each material until they start to vaporize. Each element is held within a separate cell, called an effusion cell, and the atomic vapor passes down the cells and exits from a port at the far end into a chamber where the substrate sits under a high vacuum. The flux of material out of the effusion cells is monitored very accurately, and can be controlled by varying the amount, temperature, and position of the source material within the cell and by shutters at the output ports. The composition of the films deposited on the substrate is determined by the relative sizes of the fluxes of different elemental beams (fig. 1.12*b*).

Both of these approaches have their strong points and their drawbacks. In vapor-phase epitaxy the fact that there is a relatively high pressure of gases in the deposition chamber creates problems. First, it prevents one from being able to monitor the structure and composition of the growing film by electron diffraction (see chapter 10), as is done in the MBE technique, because the vapor scatters the electron beam. Second, the deposition process cannot be controlled too precisely, in part because of turbulence in the vapor. On the other hand, the technique allows large-area films to be grown rapidly. Films grown by MBE are deposited much more slowly, but the high-vacuum conditions afford thinner films with very smooth interfaces between layers. Recently there have been attempts to combine the two techniques so as to share their advantages while minimizing the disadvantages, for example by depositing the kinds of compounds used in VPE using the effusion cells and high-vacuum environment of MBE.

The control over the composition and thickness of epitaxial layers offered by these techniques opens up tremendous possibilities for band-gap engineering. The band gap of very thin quantum-well structures is determined not only by the composition of the well layer but its thickness too: when charge carriers are confined within narrow wells, quantum-mechanical effects shift their energy bands. So the band gap, and hence the color of light emission, of III–V semiconductor photonic heterostructures can be varied simply by changing their dimensions. Moreover, films can be deposited with band gaps that vary smoothly in the lateral direction by gradually changing the composition of the mix of elements deposited.

In the early days of epitaxial growth of heterostructures it was thought that perfect epitaxy would be possible only when the lattice constants—the distance between neighboring atoms in the crystal structure—were the same for the substrate and the overlying layer. This lattice matching is possible for GaAs/AlGaAs heterostructures, but the lattice constant of silicon is rather different from that of the III–V alloys, suggesting that films of the latter would have to incorporate defects to make up for the mismatch, when grown on silicon. This would pose a serious problem for the integration of III–V semiconductor devices with silicon technology. But advances in epitaxial growth technology have now made it possible to marry layers with mismatched lattice constants in a way that enables them to take up the strain of the mismatch without suffering badly from defects. Gallium arsenide, for example, can now be grown on silicon without too great a proliferation of defects, although some problems remain to be overcome before such composite sandwiches become technologically useful.

Once layers of semiconductors have been deposited on a substrate, they need to be cut up, sandwich-style, into discrete devices. For both photonic and electronic devices, the cutting is done by various kinds of etching agent, and is restricted to selected areas by coating the topmost layer of the flat multilayer films with a patterned "resist" that prevents etching of the regions below. This resist is generally a photosensitive polymer containing chemical groups that form cross-linking bonds when excited by light. When a uniform polymer film spread on the top surface is irradiated with light through a mask patterned as a "negative" of the circuit structure that one wants to etch, those parts of the polymer film that are irradiated become cross-linked, making them insoluble; the rest of the film is washed away with a solvent. This leaves a patterned polymer film on the sandwich surface, which acts as a protective coat against the etching agent (fig. 1.13).

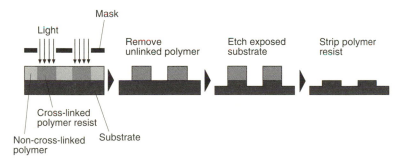

FIGURE 1.13 Device structures are carved into semiconductors by means of photolithography. A polymer film is deposited on the surface of the semiconductor, and illumination through a patterned mask fixes the pattern into the film—either by cross-linking of the irradiated parts of the polymer to form a robust "negative resist" or by breaking bonds in the irradiated parts, leaving behind a "positive resist." The non-cross-linked parts of the polymer film are then washed away, and the patterned resist confers protection to the semiconductor below from an etching process, which commonly involves exposure to a reactive plasma. After etching is completed, the resist is broken down chemically.

An agent such as a plasma of oxygen ions is then used to scour away the exposed parts of the material, leaving behind the patterned heterostructure capped with the polymer photoresist, which is subsequently removed by chemical means. This approach is said to involve a negative resist, because the pattern of the resist is the negative of that cut into the mask. Positive resists are also used, in which exposure to light induces a chemical reaction that makes the film *more* soluble, leaving behind a replica of the mask. These procedures, collectively called photolithography, can be used to inscribe extremely narrow surface patterns: the smallest features presently achieved on commercial chips are just 350 nanometers across. The lower limit to this resolution is set by the wavelength of the light used to irradiate the resist: by using ultraviolet light, researchers at the Massachusetts Institute of Technology have developed a photolithographic process that can inscribe features 200 nanometers across. But as miniaturization of integrated electronic circuits continues apace, even this is not enough: the Intel Corporation hopes by the year 2001 to be able to make silicon chips whose smallest features measure just 180 nanometers, while the dream for the further future is to push this limit to 100 nanometers. To do so, X rays rather than UV light will have to be used to irradiate the films, and that requires components for X-ray optics, such as lenses and lasers, that are still in the early stages of development.

Small Is Beautiful

Despite their proven usefulness in optoelectronic communications systems, laser diodes have a drawback—they are big beasts. To fully integrate photonics with electronics, it will be necessary to fit semiconductor lasers on a chip. In very recent years, new kinds of ultrasmall semiconductor lasers called microlasers have been constructed. The crucial aspect of most of these devices is that they emit laser light perpendicular to the layers of the sandwich structure, rather than parallel to the layers like conventional laser diodes. This difference means that the laser structure occupies less surface area, so that many more can be packed onto a chip. The development of surface-emitting structures was pioneered by Kenichi Iga and coworkers at the Tokyo Institute of Technology in the late 1970s, and led to efficient surface-emitting microlasers in the 1980s.

Typically, a surface-emitting microlaser is a cylindrical structure in which many disk-shaped layers of III–V semiconductors are stacked like coins (fig. 1.14). The principles of the laser's operation are much the same as those of a laser diode: the active region is a quantum well sandwiched between layers that inject electrons and holes into the well. But in the surface-emitting microlaser there are commonly several such wells stacked on top of each other in the active region. And most importantly, the wells are very thin—perhaps just 10 nanometers across—to keep the laser's power requirements as small as possible. This is important if they are to be used in their millions as vast emitting arrays. It means, however, that for lasing action to be achieved, the photons emitted in the active layers must bounce back and forth within them many more times than in conventional laser diodes to induce stimulated emission. And that in turn means that the

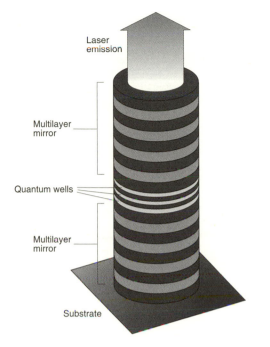

Laser emission

Multilayer mirror

Quantum wells

Multilayer mirror

Substrate

FIGURE 1.14 Microlasers are much smaller than the conventional laser diodes in commercial use. A prominent design is the surface-emitting laser, a stack of hundreds of disks of photonic semi-conductors. The central part of the stack contains an array of quantum wells, within which the laser emission originates. The laser cavity is defined by multilayer heterostructures above and below, which reflect around 99 percent of the light that falls on them. The laser beam emerges from the top of the device, perpendicular to the layers.

mirrors at the top and bottom of the laser cavity must be much more reflective—99 percent or more.

These mirrors in fact make up the bulk of the laser device. They consist of hundreds of thin, alternating layers of two materials with different refractive indices, such as gallium arsenide and aluminum arsenide. The difference in refractive index between these materials is sufficient to reflect only about 0.6 percent of the light that falls on each interface; but if there are enough layers, their combined effects add to give the required reflectivity of at least 99 percent.

So these surface-emitting microlasers are extraordinarily elaborate constructions, containing typically five hundred or more different layers in the stacked column. That makes for an elaborate fabrication process, but this is offset by the fact that you can make a million of the devices as easily as you can make one, simply by depositing a large-area sandwich of the many layers and then using photolithography to cut it up into individual lasers, just as one might cut a cucumber sandwich into petite squares (fig. 1.15). In 1989 Jack Jewell of AT&T Laboratories and coworkers from Bellcore in New Jersey fabricated an array of over one million microlasers, each between one to five micrometers across, on a single semiconductor chip of seven by eight millimeters.

An attractive feature of these microlasers, aside from their size, is that the columnar structure acts as a kind of "waveguide" for focusing the laser beam into a narrow beam—unlike the edge-emitting laser diodes, whose flat beams have a tendency to spread. Surface-emitting microlasers have mostly used gallium arsenide as the active material, emitting in the near infrared (wavelengths of around

FIGURE 1.15 Vast arrays of microlasers can be fabricated by using lithography to cut large multilayer sheets into individual stacks with circular or square cross-sections. Each microlaser in the array could be used to send a photonic signal. The two-dimensional checkerboard array shown here was made at Sandia National Laboratories; each square laser pixel is 5 × 5 micrometers. (Photograph courtesy of Paul Gourley, Sandia National Laboratories.)

800 to 1,000 nanometers), but red-light devices using aluminum gallium indium phosphide have now also been demonstrated. They are still in the early stages of development, but it is hoped that as vast two-dimensional arrays they might be used to communicate millions of messages at the same time from a single chip, or to process two-dimensional images in optical form.

The End of Band-Gap Slavery

One of the attractions of photonic information processing is that it is possible to send several optical signals down the same channel at once without getting them mixed up, simply by giving them slightly different wavelengths. The principle is the same as that which allows us to tune into one of many different radio or TV channels broadcast simultaneously, by tuning the wavelength to which the receiver is responsive. This means, however, that we need to have at our disposal a range of laser wavelengths, whereas from the discussion earlier we can see that the emission wavelength of a semiconductor laser is more or less set by the band gap of the active material in the cavity. While this can be altered by, for example, doping or adjusting the atomic spacing in the emitting layer, the wavelength of semiconductor lasers is ultimately set by the optical properties of the materials from which it is made.

But a new kind of laser devised in 1994 by Federico Capasso and colleagues at AT&T Bell Laboratories has done away with this restriction. In this laser, called a quantum-cascade laser (QCL), the color of the laser light is determined by the physical dimensions of the component parts, which can be varied at will across a wide range. The device achieves this flexibility by virtue of a lasing mechanism that is fundamentally different from that of all preceding semiconductor lasers. We saw earlier that light emission in these conventional devices arises from recombination of electrons and holes injected into the lasing cavity. In the QCL, in contrast, the emission of light relies on the injection of electrons alone.

The device consists of a sequence of about five hundred layers of doped III–V semiconductors, laid down by molecular beam epitaxy. Many lasing "cells" repeat throughout this sequence, each consisting of three narrow quantum wells of indium gallium arsenide sandwiched between indium aluminum arsenide barriers (fig. 1.16). The energy levels in the quantum wells are determined by their thickness, and they form a three-step staircase in each lasing cell, down which electrons cascade (from left to right in the figure) with the consequent emission of photons. The photon is emitted during the first step, when the electrons tunnel through the barrier from the left-most well to the middle well, which has a lower energy level. To turn this photon emission into laser emission, it must be able to stimulate the decay of other electrons down this step, and this in turn relies on maintaining a population inversion between the first and second wells, so that there are more electrons in the former than in the latter. This is the function of the third, right-most well, which siphons off the electrons from the middle well because it has an energy level just a little lower.

The electrons are supplied by an electron-injecting region to the left of the first well, and can tunnel from the third well to an electron-collecting region to the right. Both of these regions are made of a "graded alloy," comprising a multilayer sequence (a superlattice) of indium aluminum arsenide and indium gallium arsenide layers doped with silicon, whose band gaps are graded such that they provide a gentle "sloping" electric field down which the electrons pass to the next cell.

The wavelength of the laser light emitted by the QCL is determined by the difference between the energy levels in the first and second wells of each cell, which varies according to their thicknesses. Capasso and colleagues have demonstrated laser emission at a wavelength of 4.2 micrometers, well into the infrared; but there is no obvious reason why, with suitable engineering of the band gaps, it should not be possible to make visible-light lasers this way. At present, the QCL needs to be cooled to around minus 170 degrees Celsius to operate, and even then it produces only pulsed light rather than the continuous laser light needed for many applications. But the crucial point is that it shows that new laser colors can be achieved not by chemical tinkering (by altering the composition, and thus the band gap, of the active medium) but by physical engineering (altering the quantum-well thickness). As Capasso has put it, we are no longer "slaves of the band gap."

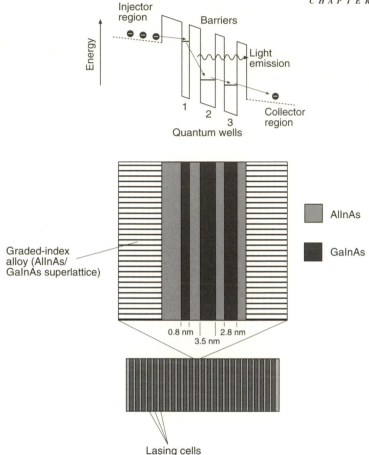

FIGURE 1.16 The quantum-cascade laser devised by Federico Capasso, Jerome Faist and colleagues at AT&T Bell Laboratories is the first solid-state laser whose emission wavelength does not depend on the chemical composition of the active lasing medium. Instead, the color of the laser light is determined by the physical dimensions of the device. The laser is a highly complex heterostructure, comprising twenty-five separate lasing cells, each containing three quantum wells of gallium indium arsenide sandwiched between barriers of aluminum indium arsenide. The whole structure has no fewer than five hundred different layers, each deposited by molecular-beam epitaxy with a precisely controlled thickness.

The quantum wells are so thin that their energy levels are influenced by the quantum-mechanical effects of confinement on charge carriers in the well. The first (left-most) well in each cell is the thinnest, and therefore has the highest energy level. Electrons injected into this well tunnel through to the lower energy level of the middle well, emitting a photon. They are then removed from the middle well by tunneling to the right-most well, which has a very slightly lower energy level—this maintains the population inversion between the first and second wells that is necessary for laser action. The electrons tunnel from the third well into a "graded alloy" superlattice of AlInAs and GaInAs to the right, through which they travel to the next cell; the band gap of this graded alloy varies smoothly so that the electrons reach the next cell with the right energy to tumble into the first (narrow) well. Thus the electrons cascade down a "staircase" in energy as they pass from left to right through the device.

Out of Step

Not all optical telecommunications systems rely on laser light, in which all of the photons are synchronized. The spontaneous emission that occurs in gallium arsenide/aluminum gallium arsenide heterostructures can be used "as is," without enclosing the device in a lasing cavity to bring about stimulated emission. Rather than a laser diode, the device is then simply a light-emitting diode (LED), and the light that it emits is incoherent—the photons are out of step.

Semiconductor LEDs are generally cheap to make and easy to switch on and off (to modulate), but they have neither the high light intensity nor the well-defined emission wavelengths of laser diodes. LEDs that emit in the infrared region, based on AlGaAs/GaAs and indium phosphide/indium gallium arsenide phosphide heterostructures, are used for short-distance optical-fiber communications links, while visible-light LEDs made from gallium arsenide phosphide find applications in display devices and as illumination sources. LEDs based on these materials are now commercially available and cover virtually the entire wavelength range from the near infrared (around 1.6 micrometers) to the green part of the spectrum. In their simplest form they consist of an active layer (say, GaAs) sandwiched between an n-doped and a p-doped layer, which inject electrons and holes respectively when a voltage is applied—just as was shown in figure 1.9. Like laser diodes, these LEDs come in several shapes or forms: some emit their light from the surface (perpendicular to the semiconductor layers), while others are edge emitters (fig. 1.17).

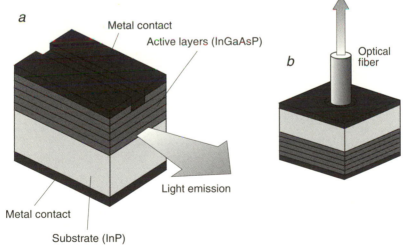

FIGURE 1.17 Photon emission due to recombination of charge carriers within a photonic semiconductor is exploited in light-emitting diodes (LEDs). Despite their low brightness and the incoherent nature of the light they emit, semiconductor LEDs are widely used in communications and display devices. The former, which emit in the infrared, commonly use indium gallium arsenide phosphide as the emitting layer. Both edge- and surface-emitting LEDs (*a* and *b*, respectively) have been fabricated; in the latter, the light can be fed efficiently into an optical fiber glued to a cavity in the surface.

SEEING THE LIGHT

At present it is necessary to convert the light-encoded information borne by optical-fiber transmission lines back into electronic form so that microelectronic circuitry at the other end can process it. This is a job performed by photodetectors. The simplest of these use photoconductive materials, in which absorbed photons excite charge carriers into mobile states so that they can carry an electrical current when a voltage is applied across the material. In so-called intrinsic photoconductors this excitation involves kicking an electron up from the material's valence band to the conduction band; in extrinsic photoconductors the electrons are excited to or from energy states of dopants. Although there are several semiconductor materials that show good photoconductivity at infrared wavelengths, photoconductive detectors are not used very much in optical telecommunications because their light sensitivity is rather low and their response is slow.

Instead, most photodetectors in information technology are photodiodes, which consist of layers of p-doped and n-doped materials. In its simplest manifestation, a photodiode is a plain old p–n junction across which a "reverse" voltage is applied: the p-doped region is given a negative voltage and the n-doped region a positive one. The result is that holes in the former and electrons in the latter are attracted away from the p–n interface, preventing recombination and creating a region depleted in mobile charge carriers on either side of the interface—this is called a "space charge" region, since the immobile countercharges that are left behind set up an electric field (fig. 1.18*a*). Provided that their energy is at least as

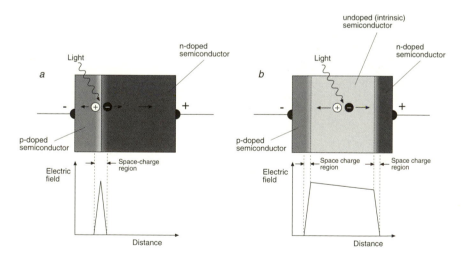

FIGURE 1.18 (*a*), Photodetectors can be constructed from p–n junctions. When a reverse bias is applied (so that the negative terminal is attached to the p-type layer and vice versa), charge carriers are drawn away from the p–n junction and a depleted region called the space-charge region is set up. When electron–hole pairs are generated in this region by absorption of a photon, they are rapidly swept outward toward the two terminals by the electric field created in the

great as the band gap, photons falling onto the material create electron–hole pairs by exciting electrons to the conduction band, and the electric field in the space-charge region will rapidly sweep the holes in one direction and the electrons in the other. So an electrical current—a photocurrent—flows in response to the light.

To maximize this photocurrent, all of the light should be absorbed in the space-charge region, which means making this region as large as possible. One way of doing so is simply to apply a large reverse voltage, but a better way is to leave a wide *un*doped ("intrinsic") region between the p and n materials: this creates a p–i–n structure (fig. 1.18*b*). Photoexcited charge carriers in the undoped region pass rapidly down the electric-field gradient set up by the space-charge regions at either end. These p–i–n photodiodes form the basis of just about all detector systems in optical telecommunications: for wavelengths from 800 to 900 nanometers (the near-infrared), doped silicon is used, while indium gallium arsenide, sometimes sandwiched between p- and n-doped indium phosphide in a heterostructure, provides photodetectors for longer wavelengths of around 1 to 1.6 micrometers.

Putting It All Together

With semiconductor-based light sources, optical fibers and photodetectors, we have all that we need for transmitting information with light. Laser diodes and LEDs provide light sources for sending the data, optical fibers are the wires that carry it, and photodetectors read out the data at the other end. At present, these are the only operations carried out with light in information technology—the difficult job of interpretation and processing the signals is left to electronic devices. This marriage of optical and electronic technology is called optoelectronics, in which light effects the communication between "thinking" electronics. In the present generation of optoelectronic telecommunications systems, the optical components are linked up to microelectronic information-processing devices in *hybrid* circuits, in which the electronic devices are mounted on silicon chips but the optical devices remain separate. This separation is not ideal—not only does it make the circuits less compact and harder to package, but it also degrades their performance (the interconnections cause signal losses) and their durability. The next challenge is to combine electronic and optical devices on a single chip.

---

space-charge region. (*b*), The standard photodetector in fiber-optic telecommunications technology is the p–i–n photodiode, consisting of a thick layer of an undoped semiconductor sandwiched between p- and n-doped materials. The device is again reverse biased. When light is absorbed in the undoped region, electron–hole pairs are created, and the two charge carriers are pulled in opposite directions by the applied electric field, and are again rapidly swept up when they enter the depletion regions, where the field varies sharply. Silicon is used for photodetectors sensitive to infrared wavelengths of 0.8 to 0.9 micrometers—these devices have relatively low light-to-electricity conversion efficiencies, but they are cheap. For longer wavelengths, indium gallium arsenide is used—these devices are more expensive but also more efficient than silicon-based ones.

This idea for an *integrated* optoelectronic technology—for handling light on a chip—was first proposed in 1969 by S. E. Miller of Bell Laboratories. But the technical difficulties associated with fabricating devices made of the semiconductor alloys needed for photonics on the substrates used for electronics meant that this idea was not realized until almost ten years later. The first integrated optoelectronic circuit was demonstrated in 1978 by a group from the California Institute of Technology that fabricated a gallium arsenide laser on a chip. The following year, the same group made an optical "repeater" circuit, in which a laser was switched on and off by an optical signal that controlled a transistor. These devices used gallium arsenide rather than silicon as the chip's base layer (the substrate), to ensure lattice matching with (and therefore easy fabrication of) the photonic heterostructure devices, which were built from gallium arsenide and gallium aluminum arsenide. The drawback, however, was that by using the same material for both the substrate and the active medium of the optical devices, the substrate could absorb some of the optical signal and would then start radiating this to other chips, resulting in cross talk. It was rather like trying to make electronic circuits on metal substrates.

One of the functions in most immediate demand for integrated optoelectronic circuits is that of a photoreceiver: a circuit that receives an optical signal and converts it into electronic form. This is of course precisely what a photodetector does on its own, but in practice the signal from the detector then has to be amplified to a high enough level for use in further signal processing. Amplification can be carried out by a conventional microelectronic transistor such as the field-effect transistor. The first integrated (that is, chip-mounted) photoreceiver was made in 1980 by Martin Pollack and colleagues at AT&T Bell Laboratories, using a p–i–n photodiode and a field-effect transistor in both of which gallium indium arsenide was the active material. They deposited these heterostructure devices on a substrate of indium phosphide (fig. 1.19).

The other type of optoelectronic circuit most needed for optical telecommunications is a transmitter, which does the opposite task of converting an electronic signal to an optical one. The transmitters used by present optical communications systems are semiconductor lasers and LEDs driven electronically in hybrid circuits. There are various practical reasons why integration of such circuits is tricky. First, the physical size of the laser determines that of the chip, because the two ends of the lasing cavity must be aligned with the edges of the chip. This leaves little room on a chip for any electronic driving components. Second, the large driving currents needed by the laser mean that any sources of electrical leakage from the chip stand to waste a lot of power, placing high demands on fabrication standards.

Although problems such as these continue to pose a challenge to the fabrication of integrated optoelectronic circuits, the potential benefits of integration as opposed to the use of hybrid circuits make the goal worth striving for. As it becomes possible to combine more and more electronic and optical devices on a single chip, the range of functions that can be obtained from them will expand. Much of the work on optoelectronic integrated circuits in recent years has focused on com-

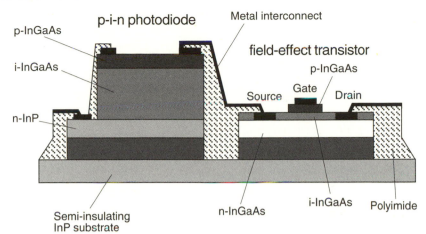

FIGURE 1.19 An optoelectronic integrated circuit (OEIC) has photonic and electronic devices coupled together on a single chip. One of the most common kinds of prototype OEICs is a photoreceiver, in which a photodetector (such as a p–i–n photodiode) is linked to an amplifying device such as a field-effect transistor. Here I show a simplified illustration of such a device, fabricated on an indium phosphide chip (generally doped with iron). In general, each of the semiconductor layers is deposited over the entire chip, and then the devices are carved out of the sandwich using lithographic methods. But more recently much effort has gone into selected-area epitaxy, in which thin films are deposited only on specific areas of the chip. The devices are "packaged" with a polymer coating (here a polyimide) to protect them from corrosion and mechanical wear.

bining optical devices based on indium phosphide (such as lasers and LEDs) with gallium arsenide chips, and gallium arsenide devices (such as photodiodes) with silicon chips. At present, none of these optoelectronic integrated circuits is commercially competitive with their hybrid counterparts, and only the latter are used for optical telecommunications. But once integration of optoelectronics becomes commercially viable, we can expect to see consequences as profound as those that accompanied the integration of microelectronic devices onto silicon chips.

Going All Optical

In addition to increasing the degree of integration, there is also a move toward increasing the range of operations that can be performed using purely optical devices. The incorporation of optical devices that can actually *do* something with data rather than just send and receive it would eliminate the need to fill up the chip with devices for converting optical signals to electrical ones and vice versa, with attendant amplifiers to accommodate for the losses that this conversion entails. Fully optical integrated circuits are called *photonic integrated circuits* (this term is, however, also sometimes applied to circuits in which a few tasks are still carried out by electronic components).

If optical logic devices are going to talk to each other on a chip, we need to be able to wire them up. What that means is that there is a need for a way of guiding light signals around on a chip—a microscopic equivalent of the optical-fiber technology used for long-distance communications. One possibility that exists for light but not for electronics is simply to send the signals through free space—to shine a laser beam from one component to the next through air. This solution has the advantage that there is extremely little absorption of the light by the medium through which it travels; but it requires very precise alignment of components and can take the signal from place to place only in straight lines. Free-space optical signaling is very attractive for getting one integrated chip to communicate with another, because it means that there need be no physical connection between the chips.

For directing the flow of light on a chip, however, it is more usual to contain it within a "waveguide," a kind of light wire that is really nothing more than a miniaturized optical fiber. In general, waveguides consist of a transparent material with a refractive index higher than that of the surrounding medium. Just as in optical fibers, this difference in refractive index allows for total reflection at the walls of the waveguide, confining the light within. Such waveguides are commonly thin strips of transparent material deposited on a substrate; the surrounding medium is then the substrate on one face and air on the others (fig. 1.20). Wave-

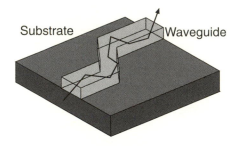

FIGURE 1.20 Waveguides on a chip are transparent channels whose refractive index differs from that of the materials on all sides. They confine light in the same way as optical fibers.

guides can also be made from strips buried beneath layers of lower-refractive-index material. They are made by depositing an entire flat layer of waveguide material and then using photolithography to etch away part of the layer, leaving only the strips in the desired pattern. This is just the same process as is used to make copper wiring on microelectronic circuit boards. The favored waveguide materials are silica and silicate glasses, the III–V semiconductors gallium arsenide and indium phosphide, and crystalline lithium niobate doped with titanium or hydrogen ions. There is now increasing interest in developing organic polymers for optical waveguides, partly because these can conceivably be laid down using much simpler technologies such as injection molding and embossing.

Nonlinear Thinking

To make photonic circuits, one needs devices that can perform all of the tasks of which electronic devices are capable. In general, devices that manipulate and process data are controlled or driven by a signal separate from that which carries the data; these are therefore active devices, and their need for a driving signal means that they consume power.

Just about every kind of active photonic device makes use of a material property called *optical nonlinearity*. "Nonlinear" can generally be regarded as a kind of technical term for "not obvious," in the sense that a nonlinear response of any sort is simply one that does not vary in direct proportion to the stimulus. Linear phenomena are easier to understand intuitively: the bigger the cause, the bigger the effect. If you increase the power supplied to a light bulb by turning up a dimmer switch, it gets proportionally brighter. Nonlinear responses embrace just about any other alternative to this kind of simple behavior: for example, if, as you turned up the dimmer switch, the light flashed alternately on and off, or if it got abruptly brighter and then stayed that way, or if it blew, those would be examples of a nonlinear response. All physical systems exhibit nonlinear behavior when driven hard enough: commonly the nonlinear response takes the form of saturation (where further increases in the driving force make no more difference to the output) or breakdown (where the system simply fails above a certain threshold).

Nonlinear behavior is often something to be avoided—for example, in the case of an audio amplifier, where a signal starts to distort when amplified too far. But just as rock guitarists have found this effect useful for their own purposes, so nonlinear effects in other situations can be put to (some might say more desirable) use. A diode is an example of a nonlinear electronic component: its output current remains negligibly low until the driving voltage reaches a threshold value, whereupon the output current increases sharply. This makes the diode a kind of switch, which is "off" when driven by voltages lower than the threshold and "on" for above-threshold voltages. A transistor can act as a more advanced version of a diode: a switch that also amplifies the signal through it.

A nonlinear *optical* response means that the amount of light transmitted through a substance is not proportional to the amount of incident light: doubling the latter will not necessarily double the former. Instead, the incident light actually changes the way that the material responds to light—for example, by altering the material's transparency or refractive index. Not only does this mean that the light that comes out might be rather different, in frequency for example, from that which went in; it also means that one light beam can exert a strong influence on a second as they pass through the material. In this way, nonlinear optical (NLO) materials may act as switches, in which one light beam can be used to control and direct the path of another.

At root, this nonlinear optical behavior is a consequence of the ease with which charges in the material can be shifted around (polarized) by an electric field.

These charges—electrons or ions in the NLO material—can be displaced by the oscillating electric field of a photon. Because of this rearrangement, the electric fields of subsequent photons interact with the material in a different way.

One example of such behavior is called second-harmonic generation (SHG). When light of a sufficiently high intensity is shined on a material that exhibits SHG, the transmitted beam takes on a different color from the incident beam. This may not sound so surprising at first: after all, stained-glass windows produce a kaleidoscope of colors even though illuminated with plain white light. But all that is happening here is that the glass is extracting some of the colors from the white light by absorbing them: those that are not absorbed pass through the glass and give it its color. Thus, the glass is not actually generating any colors that were not present already in the incident light. In nonlinear optical materials that exhibit SHG, on the other hand, light of one pure color can be shined onto the material, and light of a different pure color can emerge at the other side. Infrared light, for instance, can be converted to red or blue. The incident photons are converted into photons with a frequency twice as high (and thus a wavelength half as long). This frequency is the "second harmonic" of the oscillating electromagnetic field set up in the material by the incident light, just as higher harmonics can be heard in the note emitted by a guitar string or an organ pipe. In some materials a strong third harmonic can be excited instead, and the frequency of the light is then tripled.

Materials that show second-harmonic generation are very useful for expanding the range of available colors from laser light, and in particular for obtaining green and blue laser light. For many years attempts to make blue-light diode lasers were impaired by the lack of suitable materials. Although, as I mentioned earlier, II–VI alloys (particularly zinc selenide) and more recently nitride III–V materials are now filling this gap, second-harmonic generation provides a shortcut to blue laser light. Pass the infrared light (of wavelength around 860 nanometers) of a standard diode laser through a frequency-doubling material and out comes coherent blue light of twice the frequency, and thus half the wavelength. In the first demonstration of frequency doubling by SHG in 1961, a quartz crystal was used to turn red light from a ruby laser into ultraviolet. Subsequent studies of SHG at first used the natural mineral potassium dihydrogen phosphate; but today inorganic niobates, particularly lithium niobate, are the most commonly used frequency doublers, and can be grown in perfect crystals several centimeters across.

In 1995, the first commercial blue laser appeared on the market. Developed by Coherent Inc. in Santa Clara, California, in collaboration with IBM's Almaden research center in San Jose, it uses crystalline potassium niobate to double the frequency of an infrared laser. By employing a special trick called resonant doubling, the Coherent device achieves the high-intensity output needed for practical applications such as reading optical disks. Its high cost is sure to curtail widespread use, however, and most researchers now see these frequency-doubling measures as a stopgap before intrinsic blue-light lasers, perhaps based on gallium nitride, become commercially available.

Switching On

Probably the most important active component in any information-processing circuit is the switch, one of the roles played by the transistor in electronics. To achieve electronic switching, electrons can be rerouted along a conducting pathway by an applied voltage—in other words, one electronic signal is used to control another. In a truly photonic switch, one light beam will control another. But a kind of halfway house is a switch in which an electric field directs the path of a light beam. This kind of switch is an optoelectronic device, and can be fabricated by making use of the nonlinear optical effect known as the electro-optic effect, whereby the refractive index of a material is changed by an applied electric field.

A typical electro-optic switch comprises two parallel waveguides of a nonlinear optical material such as crystalline lithium niobate, which come together at a bottleneck (fig. 1.21). The waveguides are defined in the crystalline film by

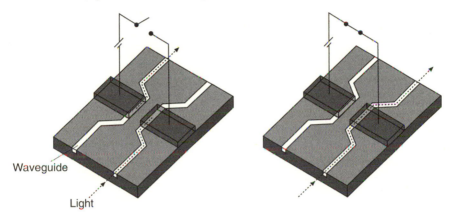

Waveguide

Light

FIGURE 1.21 An electro-optic switch controls the path of a light signal through a waveguide. Light traveling through the lower waveguide interacts with the upper waveguide in a nonlinear way at the bottleneck, causing the light to switch channels. But if an electric field is applied across the bottleneck, the nonlinear interaction is modified and the light stays in the lower channel.

doping the channels (commonly with titanium) to change their refractive index relative to the surrounding material. The light in one channel can be switched back and forth between the other channel by applying a voltage across the bottleneck—this electric field polarizes the NLO material and alters the refractive indices of the two waveguides. The application of a voltage does not, as you might think, cause the light to switch channels; rather, this happens in the *absence* of an applied field, because of the nonlinear way in which the electric field of the light in one channel interacts with the charges in the other, "empty" channel at the bottleneck. The applied voltage, meanwhile, keeps the light in the same channel as it traverses the neck. Electro-optic devices of this sort are now produced commercially for integrated optoelectronic circuits.

The electro-optic effect is also exploited in devices for modulating light signals—for chopping them up into discrete, "digital" pulses or for altering the frequency of a pulsed signal. These modulators again take advantage of the effect of an electric field on the refractive index of an NLO material. In the modulators known as Mach–Zehnder interferometers, an incoming laser beam is split into two beams, each of which is directed along a separate waveguide in the NLO material before they are merged again. When a voltage is applied across one of the branches, its refractive index is altered and the light beam along that branch is slowed down slightly. As a result, when the two beams merge again their peaks may no longer coincide and so they interfere destructively. If the refractive index of one channel is changed just enough to slow the beam it carries until the peaks coincide with the troughs of the other, unperturbed beam, there will be complete destructive interference when the two beams are reunited, and the light signal will be switched "off." So by modulating the applied voltage, the light beam can be modulated with the same frequency.

These electro-optic modulators, which are again optoelectronic devices, can achieve extremely high modulation frequencies—they can switch the light beam on and off very rapidly. This means that the modulated light beam can ferry a lot of data very quickly (each pulse can be regarded as a "bit" of information). This illustrates one of the fundamental attractions of photonic computing—the speed of data transmission and processing. Commercial electro-optic modulators, which use lithium niobate as the NLO material, can reach switching rates of twenty billion times per second, while laboratory prototypes have been made that are almost four times faster than this.

But these switches are not cheap (their cost can run to tens of thousands of dollars), primarily because the crystalline NLO materials are difficult to grow, and hence expensive. This is one of the main reasons why there is so much interest in developing organic polymers that carry out the same function as the NLO crystals. Some of the polymers that have been successfully used for electro-optic switching and other photonic functions derived from nonlinear optical behavior have their NLO properties "built in" to the polymer chains, which have rather sloppy (highly polarizable) electron clouds. Several polyimides fall into this category (fig. 1.22a shows one such). A polyimide called Pyralin 2611D, developed by Du Pont, has been used in an electro-optic modulator that provides modulation frequencies of up to 20 billion times per second—equal to the commercial lithium niobate devices. Other NLO polymers have "optically inert" backbones (such as polyacrylates and polyurethanes) to which optically responsive organic side groups are attached (fig. 1.22b). Pilot-scale commercial production of modulators made from these materials is underway. But the advantages of low cost and easy processing of polymers are somewhat offset by question marks over their long-term stability—these materials may start to break down after prolonged exposure to intense light.

The nearest thing to a commercial all-optical switch at present is a device that

a   b

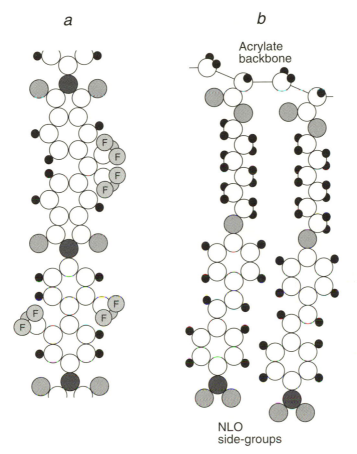

FIGURE 1.22 Organic polymers are promising candidate materials for nonlinear optical devices. Fluorinated polyimides (*a*) and polyacrylates with nonlinear optical side groups (*b*) are both being explored for use in electro-optic modulators and other optoelectronic devices.

uses optical nonlinearity in conventional glass fibers. Nonlinear optical effects in these fibers are very small—indeed, it would not be possible to send optical data reliably through hundreds of kilometers of fiber if they were not. But because the fibers are also highly transparent, one can let these small NLO effects build up slowly to significant levels between two intense, interacting beams as they travel through long lengths of coiled fibers, without having to worry about loss of signal.

One of the most common of these fiber-based switches, a nonlinear optical loop mirror (NOLM), works rather like the Mach–Zehnder modulators described above: the signal beam is split into two components, sent around a fiber-optic loop several kilometers in length, and reunited. But the two beams are sent around the coiled fiber in opposite directions, so that they reunite at the same point where

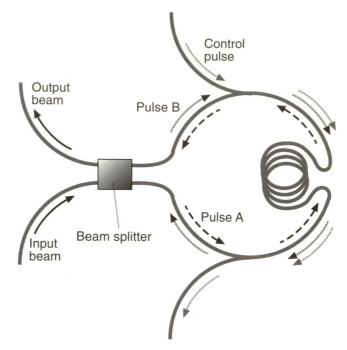

FIGURE 1.23 A nonlinear optical loop mirror (NOLM) is a fiber-optic device that allows the path of one light beam to be switched by another. The pulsed input beam is split into two (pulses A and B) by a beam splitter, and these are routed around a long optical-fiber loop in opposite directions. Switching is induced by a control pulse (gray), which is introduced into the loop so that it travels alongside one of the split signal pulses (here pulse B). The combined nonlinear optical effect of the control pulse and pulse B causes a refractive index change in the fiber along B's path, slowing down pulse B relative to A. The effect is very small, but because the loop is so long, pulses A and B are out of phase when they finally recombine after a circuit. This phase difference causes the beam splitter to direct the recombined signal beam down a particular output fiber; in the absence of a phase difference (that is, if no control pulse were introduced), the output signal would be sent down a different fiber. Signal pulses at particular intervals in a data stream can be redirected by sending a control pulse into the loop at those intervals. In this way, several pulsed data streams can be interdigitated (multiplexed), sent down a single fiber, and unraveled (demultiplexed) by a NOLM at the other end.

they were split. If at this point they are in phase, the recombined beam is sent in one direction; if they are out of phase, they are routed down a different optical-fiber channel. This switching is induced by a control beam, which is injected alongside one of the split signal beams. As it travels alongside, it causes a change in the refractive index of the glass fiber owing to the fiber's small NLO response. This refractive-index change alters the phase relationship between the two split-signal pulses (fig. 1.23).

These fiber-optic devices can also be used for multiplexing optical data: for combining two or more pulsed signals into a single stream of pulses for transmission down a fiber-optic network. At the other end, the signal must be demultiplexed—separated back into the original streams of data—and a NOLM can perform this separation too. For instance, the demultiplexing NOLM might pick out every fourth data bit from a stream of optical pulses carrying forty billion bits per second, thus converting this single signal into one that carries ten billion bits per second and one that carries thirty billion. This process can be regarded as the *selective* optical switching of every fourth bit in the input data stream: the control beam consists of pulses fired off at a rate of ten billion per second, so that a control pulse accompanies every fourth input bit and reroutes it.

But fiber-based all-optical switches and multiplexers have not yet left the laboratory. One key reason for this is their size: it is no easy matter to pack kilometers of fiber into a shoebox, let alone reduce them to the proportions that would fit on a chip. Researchers are now striving to reproduce the same kind of effect in microscopic semiconductor devices in which the coiled fiber is replaced by waveguides etched into photonic materials such as indium gallium arsenide phosphide. Demultiplexing in these miniaturized semiconductor devices has already been demonstrated in the laboratory.

Toward the All-Optical Computer

The battle between electronics and photonics is being waged in two major arenas. One is in information transmission, and here the photons are already starting to taste victory. For the long-distance cables that carry the data, there is no longer any doubt that optical fibers have outmatched copper wires; but if optical data transmission is to achieve its full potential, these transmission networks will have to become *all*-optical, dispensing with any form of electronic mediation to modulate, direct, or amplify the signals between the sending and receiving hardware. Only then will the kinds of transmission speeds that photonics makes possible be realized. We've seen how progress is being made on all of these fronts. The question of multiplexing of optical signals—sending several independent signals simultaneously down the same fiber—is particularly important, because it is here, as much as in the issue of modulation speed, that light-based transmission holds such promise.

The most well-developed kind of multiplexing at present is that in which each signal is assigned a different wavelength band—this is equivalent to the way in which radio broadcasting is divided into bands, although in that case the signal wavelengths are in the radio-wave part of the spectrum rather than the optical or infrared. By tuning the receiver appropriately, one particular band can be picked up without interference from the others, provided that their wavelength ranges do not overlap. A small-scale all-optical network that carries twenty wavelength-multiplexed signals, each at a different wavelength and each bearing up to ten billion bits of data per second, has been constructed in eastern Massachusetts by a collaboration involving AT&T Bell Laboratories, the Massachusetts Institute of Technology, and the Digital Equipment Corporation, and was demonstrated successfully in 1995. Other all-optical wavelength-multiplexing projects have been launched in the United States, Europe, and Japan. These projects serve to demonstrate that the technologies needed for all-optical networks are already at hand.

The second arena in which electronics and photonics compete lies at either end of the transmission network: the signal-processing systems, and in particular the computer. Here electronics still has the upper hand; indeed, the challenge posed by photonics remains rather feeble at present. The all-optical computer, in which logical data processing is carried out on photonic integrated circuits which need electricity for little more than just driving the semiconductor microlasers—this is something that remains a distant dream. For as we have seen, the photonic integrated circuit, which processes light on a chip, is by far the most immature of the new optical technologies. But when it arrives—maybe in the next five years, maybe in the next fifty—we will see computers change *qualitatively*. Not only will they be faster, but entirely new kinds of computer architecture should become possible. In others words, we will discover new ways to make machines think.

THE SILICON GLOW

While there is no lack of photonic materials for making light-emitting devices that can be miniaturized and coupled to silicon circuitry in hybrid structures, integration of photonic devices onto silicon chips has been hampered by the difficulties of growing the photonic materials, such as gallium arsenide, on silicon surfaces. Photodetectors, waveguides, and light modulators can all be made from silicon-based materials instead, but the light sources—semiconductor lasers and light-emitting diodes (LEDS)—cannot, because silicon does not emit light efficiently. Overcoming this obstacle would revolutionize optoelectronics by making it an all-silicon technology.

Like gallium arsenide, silicon can be made to produce photons—to luminesce—by creating electron–hole pairs which then recombine. In a simple light-emitting diode, the electrons and holes are injected by applying a voltage across the material, and the resulting light emission is known as electroluminescence. A good guide to the electroluminescent potential of a material is its efficiency of photoluminescence, in which the pair of charge carriers is excited by light absorption. In general, the energy given out by recombination is slightly less than the energy taken in by initial light absorption, so the wavelength of photoluminescence is somewhat longer (the photon is less energetic) than that of the light used to stimulate it.

Measurements of the photoluminescence of silicon in the 1950s appeared to write off the possibility of silicon-based photonics: if irradiated with visible light, silicon emitted radiation in the infrared part of the spectrum, but only very weakly. For every photon of light emitted, about a million photons had to be absorbed. This is because there are several alternative ways in which the electron–hole pair can recombine *without* emitting light. Since in silicon these processes are much more rapid than radiative recombination, most of the pairs get squandered. Radiative recombination is so slow in silicon because a peculiarity of its electronic band structure means that the net momentum of the recombined pair is different from that of the excited pair (its band gap is "indirect"). So the principle of conservation of momentum requires that some of the energy of recombination be carried away by lattice vibrations, which make up the difference in momentum. In a sense, electrons and holes have to wait for a suitable lattice vibration to pass by before they can recombine radiatively.

But this conservation rule is relaxed if the material is not perfectly crystalline, which has led to the exploration of disordered forms of silicon as photonic materials. Disordered silicon-based materials called polysilanes, which contain polymers of silicon and hydrogen, do indeed show good photoluminescence, but at the expense of the good electrical conductivity that is also central to silicon's use for microelectronics. Another way to introduce disorder without sacrificing crystallinity altogether is to introduce defects into the silicon crystal. But these must be chosen with care, since most defects actually enhance the ability of the electron–hole pair to recombine without emitting a photon. Certain defects, however, make the light-emitting transition easier.

Among them are defect atoms that have the same electronic makeup as silicon itself: carbon, germanium, and tin, which, like silicon, all have four electrons available for chemical bonding. When introduced into the silicon crystal lattice in place of a silicon atom, these are known as isoelectronic ("same-electron") defects. Efficient electroluminescence can be achieved in carbon-rich silicon, but only at the temperature of liquid nitrogen (minus 196 degrees Celsius); good electroluminescence at room temperature by isoelectronic defect doping of silicon has yet to be achieved. Another approach is to introduce defect atoms that themselves have strong luminescent properties, the idea being that the electron–hole pair will find its way to the defect and there recombine with efficient photon emission. Rare-earth metals such as erbium have been shown to enhance the luminescence properties of silicon in this way, but again the practical value of the approach remains limited.

These approaches show much promise but remain very far from providing a silicon-based device that will emit light efficiently at room temperature. This is why the discovery made by Leigh Canham of the Defence Research Agency in Malvern, U.K., has excited such interest. Canham announced in 1990 that silicon can be turned into a good emitter of light by cutting most of it away—specifically, by using an electrochemical method to etch channels throughout the material, leaving a ramified array of thin silicon filaments, like a sponge. When the husband-and-wife team of A. and I. Ulhir, working at Bell Laboratories in New Jersey, first found in 1956 that electrochemical etching of silicon produced this result, they were far from pleased, because it undermined their attempts to make smooth, polished silicon surfaces for microelectronics applications. But four decades later, this porous form of silicon may prove to be the optical material that photonic engineers have been dreaming of. A disk of porous silicon illuminated by ultraviolet light will emit an orange-red glow, without any need for careful surface treatments or sophisticated defect doping (plate 2). Indeed, the material is produced by a technology so simple that Canham can do it in a bucket.

The etching process used first by the Ulhirs and later by Canham involves immersing a silicon wafer in a bath of hydrofluoric acid and passing an electrical current through the solution, with the silicon acting as the positive electrode. Under these conditions the acid eats away at the silicon in such a way as to carve out tiny channels. As more and more of the material is eroded away, what is left is a delicate filigree of silicon consisting mostly of empty space. The remaining material forms an interconnected web of fine silicon wires, no more than a few nanometers thick (fig. 1.24). About 20,000 of these wires side by side would be no thicker than a human hair.

Porous silicon shows both photoluminescence—visible-light emission in response to irradiation—and electroluminescence, where the emission is stimulated by an applied voltage. The reason for this behavior was hotly debated when the discovery was first announced. Some researchers believed that the dissolution process was forming silicon polymers, called siloxenes, on the surfaces of the material; these compounds were already known to have luminescent properties. Others suggested that silicon hydrides at the surface were responsible. But Canham believed that the key to the effect lay with the narrowness of the silicon wires themselves. He suggested that the emission was the result of the same electron–hole recombination processes that in bulk silicon cause inefficient emission in the infrared region of the spectrum. The reason that in porous silicon the emission is both more efficient and shifted to shorter (visible) wavelengths might, Canham said, be connected with the confinement of the electrons and holes to very narrow channels.

The effects of confinement are quantum-mechanical in origin: the discrete energy levels of a quantum particle (like an electron) held in a box depend on the size of the box. Although that might seem a little strange, it can be compared with the effect on a flowing stream of water of confining it within an ever-narrower channel: as the channel gets tighter, the stream flows ever faster, and the kinetic energy of the water increases.

For a semiconductor like silicon, quantum-confinement effects bring about a change in the energies of the electronic bands: in a narrow silicon wire the conduction and valence bands are shifted farther apart, and so the band gap is increased. Consequently there is a greater energy loss in the recombination process, leading to the emission of a photon of greater energy than from the bulk material. So the luminescence shifts to shorter wavelengths as the wire gets thinner. In addition, when confined to a narrow space, electrons and holes attract each other more strongly, and so radiative recombination is able to compete more effectively with the nonradiative processes that suppress luminescence. A considerable body of experimental results now supports this interpretation.

Perhaps the most notable benefit of the quantum-confinement effect is that it allows the color of the emitted light to be tuned simply by varying the width of the silicon wires, which can be done by carrying out the etching process for different lengths of time: the longer one etches, the more silicon is eaten away and so the finer the wires are. As finer wires have a larger band gap, the color of the

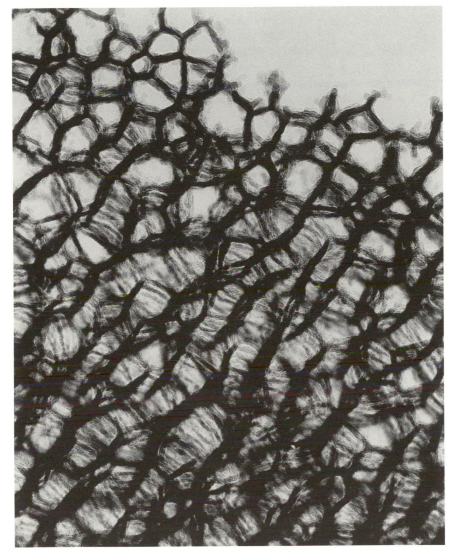

FIGURE 1.24 When etched into tiny wires no more than a few nanometers across (dark regions in this electron micrograph), silicon will emit visible light efficiently. (Photograph courtesy of Leigh Canham, Defence Research Agency, Malvern, U.K.)

luminescence moves from red to green and toward blue for more highly etched silicon.

Porous silicon that emits red, orange, yellow, and green light has now been prepared. But once the wires get too thin, they start to become so fragile that they break during the drying process, making it hard to extend the range of colors even to green. One answer to this problem is to use a process called supercritical drying, which is described in chapter 7. Canham and colleagues have used this approach to make highly porous silicon that emits uniform green photo-

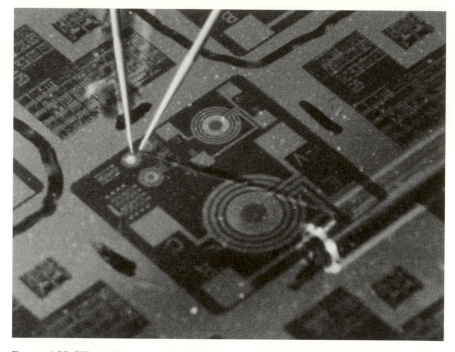

FIGURE 1.25 Silicon glows on a chip, in this porous-silicon light-emitting diode (indicated with tweezers) fabricated by researchers at Rochester University, New York. (Photograph courtesy of Philippe Fauchet, Rochester University.)

luminescence, and they anticipate that blue emission might even be within reach. But the fragility of these porous materials may pose problems when it comes to making working light-emitting diodes from it. Canham has suggested that perhaps one could fill up the empty pore space with some transparent support material to make the network more robust.

But porous silicon is also highly chemically reactive, and so it will be difficult to use it in a stable, long-lived light-emitting device. Philippe Fauchet and co-workers at the University of Rochester in New York have come up with a solution, however, which has enabled them to make the first ever integrated, all-silicon optoelectronic cirucit. Fauchet's group has found that porous silicon can be partially oxidized to produce a material called silicon-rich silicon oxide (SRSO), which retains the visible-light-emitting behavior of porous silicon itself without its instability. In 1996 they fabricated an integrated light-emitting diode on a silicon chip in which the active, light-emitting material was SRSO (fig. 1.25). The Rochester team was able to construct addressable arrays of these devices in which each one could be switched on and off individually. So just six years after the discovery of glowing silicon sponges, they seem on the verge of realizing at last the dream of an all-silicon optoelectronic technology.

Total Recall

MATERIALS FOR INFORMATION STORAGE

> The ages have been inundated with vast oceans of words. We have
> been virtually drowned in them.
>
> —Ben Okri, *Beyond Words*

The information age is producing a data glut: we are on the verge of generating more data than we can hold on to. More capacious yet more compact memory devices are urgently needed. These demands will eventually push conventional magnetic data storage to its limit, and are motivating the introduction of new kinds of light-based memories.

THE WORLD is awash with information in transit. Conversations go speeding down subsea cables, or bounce from satellite to satellite. The air is filled with invisible radio and TV shows en route to a billion homes. Computers talk to each other day and night from one side of the globe to the other. But all of this information is as nothing compared with that which lies in storage. While some transmitted signals carry only ephemeral, fleeting messages—the idle chatter of friends and lovers—many more bear information that someone, somewhere, has decided they need to hold and to peruse. Information processing technology is largely occupied with some tiny fraction of a vast data dump, on its way to another data dump. So although the information revolution has placed unprecedented demands on the capacity of communications networks, this is nothing compared with the demands on data storage systems. Some of mankind's most ambitious projects, such as attempts to read the entire genetic blueprint of humans and other animals, face limitations not just from the speed at which the information can be obtained but the capacity with which it can be stored.

And so the advances that are taking place in ultrafast computation place ever greater demands on the technologies that hold data for posterity. Put simply, the faster we are able to generate data, the more room we need to make for it. In practical terms, this means that we need to find ways of squeezing more and more information into smaller and smaller spaces: to obtain higher data storage *densities*.

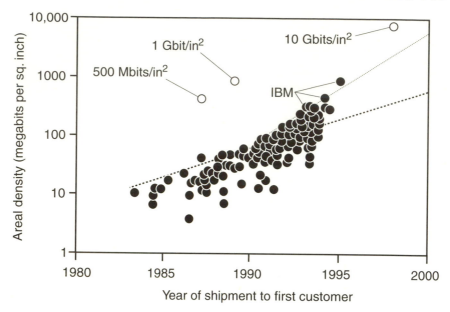

FIGURE 2.1 Over the past decade, the areal (two-dimensional) data storage density of commercial disk drives for computers has roughly doubled every three years. The bold dashed line represents a conservative estimate of the upward trend; over the last five years or so there are signs that this trend could begin to rise even more steeply (thin dashed line). The points in white represent demonstration models rather than commercial devices.

Today's memory banks, whether they be on cassette tape, on disk, on videotape, or on hard drives, are almost all magnetic: the data is recorded as a pattern in a magnetic medium. And magnetic recording technology is taking giant strides forward: to ever higher storage densities (the amount of information in a given volume of material—or, for media in which the information is stored two-dimensionally, with items of data side-by-side like words on a page, the amount in a given area), and to ever greater fidelity, better durability, and reliability. One reads, from time to time, the suggestion that the magnetic recording industry has reached the limit of its potential, that it has no further scope for improvement. This is, of course, an enticing line if one is in the business of singing the praises of a new technology for data storage (examples of which we shall encounter in this chapter); but it is scarcely accurate.

A glance at figure 2.1 should persuade you that the magnetic storage industry is anything but moribund. This figure shows the way in which the areal storage density of commercial disk drives has been changing over the past decade. Analysts argue about how steeply the curve is rising, but there is no question that the trend is upwards—to ever more amounts of data packed into an ever smaller area of disk. The ballooning of the home computer business has been powered by this trend, which allows magnetic hard drives to hold immensely sophisticated software without the need for battalions of add-on devices to expand the memory.

This is the sort of graph that industrialists like to present to show just how rosy the future looks. But its upward slope will persist only so long as fundamental research continues to push at the envelope. Magnetic recording has existed as a burgeoning commercial enterprise since the 1950s, but the advances have not been a matter of refining old ideas; instead, new ways of imprinting data have been devised, and new materials developed to contain the information revolution.

Are there *fundamental* limits to this growth? Is there anything in the physics of magnetic recording, or in the properties of the magnetic media used for storage, that will ultimately limit the rise, in the same way that some say physical laws will ultimately curtail the miniaturization and increase in speed of computers? So far, there seems to be little prospect of that. On the contrary, the potential for growth looks ever brighter, as recent years have produced entirely new approaches to data storage that could accelerate the trend even further if their technological potential can be realized.

But magnetic storage, unchallenged for decades, now has a rival. Just as photons threaten to usurp electrons for data processing, so too does light promise to become the memory medium of the future. Optical memories claim to offer not only potentially higher storage densities, but also entirely new kinds of memory, some of which operate more like our own minds than like the linear data banks of magnetic tape. They are, at present, the David to magnetism's Goliath, but the story may yet have the biblical conclusion.

GETTING IT TAPED

Like many technologies that today we take for granted—the fax machine is another example—magnetic recording was an idea that had to wait a long time for applied science to catch up with the theory. It was mooted in the nineteenth century, and was first put into practice by the Danish engineer Valdemar Poulsen, who in 1898 invented the Telegraphone, a sound-recording device that employed steel wires as the magnetic medium. Magnetic recording became commercial only in the 1930s, however, when the German company AEG Telefunken created the Magnetophon, a device that recorded sound signals on magnetic tape developed by the German company BASF. Versions of this device found their way into German broadcasting during the Second World War, and after the war similar sound-recording devices were developed outside of Germany, notably by the Ampex company in California. The sound quality improved steadily, and domestic tape recording was given a new impetus by the invention in 1964 of the compact cassette, a tape reel enclosed in a plastic box for ease of handling.

Pointing the Way

Although all of these early magnetic-tape recorders were analog devices (in which the input signal is modulated continuously), the principles of magnetic recording are more easily understood for digital data—increasingly the prevalent means of encoding data in today's information age. Digital information—whether it represents a sound, a picture, or a stream of numbers or words—is recorded as a sequence of pulses (bits), each of which takes one of two values. This sequence of data pulses can be written as a series of 1's and 0's; in practical terms these bits are stored in a memory that can be thought of as comprising an array of switches, which are flicked one way ("on," say) to represent a "1" and the other ("off") to represent a "0."

A string of data—the words on this page, for instance—can be encoded using just 1's and 0's by creating a code in which a string of several bits represents a single character (the letter "a," for example, or a space or a comma). As there are thirty-two ways of arranging a sequence of five 1's and 0's, and sixty-four ways of arranging six, it is clear that one quickly acquires enough scope to include all of the characters on a keyboard in a binary code of just a few bits per character. To store a more complicated kind of data set—say, the information in a photograph—requires more ingenuity, but only a little. One can break the image down into tiny squares, like the pixels of a television screen, and for each pixel one can use a binary code to record the relative intensity of the primary colors red, blue, and green in that pixel.

An information storage device working on this basis needs three basic components: a medium for storing the binary data (this will comprise, in effect, an array of switches that can be flipped one way or the other, or a substance that can locally exist in one of two states), a means of writing the information into the storage medium (recording), and a means of getting the information back out (readout or playback). In many storage systems it is also desirable to erase information (to reset the switches) and to overwrite old information with new (which is essentially the same operation as erasing). Magnetic audiotape can of course be erased and over-written; but CD-ROM storage media (which are optical memory devices, described later) cannot (ROM stands for "read-only memory," which means that once the information is written onto the disk, it can be read but not erased).

Magnetic materials are ideal candidates for storage media because magnets are rather like switches—they can be made to point in one direction or another if one places them in an external magnetic field. The extent to which this simple fact changed the world is not always sufficiently appreciated. After the invention in ancient China, nearly two millennia ago, of the floating compass—a magnetic needle that aligned itself with the Earth's magnetic field—the magnetic iron oxide mineral magnetite became known as the lodestone (the "leading stone"), and ensured that seafarers were no longer at the mercy of the stars for accurate navigation. Without this navigational device, European explorers, and later settlers, in the sixteenth century might never have found their way to the New World.

These discoveries of the properties of natural magnets were very much of an empirical nature; just as the earliest navigators had no real conception of *why* the stars showed them the way, so the mysterious rotation of the lodestone was unexplained for many centuries. It was only in 1600 that William Gilbert, a scientific genius of the order of Isaac Newton, suggested in his book *De Magnete* that perhaps the Earth itself is a giant magnet.

In the nineteenth century, Michael Faraday showed that magnetism and electricity are like opposite sides of a coin—the flow of electricity induces a magnetic field, and a changing magnetic field induces an electrical current. These two fields, electric and magnetic, can be self-supporting in the sense that a varying electric field can propagate out into empty space with a varying magnetic field propagating perpendicular to it. These propagating fields constitute electromagnetic radiation—radio waves, light, X rays, and all the rest.

The unified theory of electricity and magnetism, developed by James Clerk Maxwell in the 1880s, provides one of the central ideas in physical science. It is not an easy theory, but for the purposes of understanding magnetic recording we need know nothing more than Faraday's principle of electromagnetic induction: a (changing) electric field can induce a magnetic field, and vice versa. We will see why this is useful shortly.

A magnet, then, is a potential switch: place it in a magnetic field, and it will tend to align its poles with those of the field (fig. 2.2a). Reverse the direction of the field (so that the north pole becomes the south), and the magnet will switch to point in the other direction. Here we have a simple system for storing binary information: the magnet represents a "1" when pointing in one direction, and a "0" when pointing in the other. A magnetic memory can be constructed from a whole array of magnets, in which the binary data can be encoded by applying an external magnetic field to each magnet in turn, the direction of which determines whether we write a "1" or a "0" (fig. 2.2b).

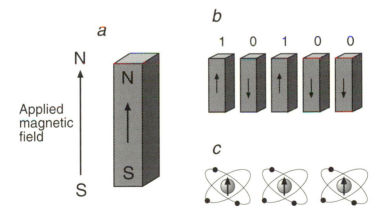

FIGURE 2.2 (a), A magnet that is free to rotate will align itself in an external magnetic field so that its poles point in the same direction as those of the field. (b), By using an applied field to orient magnets in one direction or the other, one can record binary data in a magnetic array. (c), Some atoms behave like tiny magnets, possessing a magnetic dipole (a "magnetic moment") which can be oriented by an external field.

Just a Moment

This is all very well, but if we want to have a realistic storage density, then our magnets will have to be very small. In principle that is simple enough, since certain individual *atoms* act as magnets. That is to say, these atoms (iron is the typical example) can be thought of as tiny magnets possessing a north and south pole, which can be flipped in one direction or another in a magnetic field.

This is not a property that all atoms possess. Rather, it is specific, in general, to atoms that have a certain arrangement of electrons. Electrons are negatively charged particles that surround the central, positively charged nucleus of an atom. They can be considered (in the crude so-called classical picture of an atom, which retains a great deal of value even though it has been superseded by quantum mechanics) to be circulating the nuclei in orbits. Now, a circulating, negatively charged electron is really nothing more than a circulating electrical current, which, because of electromagnetic induction, gives rise to a magnetic field. In many atoms electrons are paired up in each orbit; crudely speaking, they can be considered to be circulating in opposite directions, giving rise to opposing magnetic fields which therefore cancel out.

But when electrons are not paired up in this way, they can generate a net magnetic field. This contributes to the magnetic field of the atom, but that is not the whole story. Electrons also possess a property called *spin*. This is a quantum-mechanical property; it is not really so simple as the spinning of the electron about an axis (in the same way as the Earth rotates on an axis as it orbits the Sun), but that again is not a totally inept analogy. Electron spin also generates a magnetic field, and the overall contribution of an unpaired electron to the magnetism of an atom results from the combination of its orbital and spin components. An atom that possesses net magnetism is said to have a magnetic moment—a kind of magnetic dipole that points in a certain direction (fig. 2.2c). Metal atoms commonly have unpaired electrons, both in the pure metal and in some compounds (such as natural oxide minerals). These substances are therefore potential magnets.

Only "potential" magnets, mind you, because the existence of magnetic atoms in a substance does not in itself guarantee that the substance will be magnetic in the sense that we generally recognize it—picking up iron filings and so forth. Scientific techniques for revealing such things will show that an iron nail is made up of atoms possessing magnetic moments, yet the nail will not rotate to point north when hung on a string. Why not?

In iron, the magnetic moments on each atom have a tendency to line up so that they all point in the same direction. At first glance this seems intuitively natural, but it is in fact less obvious. If you place two bar magnets side by side, with the north poles of each facing the same direction, they will certainly *not* be happy to maintain this arrangement; they will rotate so that the like poles are not adjacent. If they can rotate freely, they will tend to align in opposite directions, and indeed may pull themselves together with the north pole of one facing the south pole of the other and vice versa. As in the case of electric charge, one can say that opposites attract.

One might therefore expect the magnetic moments in iron to line up in an alternating fashion, pointing first north and then south as one passes down a row. That this does not happen is because the way in which the moments interact is subtle, involving quantum-mechanical effects that are determined by the nature of

Ferromagnet   Antiferromagnet

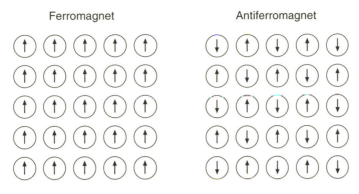

FIGURE 2.3 In a ferromagnet, the magnetic moments of all the atoms in a crystal point in the same direction. The bulk material is then a permanent magnet, and is said to have a net magnetization. In an antiferromagnet, the moments on adjacent atoms prefer to point in opposite directions. The magnetic moments then cancel each other out, and there is no net magnetization.

the electron orbits responsible for the moment, and the way in which these orbits overlap with one another between neighboring atoms. This subtlety means that to deduce how one atomic magnetic moment will influence that of its neighbor, one must carry out some rather careful and sophisticated calculations. In iron and nickel, it so happens that alignment in the same direction is preferred. In other magnetic metals, such as chromium and manganese, alignment in opposite directions is found instead (fig. 2.3). The former situation is known as *ferromagnetism* (deriving from Latin *ferrum*, for iron); the latter is called, reasonably enough, *antiferromagnetism*.

Both of these types of magnetism derive from the same principles, but the way in which they become manifest in the bulk material is very different. In a ferromagnet, the fact that all of the atomic magnetic moments are aligned means that they all add up to give the material an overall magnetization—it is "magnetic" in the colloquial sense. But in an antiferromagnet the magnetic moments cancel those of their neighbors, so the net effect is to leave the material devoid of net magnetization.

What, then, of our iron nail, in which all the magnetic moments are aligned? The reality is that, although this is the ideal situation, it is not the one that occurs in practice. Rather, the alignment is not sustained throughout the whole object. The metal is divided up into a mosaic of tiny *domains*, each one typically no more than a few micrometers across. Within the domains all the magnetic moments are aligned, but those in one domain point in a different direction to those in its neighbors (fig. 2.4). At domain boundaries we find an unhappy juncture of unaligned moments; but to correct this misalignment would require that *all* of the moments in one or other of the domains be reoriented. That is something that costs a lot of energy, so it will not happen spontaneously. Thus, even though the magnetic moments on all atoms would prefer to be aligned, the misalignments at domain boundaries are frozen in. The result of this patchwork of domains with

Domain wall

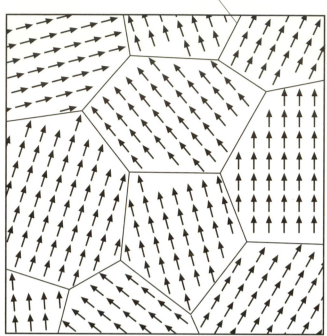

FIGURE 2.4 Iron is a ferromagnet, but an iron nail does not display a spontaneous magnetization because it consists of many microscopic domains, within which all the magnetic moments of the iron atoms are aligned but between which there is no net alignment. When averaged over the entire bulk material, the moments of the domains cancel one another out. But the magnetization of every domain can be oriented in the same direction by an applied magnetic field—the nail will then become magnetized.

different, random magnetic orientations is that they cancel each other's magnetic moments out on average, and the material has no net magnetization.

That imperfect arrangement can be rectified, however. If it is placed in a strong magnetic field, an iron nail can be magnetized. This is because, if the field is strong enough, the magnetic moments on each of the iron atoms will tend to become aligned with the direction of the field. The domain boundaries will thus be destroyed, and all of the atomic moments will point in a single direction. When the external magnetic field is switched off, the nail will have acquired the ideal, fully aligned state, so it will have no tendency to revert to the domain-structured state: it will have become a permanent magnet.

Experience teaches us that this is not the full story, however. An object magnetized in this way will generally tend to lose its magnetism over time. The problem is that alignment of magnetic moments is constantly fighting a battle with the randomizing effect of thermal motions of the atoms. Heat has the effect of destroying this alignment, and at room temperature a magnetized nail will gradually find its perfect alignment of atomic moments disrupted by thermal jostling. This disruption increases as the temperature increases; at 770 degrees Celsius thermal effects are so great that a piece of iron loses its ability to align the moments of neighboring atoms entirely in the absence of a magnetic field. Every ferromagnet has such a point, called the Curie temperature after the French physicist Pierre Curie who first explained this property.

Writing with Magnetism

The ability to influence the alignment of magnetic moments by applying a magnetic field forms the basis of magnetic recording. By applying a strong, localized magnetic field to one region of a thin layer of a ferromagnetic material such as iron or its magnetic oxide magnetite, we can create a domain in which all of the magnetic moments of the iron atoms are aligned (provided that the temperature is below the magnetic medium's Curie temperature). Applying the field elsewhere in the opposite direction, we can form a domain of opposite alignment. If we take one alignment direction to represent the binary digit "1," and the opposite alignment to represent "0," we can write into the magnetic layer a pattern of 1's and 0's—binary data, in other words. For practical reasons that will become clear shortly, in current magnetic storage media based on thin magnetic films the two different alignment directions are both parallel to the plane of the film. Alignment perpendicular to the plane is also possible in principle, and has potential advantages in terms of storage density, as we shall see; but it is difficult to achieve for technical reasons.

The earliest thin-film magnetic recording media, developed in the 1930s, used films of iron particles stuck to paper tape. But iron oxide soon became the preferred material—not only does pure iron go rusty, but in the form of a fine powder it is highly flammable. The early tapes developed by BASF used the plastic cellulose acetate in place of paper; today other plastics are used as the support for the magnetic particles. The information is written onto the tape by a recording head, which generates a magnetic field that varies the direction of alignment of the magnetic moments within the particles (the magnetized domains are bigger than the size of the particles themselves).

This writing process uses the phenomenon of electromagnetic induction. The information is fed into the recording head in electronic form—it might be, for example, audio information from a microphone, or visual (video) information from a television receiver. In most recording devices currently in use, the recording head contains an induction coil: an electromagnet, in which a coil of wire surrounds a core of magnetic material such as iron. The electric current circulating around the coil induces a magnetic field in the core; reversing the direction in which the current flows reverses the orientation of the induced magnetic field.

The core consists of a ring with a small gap. When the core is magnetized, a magnetic field bridges the gap. This field can be thought of as a series of *field lines*, or lines of *magnetic flux*, which can be made visible around a bar magnet by scattering iron filings around it—the particles become arranged into lines running from one pole to the other. The flux lines spread out somewhat as they pass from one side of the gap to the other, and so they can pass through a magnetic tape or disk lying just below the gap (fig. 2.5). The direction of the magnetic moments in the recording medium will align themselves with the direction of the field across the gap in the recording head. You can now see from figure 2.5 why the direction of alignment is parallel to the plane of the recording medium.

Thus, as the recording medium is pulled past the recording head, a series of current pulses in one direction or another will write into the medium a series of magnetized domains of differing orientation. To read this information back out, the same process is used in reverse. The magnetized regions of the recording medium also have associated magnetic flux lines which pass from one "pole" to the other (fig. 2.6). (Notice that the direction of the field lines of this so-called *demagnetizing field* is opposite to the direction of the magnetization.) These flux lines can induce a magnetization in the core of a readout head more or less identical to the recording head. Every time the direction of this induced magnetization changes (that is, every time the boundary between one magnetized region of the

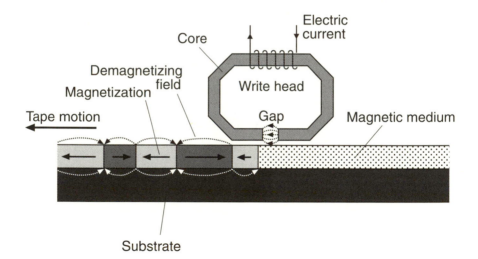

FIGURE 2.5 In conventional magnetic recording, regions within a magnetic film deposited on a substrate (a rigid disk for magnetic hard disks, a flexible plastic tape for magnetic tape) are magnetized in different directions by an electromagnetic head. The direction of magnetization between the poles of a gap in the head is determined by the direction of the current that flows through the electromagnetic induction coil wound around the head's core. This in turn determines the direction of magnetization in the storage medium, because the magnetic flux lines across the gap penetrate this medium. Data is thus encoded in binary form as a series of domains of differently oriented magnetization. A "demagnetizing field" that opposes the magnetization of a given domain is set up between the oppositely oriented domains on either side (see text); unless the resistance to reorientation (the coercivity) of the magnetic medium is large enough, this field will flip the orientation of a domain so that it is aligned with its neighbors.

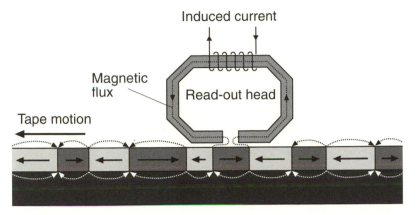

FIGURE 2.6 The readout process is essentially the reverse of the write process. As the tape passes below the readout head, the core experiences a changing magnetic flux each time a boundary between domains passes beneath the gap, owing to the demagnetizing fields that extend beyond the magnetic medium itself. This changing flux induces an electric current in the induction coil of the head.

recording medium and another of opposite orientation passes by the readout head), a current is induced in a coil of wire wound around the core of the head. Note that this induction of electric current requires a *change* in the magnetic field—the mere presence of a constant magnetic field in the core will not have this effect.

For magnetic tape devices, the tape is pulled past the recording head. But storing information in the essentially one-dimensional format of a magnetic tape has the drawback that one has to trawl through long stretches of tape in order to find a specific piece of information. The two-dimensional format of a disk allows for much more efficient writing and retrieval—the write or readout head can be moved rapidly to the relevant part of the disk. In modern computer technology, the magnetic disk has all but entirely replaced the magnetic tape. In disk drives, information is written and located both by spinning the disk and by moving the recording head across its surface. But the principles of the recording process are the same.

These principles involve physics that was well known at the end of the nineteenth century. But the practical requirements of today's top-range recording devices are awesome and are realized only thanks to the phenomenal capacity of modern electronic and mechanical engineering. Commercially available hard-disk storage devices can hold several billion bits of information in a square inch of magnetic recording medium (equivalent to a stack of typed pages about 100 meters tall). This density of information is useful only if it can be accessed on a realistic timescale, and modern readout devices can retrieve several million characters per second. This requires that the head skim over the surface of the disk at speeds of around 100 miles per hour, with jarring halts and reversals in direction. And all the while, the head sits poised only ten millionths of an inch above the recording medium. If this were a fairground ride, you wouldn't want to be on board!

Denser and Denser

My computer came supplied with an encyclopedia on CD-ROM that staggers me. It seems to have a limitless capacity to supply me with more than I need to know about every topic that I ask of it, and provides pictures and sometimes sound and even movies to boot. (The CD-ROM drive is in fact an optical, not a magnetic system, but the computer does not balk as I load this mountain of information into its magnetic hard drive.)

And yet I know that this is nothing compared with what will be available in just a few years, as storage densities and memory capacities become ever greater. What is it that is bringing about this explosion in data storage?

We saw above that data is stored in magnetic media as a series of regions magnetized in different directions. In principle, the storage density can be increased simply by making these regions smaller. But there is a limit to how far this can be taken, which is determined in large part by the detailed nature of the boundary between one region and the next. It is these boundaries that play the crucial part in data storage because, as explained above, the readout head produces a pulse of electricity only when the magnetization of the recording medium beneath it *changes*: in other words, when we pass from one magnetized region to the next.

The boundaries are not abrupt; they have a finite width, called the transition region. One can imagine increasing the storage density of the medium by reducing the width of the transition region, so that more magnetized regions can be fitted within a given area of material.

We can think of the transition regions as a kind of buffer between the magnetic poles of the magnetized regions. Because the poles of adjacent regions are oriented in different directions, poles that are alike abut one another on either side of the transition region (fig. 2.7). But there lies the rub: for, as is well known from experience with bar magnets, like poles repel one another. There is consequently a driving force for any given magnetized region to reverse its direction, so that its poles are aligned with those of its neighbors and opposite poles meet at the boundaries. In effect, one can think of the two neighbors on either side of a magnetized region in a track as setting up a magnetic field that opposes that of the central region and which thus has a tendency to reverse it. This is the demagnetizing field mentioned earlier.

If the transition region is made narrower, the poles of neighboring magnetized regions become closer together and the driving force for reversing the direction of magnetization becomes stronger. Whether or not this reversal actually occurs (thereby erasing data) depends on a property of the magnetic medium called its *coercivity*, which is a measure of the strength of the magnetic field required to flip the direction of magnetization. A material with a high coercivity will be more able

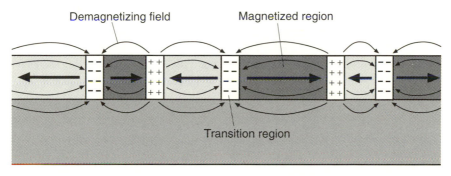

Figure 2.7 Between each magnetized domain in a magnetic storage medium there is a transition region in which the flux lines of the demagnetizing fields start and end. This transition region is a kind of buffer between the like poles of adjacent domains, which repel each other. The storage density of the medium is limited by the width of this transition region: if it is made too narrow, the repulsion between like poles can become so great that the demagnetizing field flips the orientation of a magnetized region to align it with those regions on either side.

to resist flipping, and so will be able to support narrower transition regions and higher storage densities.

The search for materials for high-density magnetic storage is therefore in part a search for materials with high coercivity. The coercivity of iron oxide (specifically, the form denoted $\gamma\text{-Fe}_2\text{O}_3$) is sufficient to make it still the mainstay of magnetic recording, but other, superior materials are now available. Chromium dioxide is used for improved sound fidelity on commercial audio cassettes, because it has a higher coercivity than iron oxide and therefore captures a more accurate record of the audio signal. If you really want exceptional sound quality, however, you can resort to metal tapes, which use fine particles of pure iron as the recording medium. The coercivity of pure iron is considerably greater than that of its oxide. The manufacture of metal tapes is a delicate process because, as indicated earlier, the metal particles are highly susceptible to oxidation (basically, to rusting, although this can manifest itself as inflammability for very small particles) and are hard to disperse evenly on the tape.

While these are the materials most commonly used today in magnetic disk and tape technology, more advanced media are required for significant improvements in storage density. The most promising new materials are continuous thin films of metal alloys, particularly cobalt–nickel alloys. Whereas recording media based on metal oxide particles are produced by dispersing the small particles within a binding matrix, thin-film media are made by depositing continuous films of the alloy on a rigid substrate—the resulting films are a mosaic of small crystals of the alloy, packed intimately together. These films are laid down either by a chemical plating process, in which the substrate is immersed in a bath containing the film medium, or by so-called sputtering, where high-energy ions accelerated by an electric field are used to knock atoms or clusters of atoms from a lump of the magnetic medium; these vaporized particles are then deposited on the substrate.

Because these manufacturing processes are rather complicated, continuous thin-film recording media are more expensive than particulate tapes. But the coercivity of the films is much greater than those of the iron or chromium oxides used in particulate recording media, so they offer much improved storage densities. Rigid disks made of cobalt–nickel thin films on an aluminum substrate have now captured over half of the hard-disk market.

Head Start

The properties of the magnetic recording medium are not the only factor that determines the storage density or data-handling speed of magnetic devices: better memories are also afforded by changes in device design, and in particular in the nature of the write and readout heads. As I mentioned above, the head of a hard-disk drive operates in frightening proximity to the rapidly spinning disk; a slight misalignment, or contamination by dust or smoke particles, can send the head crashing into the disk, destroying the data. Yet the size of the magnetic signal that a readout head can detect from a recording medium, or that a write head can induce in the medium, is governed by the gap between them. A smaller gap means that the readout head would register stronger signals, allowing for faster readout without the risk of error. So engineers are engaged in a delicate game of bringing the heads as close to the medium as they dare without incurring an unacceptably high risk of crashes. Often the recording medium carries an overlayer of a lubricant to reduce mechanical wear of both the medium and the head. The problem of wear also makes hardness a desirable property in the recording medium, and represents a drawback in the use of pure iron particles. Although iron metal's coercivity is higher than that of iron oxide, it is softer and more susceptible to wear.

To obtain a higher storage density, the obvious strategy is to reduce the size of the magnetized domains. As we saw above, the limit here is set by the magnetic medium's coercivity, but to approach this limit entails a reduction in the gap width of the head, so that smaller regions can be magnetized and detected. By using techniques developed for patterning semiconductor microelectronic devices, head gaps of about half a micrometer can be achieved. There is also a trend toward thinner heads themselves, allowing the width of data tracks on disks and tapes to be reduced.

A new generation of readout heads is now appearing that dispenses entirely with Faraday's principle of electromagnetic induction. Instead of relying on a change in magnetic field strength (at magnetized domain boundaries) to induce an electrical signal in the head, these devices couple magnetism to electricity via the phenomenon of *magnetoresistance*. The electrical resistivity of magnetoresistive materials varies markedly as its magnetization varies. When placed in a magnetic field (such as that created by an underlying magnetized medium), such a material experiences a change in resistivity, which can be registered as a change in the voltage across the material if a constant current flows through it. In this way, the information written into a magnetic recording medium can be decoded in the form of a variable voltage as the medium skims beneath a magnetoresistive readout head.

The advantage of a device of this sort is that magnetoresistive heads can be far more sensitive to changes in the magnetic flux of a recording medium than are inductive heads: a small change in flux can generate a rather large change in resistivity. Magnetoresistive heads can be made very thin, and so can be combined with inductive write heads in a single device by placing the magnetoresistive medium in the gap of the inductive loop.

Particularly promising for this new class of readout heads are the composite materials known as *magnetic multilayers,* in which alternating films of two different metals, just a few nanometers thick, are deposited on top of one another. These stacks of thin layers are generally called superlattices. Magnetic multilayers made from iron and chromium, and from cobalt and copper, show extremely large magnetoresistance—so large, in fact, that it is distinguished by the technical term "giant magnetoresistance." Iron–chromium superlattices can be made whose resistance in the presence of a strong magnetic field is just half that in the absence of a field (fig. 2.8). This behavior is a consequence of subtle magnetic interactions between the ferromagnetic iron layers: it turns out that the magnetic coupling between adjacent layers can be like that in an *anti*ferromagnet. This coupling, which is very sensitive to the thickness of the intervening chromium layer, leads to changes in the ability of mobile charge carriers to pass through the metal superlattice, and thus to changes in the electrical resistance.

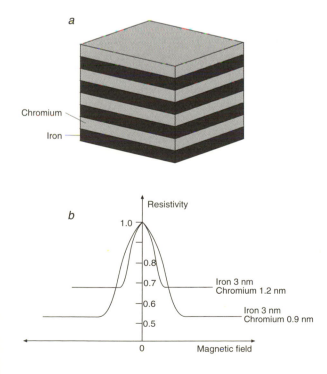

FIGURE 2.8 Superlattices of thin, alternating layers of iron and chromium (*a*) display the property of giant magnetoresistance: their electrical resistivity decreases dramatically in an applied magnetic field (*b*). The magnitude of the effect depends on the thickness and number of the layers. For a thirty-five-layer stack of 3-nanometer-thick iron films and 1.2-nanometer-thick chromium films, the resistivity decreases to around 67 percent of its value when no magnetic field is applied; for a sixty-layer stack with 0.9-nanometer chromium layers, the decrease is to about 54 percent.

Stand-Up Recording

The storage density of a magnetic medium is set by the point at which its coercivity is no longer large enough to prevent reorientation of recording domains by the demagnetizing fields that are set up between adjacent domains. As the domains get smaller, the transition region between two domains gets thinner and the reorienting effect of the demagnetizing field increases. Although materials with higher coercivities therefore support smaller recording domains and greater storage densities, they also require stronger fields in the write head to encode the domains in the first place. An obvious solution is to do away altogether with the uncomfortable end-to-end orientation of neighboring domains, and instead to orient the magnetic moments in head-to-toe fashion, perpendicular to the plane of the medium. Then, adjacent domains of opposite magnetization have their opposite poles, not their like poles, next to one another (fig. 2.9). This means that the demagnetizing fields actually get weaker as the data density increases.

Obvious, in theory. But if you take another look at figures 2.5 and 2.6, you'll see one reason why this idea has not more readily been put into practice: any usual design for an inductive write or readout head involves magnetization of the recording medium *parallel* to the plane of the film (called longitudinal recording). Perpendicular recording requires a considerable amount of rethinking of both the device design and choice of recording medium.

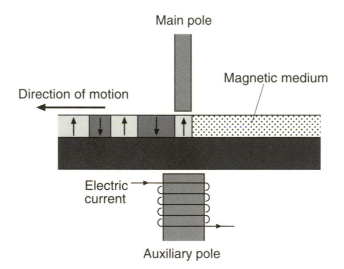

FIGURE 2.9 In perpendicular recording, the preferred direction of the magnetic moments in the magnetic storage medium is perpendicular to the plane of the film. Recording in such a medium can be achieved by means of two magnetizable heads, one on each side of the film. A signal-bearing current induces magnetization of the so-called auxiliary pole (here the lower head), and this in turn magnetizes the main pole (upper head), which lies close to the magnetic medium. The combined influence of both poles is sufficient to orient the magnetization of the storage medium.

In traditional particulate magnetic media based on iron or chromium oxide, the particles have a needlelike shape and lie with the needles parallel to the substrate surface. As the preferred orientation of the magnetization in these particles is along the needle axis, they are best suited to longitudinal recording. To make a medium for perpendicular recording, one needs to identify a material whose preferred magnetization direction is perpendicular to the substrate on which they are deposited. Polycrystalline, continuous films of a cobalt–chromium alloy have been explored for this purpose ever since the idea of perpendicular recording was first proposed by S. Iwasaki in 1975. These films, made by sputtering, contain column-shaped crystalline grains that stand up from the substrate, whose magnetization can be directed along the column axis. Much interest is also now being shown in the magnetic ceramic material barium ferrite, which forms platelike crystals whose preferred magnetization direction is perpendicular to the plates. Barium ferrite could be used as a particulate perpendicular recording medium, mixed with other materials to convey good wear resistance.

There is no unique solution to the head design for perpendicular recording. In one approach, two magnetizable heads are arranged one on each side of the recording medium. One head, called the auxiliary pole and lying on the lower side of the recording medium, is magnetized inductively by a coil of wire wound around it; this wire carries the signal that is to be recorded. The magnetic field generated by the auxiliary pole is not large enough by itself to flip the orientation of magnetization in the recording medium, but it *is* sufficient to magnetize the other head (the main pole, which lies very close to the magnetic film). Magnetic flux lines then pass from the lower auxiliary pole to the upper main pole, and the combined effect of these poles is enough to orient a domain in the recording medium (fig. 2.9). Despite the evident advantages in terms of storage density, the technical difficulties in implementing perpendicular recording have so far prevented it from finding major commercial applications.

THE NEXT WAVE

Conventional magnetic information storage is a technology designed to be compatible with electronic information processing. But when photonics replaces this means of handling data (as it surely will), information storage will have to adapt to suit the new regime. As we saw in the previous chapter, optical transmission of data has already replaced electronics, and data *processing* with light is steadily becoming a reality. If photonics is to be the information technology of the future, there will be ever more of a pressing need to develop ways of storing data optically.

One such approach has already sparked a revolution in data storage: the compact disk (CD), which is read by a laser beam. The advantages of CDs as memory banks are well rehearsed: they have high storage densities (a lot of information can be packed onto a disk), they are robust, and access to the data is very fast (because it is all laid out in front of you rather than being buried somewhere on a tape spool). One of the biggest disadvantages, however, is that it is not (yet) possible to erase and write over the information on a disk. This is not a great limitation for audio CDs—the last thing we want, after all, is to overwrite our precious recording of the Boston Symphony Orchestra or the Singing Balalaikas of the Ukraine—but for information technology at large it is a fatal flaw. A computer with no erasable memory is not a computer at all. A further drawback to CDs is that their readout speed is considerably slower than that of a magnetic system.

This is why hard and floppy disks for computers all still rely on magnetic storage media, which are erasable. But there is now a fast-developing recording and storage technology that represents a hybrid of the magnetic and the optical systems. In the magneto-optic system, data are still stored by magnetizing regions of a thin film of magnetic material, but both the write and the read operations are performed by light beams. At present this field remains in an early stage of development, commanding only a tiny fraction of the commercial market of conventional magnetic storage media. But those working on magneto-optic systems are confident that their advantages, particularly that of high storage density, will ensure that this market grows very rapidly in the years to come.

Magneto-optic storage uses perpendicular recording media, whose preferred orientation of magnetization is perpendicular to the plane of the film. The key to magneto-optic recording is the way in which the coercivity of the medium depends on temperature. As the temperature of the medium is increased, its coercivity drops, so that it becomes easier to alter the direction of magnetization with an applied field. During the writing process, the magneto-optic film is placed in a magnetic field that is too low, at room temperature, to flip the magnetization. A laser pulse from a semiconductor laser diode is focused onto a small region of the film, raising the temperature in that region far enough that the coercivity is too small to resist reorientation of the magnetization. The local temperature induced by the laser beam is typically around 300 degrees Celsius. The heated region therefore acquires a reverse magnetization (fig. 2.10), and as the film cools down after the laser irradiation ceases, this reverse magnetization becomes frozen in.

The same laser diode is used, at a much reduced intensity, to read the data back out of the magneto-optic disk. The laser light passes through a polarizing filter, which has the effect of confining the electromagnetic oscillation of the light to a single plane. When it is reflected from a magnetic medium, polarized light experiences a rotation in its plane of polarization, a phenomenon known as the Kerr effect. The direction of rotation—clockwise or anticlockwise—depends on the direction of magnetization of the magnetic medium. To deduce which way the plane of the light has been rotated, and thus whether the region on which the laser spot is focused has a magnetization that is up or down, a second polarizer is used to convert the change in polarization into a change in light intensity, which is then detected by a photodetector (fig. 2.11).

The width of the laser beam used for reading and writing can be focused very finely, to illuminate a spot on the recording medium less than one micrometer across; and the beam can also be guided very accurately onto specific locations on the disk. This means that the data can be written at high density into very thin

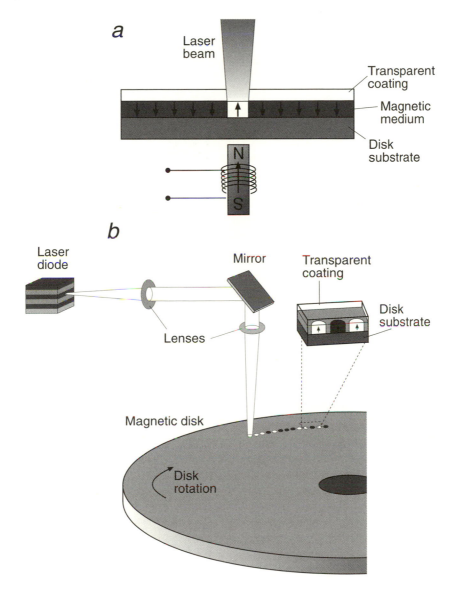

FIGURE 2.10 Magneto-optic memory devices use light to write information to, and read it from, a magnetic medium. The medium has a preferred orientation of magnetization perpendicular to the plane of the film. A magnetic field is applied in the opposite direction to the magnetization in the film, but at room temperature the coercivity of the magnetic medium is sufficient to resist reversal of the magnetization (*a*). When a laser beam induces localized heating of the medium, the coercivity is reduced and the applied film then flips the magnetization of this region. The magnetic film is deposited on a solid disk-shaped substrate, and is protected from corrosion by a transparent coating. The magnetic disk is spun beneath the laser beam, and the information is written into each radial track by a series of laser pulses (*b*).

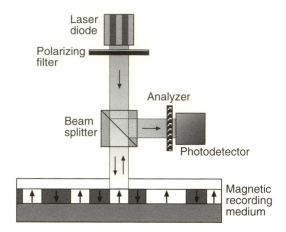

Laser diode

Polarizing filter

Analyzer

Beam splitter

Photodetector

Magnetic recording medium

FIGURE 2.11 To read out stored data in a magneto-optic memory, polarized laser light is shined onto the storage medium at low intensity. The interaction of polarized light with a magnetized material causes rotation of the plane of polarization (the Kerr magneto-optic effect). The direction of rotation depends on the direction of magnetization in the medium. A second polarizing filter (the analyzer) distinguishes between rotation in one sense and in the other in the reflected beam, converting this difference to a difference in intensity that is registered by a photodetector.

tracks on the disk, allowing magneto-optic systems to attain areal storage densities considerably larger than those of conventional magnetic media. Another advantage is that the magnetization of the recording medium is read out "remotely"—that is, via its influence on the plane of the polarized laser beam, rather than via the induction of a current pulse in an electromagnetic readout head. So the readout head of a magneto-optic system can be much further from the surface of the disk, removing the danger of crashes. This also means that the medium can be given a protective transparent coating, making it robust against degradation and allowing magneto-optic disks to be removed from the drive and handled in as cavalier a fashion as most of us treat compact disks.

On the other hand, the recording and readout head carries a lot of paraphernalia: the laser diode, the polarizers, lenses and mirrors for guiding and focusing the beam, and a mechanism that allows the lens to track the data on the disk. These parts can all be made incredibly small, but even so the head is a rather bulky affair compared with that of a conventional magnetic device. In consequence, the head cannot be shifted around so fast, and this is why the read and write speeds of magneto-optic systems are not yet competitive with those of magnetic systems. These drawbacks will surely be overcome, however, for instance by inventive head designs incorporating fiber optics.

One of the critical questions facing magneto-optic storage has been whether it can provide erasable and rewritable data storage. In principle, erasing data is straightforward: one reverses the magnetic field used for writing, and then reheats the entire magnetic disk by spinning it through the laser beam track by track; as the medium passes beneath the beam and its coercivity is lowered by the heat, it is realigned with the applied field. The restored blank disk is then ready to accept new data. But this need to erase old data before writing in new information is a serious drawback, since it means that overwriting requires two full passes of the disk below the laser beam, slowing down the process significantly.

One way of solving the problem would be to keep reversing the direction of the applied magnetic field as the data-filled disk is spun beneath the laser beam, in such a way that successive regions of the disk acquire different magnetization directions corresponding to a pattern that encodes the new data. But this requires that the applied field be modulated extremely rapidly, which is possible only by placing a small recording head very close to the surface of the magnetic disk. In doing so, one would lose one of the key advantages of magneto-optic recording, the fact that the recording head and the storage medium can be kept far apart.

A better way of introducing direct overwriting is being explored which eliminates the need for an applied magnetic field at all. Instead, it uses the demagnetizing fields between domains in the magnetic material itself to flip the magnetization direction of laser-heated regions. The recording medium then has to have either very strong demagnetizing fields or very low coercivity at high temperatures, so that the demagnetizing fields are strong enough to induce flipping without external help. To use this technique for overwriting data, a system of two laser beams is used: one to read out the existing information on the disk, and one to alter that information as and where necessary. The first beam is a conventional low-power readout beam, and the information that is read is then compared with that which is to be written in. If the medium presently encodes a "1" where the new data set requires a "0," the second beam is used to heat the disk at the appropriate point, whereupon the magnetization is then reversed by the influence of the intrinsic demagnetizing fields (fig. 2.12). This kind of approach is still under development, but it seems likely to provide magneto-optic recording with a direct overwriting facility in the near future.

Crucial to the evolution of these systems is the recording medium itself. This, as we have seen, must be a material whose coercivity varies rather markedly with temperature. The materials currently in use are alloys of transition metals, commonly cobalt and iron, with so-called rare-earth elements such as gadolinium and erbium. These alloys are deposited as amorphous (noncrystalline) thin films by the sputtering technique. The strong temperature dependence of the coercivity in these alloys is a consequence of the fact that the magnetic moments of the transition metals point in the opposite direction to those of the rare-earth metals. So there is a competition between these opposing moments; because the magnetization of the two different kinds of metal depends on the temperature to differing degrees, the balance in this competition is rather sensitive to temperature changes. At low temperatures, the magnetization direction of the alloy is governed by that of the rare-earth element, whereas at high temperatures the transition metal wins

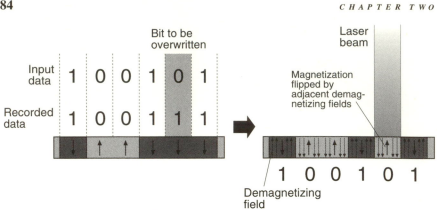

FIGURE 2.12 Direct overwriting in a magneto-optic storage device can be achieved by using the demagnetizing fields to reorient the magnetization of domains warmed by a laser pulse. Reorientation is effected selectively, by comparing the existing data pattern in the medium with the new input data stream: where these do not coincide, the medium is heated and the direction of magnetization is reversed.

out. Somewhere between these two extremes, the opposing moments balance exactly: this is called the compensation point, because the two moments are then compensated and the material has no net magnetization at all. Around this point, the coercivity becomes extremely high: it takes a huge magnetic field to reverse the magnetization direction, for the simple reason that the magnetization is so small that it scarcely interacts with the applied field. The coercivity decreases sharply to either side of the compensation temperature, which itself depends on the relative amounts of the two kinds of metal in the alloy.

Other materials are also being explored as magneto-optic recording media, particularly magnetic metal oxides such as garnets. These are iron oxides that also contain small quantities of other elements such as yttrium. They too contain magnetic atoms with opposed magnetization directions, and their compensation temperatures can be in the region of room temperature for certain compositions. Garnets doped with bismuth show much promise as recording media. Magnetic multilayers might find use here as well as in magnetoresistive recording heads. In particular, cobalt–platinum and cobalt–palladium multilayers are promising candidates for magneto-optic media: typically, the cobalt layers are just 0.4 nanometers thick, while the platinum or palladium layers are a little over twice that. These multilayers do not have opposed magnetic moments, and therefore have no compensation point; rather, they have a temperature-dependent coercivity that is simply high at room temperature and decreases as the temperature is raised.

The storage density of magneto-optic systems, like that of conventional magnetic systems, is limited by the minimum size of the magnetized domains that can be written in. But whereas in conventional magnetic storage this is determined by the read and write head design and by the intrinsic properties of the magnetic medium, in magneto-optics the limit is set largely by the extent to which the laser beam can be focused to a small point. This in turn depends on the wavelength of the laser light: the shorter the wavelength, the smaller the size of the focused spot. At present, magneto-optic systems can write data into domains of about 0.8 micrometers across, allowing a storage density of around ten million characters per square centimeter, using aluminum gallium arsenide lasers to provide infrared write beams. But it is hoped that as short-wavelength blue-light lasers become available, domain sizes half as small (and so with areal storage densities four times larger) will be possible.

Magneto-optics is still a nascent data storage technology. The latest systems can store up to 127 million bytes of data on removable three-and-a-half-inch disks, which will fit into the storage drives on standard personal computers. (For comparison, an optical compact disk can hold about 640 million bytes.) But these are still very much prototypes and command a market that is only a tiny fraction of that occupied by conventional magnetic hard disk drives. Nonetheless, this market is expected to grow rapidly in the coming years as the present limitations in magneto-optic recording are overcome and their potential advantages shine through ever more clearly.

Memories Sculpted in Light

By the time magneto-optic storage systems begin to realize their commercial potential, however, it seems likely that they will have a new competitor to contend with. Magneto-optics is, as we have seen, a kind of hybrid technology, a marriage of the new (photonics) with the old (magnetic data storage). But as the revolution in photonic information processing and transmission gathers pace, it seems increasingly natural to complete the picture by creating all-optical memories, in which light is used to write information directly into a light-sensitive medium, and to read it out again. Magneto-optics does do this in a sense, but only by exploiting the incidental fact that bright light imparts heat. In all-optical memories, in contrast, the aim is to use the fundamental interaction of electromagnetic radiation with a nonlinear optical medium to imprint a pattern of light directly into the memory material. The memory thereby retains the data as an image of bright and dark spots, like a black-and-white photograph.

One appealing aspect of optical memories is that they can be three-dimensional. Magnetic memories are flat, two-dimensional devices—disks or tapes. Not only is it possible to pack a greater density of information into three-dimensional memories; in addition one can retrieve blocks of information literally in a flash, so that for example an entire page can be read out at once rather than character by character as in conventional two-dimensional storage media. But if magneto-optics is a nascent technology, we'd have to describe 3D optical memories as embryonic. Laboratory prototypes abound, but commercial devices are still some way off. Given the potential of this new technology to boost data storage capacity one thousandfold, however, it is no surprise that computer companies such as IBM and Rockwell are now devoting a great deal of attention to it.

The 3D Memory

Most research on optical memories involves the use of holography to store information in a block of photosensitive material. The science of holograms was developed in the 1940s by the Hungarian physicist Dennis Gabor. He realized that if a beam of coherent light, like that produced in a laser, is split into two, one half is shined onto an object and the scattered beam intersects the other pristine beam, then the two will interfere to form a spatial pattern of light and dark that is a three-dimensional image of the object. If this interference pattern could be recorded in some photosensitive medium like a photographic film, Gabor pointed out that the three-dimensional image would then reappear when this medium— the hologram—is illuminated with the same kind of coherent light. In other words, all the "information" representing the three-dimensional object is stored in the hologram, ready to be deciphered by a laser beam.

But lasers were not yet invented in the 1940s, so it was not easy to test this idea experimentally. (Gabor was in any case more interested in the possibilities of the effect for improving the performance of electron microscopes.) When lasers became available in the 1960s, others set about putting Gabor's ideas into practice. The fruits of those efforts now hang in exhibition halls and adorn credit cards: holographic images in which a solid 3D object lies apparently suspended within a flat plane, with peripheral features that become visible when you look at the image from different angles.

Technologists, however, were interested in more profound applications. Pieter van Heerden, working for the Polaroid company, was the first to propose (in 1963) that holography could be used to imprint information into a three-dimensional block of photosensitive material, in the form of an interference pattern in which bright regions could represent the 1's, and dark regions the 0's, of binary data. In the early 1970s, a group at RCA Laboratories showed that the idea would work by storing five hundred simple holograms in a photosensitive crystal of lithium niobate. But holographic memories then languished for almost two decades, because memories based on semiconducting and magnetic materials looked far more promising in the short term. Interest was revived in 1991 when Fai Mok, working at the California Institute of Technology, showed that high-resolution photographic images could be stored and retrieved in holographic media. Given the impetus of the photonics revolution and the benefits of new technologies for manipulating light, holographic memories are now very much again in the limelight.

They work as follows. The interference pattern that imprints the hologram into a photosensitive medium is set up when a laser beam carrying the data (the write beam) intersects with a reference beam passing more or less at right angles through the medium. A page of data consists of a pattern of bright and dark spots imprinted into the write beam, like a checkerboard. Each bright bit of this pattern sets up an interference pattern with the reference beam throughout the whole of the holographic medium. To read the data out, the reference beam is passed back through the holographic medium at the same angle, and the interaction between this laser light and the image imprinted in the recording medium (an interaction called *diffraction*) reconstructs the pattern of light and dark spots in the original write beam, setting up an optical signal that can then be converted to electronic form by photodetectors (fig. 2.13).

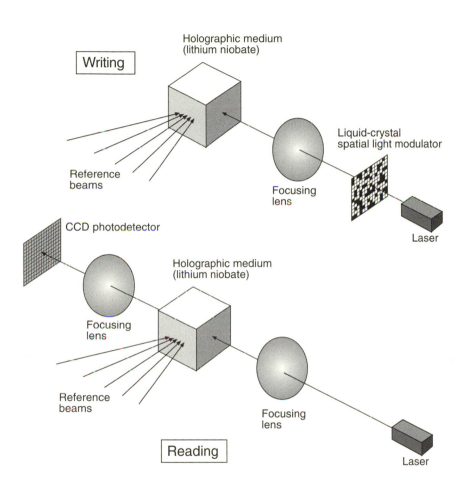

FIGURE 2.13 Holographic memories store information optically in a nonlinear optical material. The information is encoded in a pattern of light and dark regions in the input ("write") laser beam—this pattern can be imposed by a pixelated mask (a spatial light modulator) in which each pixel can be made transparent or opaque. Typically, switchable liquid-crystal devices are used to achieve this spatial modulation of the input beam. Each data image is imprinted into the storage medium by intersecting the write beam with a reference laser beam: this sets up an interference pattern, manifested as a pattern of bright and dark regions within the medium which creates a corresponding variation in the optical properties (generally the refractive index) of the medium. Many images can be superimposed in the same medium by using reference beams at different angles. To read the information back out, a read beam is directed into the medium at the same angle as the reference beam used to write that information. The interaction of this beam with the modulated optical properties of the storage medium creates an output beam with the same pattern of light and dark as the input beam; this is registered by an array of photodetectors.

For data storage, the strength of this approach lies in the fact that many such holograms can be recorded in a single block of holographic material. Simply by varying the angle between the write beam and the reference beam, holograms are stored at different orientations in the medium, and provided that the angle between them is large enough, they will not corrupt one another. Therefore, many pages of digital information can be written into the memory by rotating the angle of the reference beam at each writing step. Commonly, a difference in the angle of the reference beam of just one thousandth of a degree is enough to distinguish two holograms, making it possible to store thousands of pages in a single block of material. For readout, different pages are selected by varying the angle at which the reference beam impinges on the memory medium. The memory then becomes, by degrees, like a photographic film subjected to multiple exposures—but one in which we can pick out each image just by looking at the film from a different angle. Ultimately the number of pages that can be stored is limited by the fact that the strength with which each hologram diffracts the readout beam decreases as the number of holograms sharing the same block of photosensitive material increases. Eventually the intensity of the reconstructed beams drops so low that the pages of data cannot be retrieved reliably—the noise (small random fluctuations in the intensity) starts to overwhelm the signal. Demetri Psaltis and colleagues at the California Institute of Technology have succeeded in capturing 10,000 pages in a holographic memory, corresponding to 100 million bytes of data (roughly speaking, a byte is equivalent to a single character), with only one bit in every 100,000 being misread on readout.

Transforming holographic memories from bench-top curiosities to practical devices is a matter of solving several independent problems. In particular, interfacing a device of this sort with conventional electronic data-handling hardware is no mean task. To write data quickly into the memory, the write beam must be modulated by a mask of pixels that can be switched rapidly between "open" and "closed." The preferred way of doing this is to use liquid-crystal technology: liquid-crystal pixels can be switched from opaque to transparent by electronic means. Such pixels are incorporated into checkerboard arrays called spatial light modulators.

A corresponding pixelated array of photodetectors is used to decode the intensity pattern of the readout beam (fig. 2.13). Modern photodetectors called charge-coupled devices provide a highly sensitive means of converting an optical signal back to an electronic one—these detectors are also used in astronomical telescopes to detect extremely faint galaxies.

In 1994, Lambertus Hesselink and colleagues from Stanford University reported the first demonstration of a system like this to store and read back information from a standard computer hard drive. Hesselink used a lithium niobate crystal, scarcely bigger than a sugar cube, as the holographic storage medium—the same material as was used in the first prototype memories in the 1970s, and in just about all others since then. His system's performance was not particularly impressive in comparison with a standard floppy disk memory—it held less than a fifth of the amount of information (about 245,000 bytes), took an hour to store it, and became corrupted after only a few readout cycles. But all of that was to be expected for a first shot. The important message of the Stanford work was that holographic data storage is now a real technology, compatible with off-the-shelf computer hardware. Hesselink's team has successfully stored all manner of digital data, including visual images and video data. And considerable improvements in performance should be possible with relatively straightforward technology: improving the optics system with antireflective coatings, for instance, or using electronic means to control the position of the readout beam (rather than, as in Hesselink's prototype system, rotating the holographic crystal itself). Relative to magnetic memories, holographic storage has a lot of catching up to do. But Hesselink firmly believes that it will find its way to the market.

Thoughts in Plastic

How does a holographic medium retain the imprint of an interference pattern? I have skirted this issue so far, implying that this medium is simply rather like a block of photographic film. And indeed, flat films of photographic emulsions were used in some early prototype holographic memories (and they continue to be the medium in which the photographic holograms one can buy in art shops are recorded). But their long-term prospects for optical memories were virtually zero. For one thing, they have to be developed to reveal and to fix the holograms. Not only is this inconvenient, but it rules out the possibility of erasure and overwriting. Furthermore, only a few holographic patterns can be superimposed within them before these start to interfere with one another.

An erasable holographic memory needs a material that will store a hologram quickly and reliably, at high spatial resolution and with no need for developing. The hologram must remain stable within the medium for long periods, but it must be possible to erase it and write over it when needed. The solution to this apparently rather demanding list of requirements comes in the form of so-called *photorefractive* materials.

These are another member of the diverse class of nonlinear optical materials introduced in chapter 1. In photorefractive materials the optical nonlinearity stems from the fact that light alters the distribution of electric charge within the material. Specifically, a stationary pattern of light and dark "fringes," such as is generated when two coherent laser beams interfere to form a holographic image, gives rise to a corresponding pattern of positively and negatively charged regions within the material. This pattern can stay fixed even when the light is switched off, for a length of time that depends on the material (obviously, for holographic memories one wants to choose a material in which the pattern stays put for as long as possible).

This redistribution of charge within a photorefractive material is the consequence of several processes. First, the incident light generates mobile charge carriers—electrons and holes—by absorption of photons. The charge carriers are usually excited from dopant atoms or defects within the material: as we saw earlier, these tend to have associated with them charge carriers in localized energy

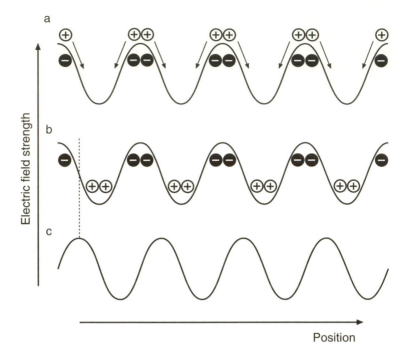

FIGURE 2.14 In photorefractive materials, the interaction of light with mobile charge carriers in the material sets up a modulation of the material's optical properties. The input signal is an interference pattern between light beams (a write and reference beam), which sets up a stationary pattern of bright and dark regions. The bright parts of the interference pattern create mobile charge carriers by photon absorption (*a*). These regions correspond to an intense electric field (recall that light consists of mutually perpendicular, oscillatory electric and magnetic fields), and the charge carriers move in this field such that opposite charges accumulate at the peaks and in the troughs of the interference pattern (*b*). This redistribution of charge sets up an internal electric field in the photorefractive medium which is a quarter of a cycle out of step with that of the optical interference pattern (*c*). If the charge carriers get trapped where they sit, for example by defects in the medium, the internal field remains when the illumination is stopped, and this creates a corresponding modulation of the refractive index of the material, which is a kind of replica of the original interference pattern. This refractive-index "grid" interacts with (diffracts) a readout beam to recreate the intensity distribution of the write beam.

levels, which can be boosted into the conduction band (for electrons) or the valence band (for holes) where they become mobile. The presence of an interference pattern formed from intersecting laser beams means that these charge carriers find themselves in an electromagnetic field that varies in intensity from place to place through the material. They will therefore flow down the slopes of this field, moving from the bright (high-field) regions to the dark (low-field) regions like water settling into the valleys of sand ripples on a beach (fig. 2.14). The charge carriers accumulate in the dark fringes, where they become trapped again (and therefore held in place) by defects in the crystal structure of the material. Thus, the interference pattern of the laser beams sets up within the material a pattern of charge that mimics it exactly.

The interference pattern therefore becomes written into the photorefractive material in the form of an internal electric field created by the migration and retrapping of charge carriers. The reason why this latter pattern is essentially a hologram is that the redistribution of charge causes a modulation of the refractive index of the material. A laser beam passing through the material will no longer see a smooth, homogeneous medium but a material whose optical properties vary from place to place, almost as if it were a composite of different materials. The effect is rather like the writing process in reverse: the laser beam is modulated by the variation in refractive index to give an output that reproduces the original write beam—a highly nonlinear optical response.

The photorefractive materials currently most favored for holographic memories are inorganic crystalline materials belonging to a class called ferroelectric oxides, which undergo changes in structure when placed in an electric field. By far the most common medium is lithium niobate, generally doped with small amounts or iron; but barium titanate has also been tried. But these inorganic materials are not ideal from a technological perspective. As I indicated earlier, they are difficult (and therefore expensive) to grow in perfect crystalline form—a single crystal of lithium niobate can cost several thousands of dollars. Because of their brittleness, subsequent processing is difficult too. Moreover, they offer little versatility: one cannot easily and systematically tailor their optical properties to specific applications. For instance, iron-doped lithium niobate is sensitive to green light, which means that it can be used only in conjunction with green or blue-green lasers. These are relatively costly and until very recently they could not be manufactured in miniaturized form from semiconductors; instead, researchers were forced to use bulky argon gas lasers for holographic recording. This particular drawback has been somewhat ameliorated recently by the introduction of lithium niobate crystals doped with both iron and cerium, which are sensitive to red light (like that produced by semiconductor laser diodes) rather than green. But there are still plenty of reasons to expand the repertoire of photorefractive materials for holography.

Probably the most promising alternatives are photosensitive organic polymers. The Du Pont company has developed a so-called photopolymer that undergoes chemical changes when exposed to light, which can be used for read-only memories (ROMs) but—because the chemical changes are irreversible—is no good for erasable data storage devices. This material entered the commercial market as a holographic medium in 1994. But more promising in the long term are polymers that show the photorefractive effect. Although the first of these was created only in 1990, some are now known that have a photorefractive performance equal to that of the best inorganic materials. It is likely that they will form the "brain" of commercial memory devices in the future.

There are several functions that an efficient photorefractive material must be able to fulfill; the beauty of the polymer approach is that one can build these functions into the material in a modular fashion. The material must first produce charge carriers—electrons and holes—in response to irradiation with light. There must be a transport medium to allow these charges to migrate through the material so as to collect in the peaks and valleys of the interference pattern. Once there, the charge carriers must be trapped so that the pattern stays fixed even when the light is switched off. And the material must exhibit the electro-optic effect, whereby the internal electric fields set up by the uneven distribution of charge causes a change in the refractive index, so that the material provides a grating to diffract a readout beam and reconstruct the hologram.

To create a polymeric material the meets all of these needs, researchers simply include separate "functional groups" in the material's molecular structure, each of which is responsible for a particular task. Groups that provide charge generation, for example, will be chosen for their ability to absorb light at the appropriate wavelength and to use the energy to spit an electron into the conduction band of the material, creating an electron–hole pair. To effect charge transport, one can either use discrete functional groups between which charge carriers can easily hop, or provide a continuous conductive pathway along the backbone of the polymer molecules. These various functional groups can be either attached to the polymer chains or interspersed as separate units between them, like peas among spaghetti (fig. 2.15). Building polymer molecules which have several different kinds of monomer groups alternating regularly or at random along the chains is now a standard practice in polymer science; such materials are called copolymers (see chapter 9).

The first photorefractive polymer, prepared in 1990 by a group led by William Moerner and Robert Twieg at IBM's Almaden Research Center in San Jose, California, employed the first of these approaches. That is to say, the polymer molecules themselves served to provide just two of the necessary properties: to supply charge carriers on irradiation, and to display the electro-optic effect once the internal electric fields had been set up. The compound was a polymeric ether to

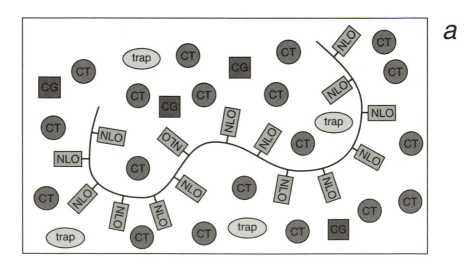

a

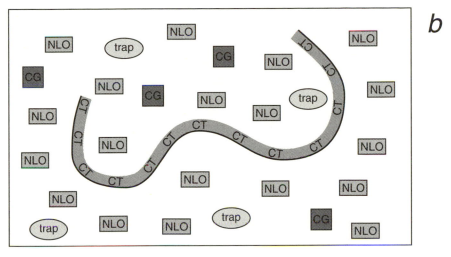

b

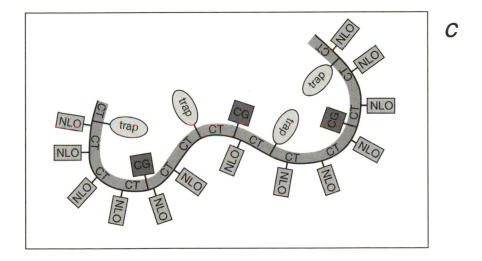

c

FIGURE 2.15 Photorefractive polymeric materials can be created in several different ways. The materials must have four components: nonlinear optical chromophores (NLO), which absorb light; charge-generation groups (CG), which produce photoexcited charge carriers when energy is transferred to them from the NLO groups; charge-transport groups (CT), which enable the charge carriers to move (this can occur either by transport along a continuous pathway or by hopping between discrete CT sites); and trapping sites (trap), which immobilize the charge carriers once they have migrated in the electric field of the illumination pattern. One approach is to disperse discrete molecules that perform some of these functions in a matrix of a polymer that performs the others. For instance, the polymer might contain NLO chromophores attached to the backbone (*a*), or it might be an electrically conducting polymer (see chapter 9) whose backbone provides the charge-transport pathways (*b*). Alternatively, all of the functional groups can be attached to the backbone of the polymer (*c*).

which were attached chromophores—chemical groups possessing the required nonlinear optical and photochemical properties. To effect charge transport, the IBM team doped this material with an organic compound used for the same purpose in photocopiers and laser printers. They hoped that trapping of the charge carriers would take care of itself, owing perhaps to inhomogeneous regions in the material. This material did indeed show photorefractive behavior, but only just. One measure of the photorefractive performance is the diffraction efficiency, which quantifies how much of the light in a readout beam is diffracted by a holographic pattern written into the material. For most applications, diffraction efficiencies of at least 10 percent are needed, whereas the IBM material gave a response one thousand times smaller than this.

Subsequent improvements to the design of the functional groups have boosted the diffraction efficiency of photorefractive polymers to well above the 10 percent threshold. One of the most promising materials is based on a polymer with an electrically conducting backbone, called poly(N-vinylcarbazole) (PVK). At the State University of New York in Buffalo, a group led by Paras Prasad used this compound as the basis for a material whose photorefractive behavior could be controlled with an electric field. They attached to the PVK chain chromophores that provided electro-optic activity, and then doped the polymer with C_{60}, soccer-ball-shaped molecules made from sixty carbon atoms linked into a hollow sphere. These footballs provided the light-generated charge carriers. In most polymeric materials possessing chromophores with nonlinear optical behavior, the chromophores must be aligned during processing (while the polymer chains are still flexible) by applying an electric field. This alignment, called poling, is necessary to give the nonlinear optical behavior and is generally frozen in once processing is completed. But Prasad's team added to their material a compound that acted as a plasticizer, making the material soft even at room temperature. This left the chromophores on the PVK backbone free to wave around and randomize their orientation, so that the material did not show nonlinear optical behavior. Applying an electric field, however, caused all of the chromophores to line up in the field direction, and the material then became highly photorefractive, with a diffraction efficiency of more than 50 percent. The ability to control the photorefractive behavior in this way might be useful in light-based data processing as well as data storage.

A further improvement in performance was reported by Nasser Peyghambarian and colleagues at the University of Arizona in 1994. They used a PVK backbone and a different selection of chromophores and charge generators to make a polymeric material that could be poled at room temperature and which possessed a diffraction efficiency of 86 percent. This value was virtually as much as could be hoped for, because the remaining 14 percent of the light was lost not through inefficient photorefractive behavior but through absorption and reflection of the incident light.

The use of polymers to which *all* of the necessary functional groups are attached (fig. 2.15c) is being explored by Luping Yu and colleagues at the University of Chicago. This approach has the advantage that one can put all of the groups exactly where one wants them, and know that they will stay there; in contrast, adding some of these groups as discrete dopants can run into the problem that the various components of the system tend to separate from each other and aggregate with more of their own kind, a form of separatism known as phase separation. Yu's group has fabricated several multifunctional polymers that show photorefractive behavior, but sadly these materials also tend to absorb light rather strongly, thus reducing the diffraction efficiency.

One of the main problems still to be overcome in using photorefractive polymers in optical memories is that they are not particularly quick-witted: it takes too long to write holograms into the material, a critical factor in a technology for which speed is of the essence. Nonetheless, the advances that have been made in only a very few years bode well for what polymers have to offer in the future; already IBM has produced prototype plastic memories, and it seems likely that thin, flexible devices will soon be on the market.

A SALTY STORY

There is fierce competition among scientists and engineers to create optical memory devices that can store large amounts of information in a polymeric organic material, and this is sure to bring about rapid improvements in the optical properties of polymers. But one of the most sophisticated optical polymers known has been developed in an arena far fiercer: that of the natural world, where the pressures of natural selection present exigencies that are a matter of life or death. This material is a protein called bacteriorhodopsin, which enables certain kinds of bacteria to survive by extracting energy from sunlight.

Bacteriorhodopsin is found in the cell walls of bacteria of the species *Halobacterium halobium*, which grow in salt marshes where the concentration of salt is six times greater than that of sea water. In these environments, the temperature of the waters can sometimes reach as high as 66 degrees Celsius. Such warm, salty water is considerably corrosive, and the bacteria have had to develop a very robust set of tools for survival. They are also faced with the problem that the oxygen content of their surroundings can vary markedly, so that they have to adapt to both low-oxygen and high-oxygen situations. In the latter situation the bacteria use oxygen for respiration, but when oxygen is low they have to switch to photosynthesis, which is how plants obtain their energy. Bacteriorhodopsin is a protein that can perform the central process underlying photosynthesis—the conversion of light energy to chemical energy—under these harsh conditions. When irradiated, it pumps hydrogen ions across the bacterial membrane, setting up a chemical gradient that can then be tapped for energy. Within the bacterial cell wall the protein folds up into seven helical coils that pass back and forth across the membrane, providing a channel for hydrogen-ion transport (fig. 2.16).

Ever since the 1970s, when the optical properties of bacteriorhodopsin were first elucidated, it has been appreciated that these might be exploited in optical

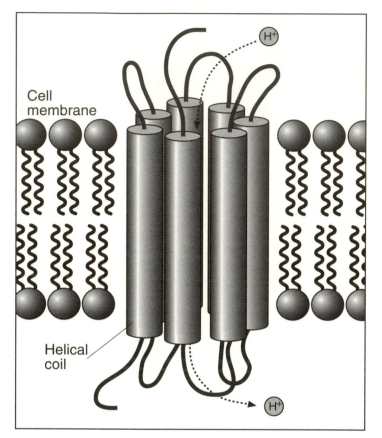

Figure 2.16 Bacteriorhodopsin is a protein molecule that effects transport of hydrogen ions across the cell walls of certain bacteria. The protein contains seven helical coils joined end to end, which fold up into a seven-helix bundle within the cell wall. The protein absorbs green light strongly, and uses the energy to pump a hydrogen ion across the membrane.

devices. And because the protein comes from such a tough neighborhood, it will not suffer from the problem of fragility that hinders technological applications of many other biological materials. During the 1970s, Soviet scientists led by Yuri Ovchinnikov in Moscow worked under great secrecy to develop optoelectronic devices using bacteriorhodopsin. Many of the details of this initiative, given the splendidly Cold War–style name of Project Rhodopsin, remain murky—if not in fact still classified as military secrets—but it seems that the fruits included a microfiche film called Biochrome and optical data processors. In these thankfully less paranoid times, bioelectronic devices made from bacteriorhodopsin find their way to the light of day.

    Underlying all of these devices is the idea that bacteriorhodopsin can act as a two-state switch, since by absorbing light it can be transformed into a new, long-lived state with different optical properties. The pressures of life have allowed the protein to evolve into a highly efficient light absorber over a wide range of wavelengths. The protein undergoes a series of changes in electronic structure—a so-called photocycle—when it absorbs light. The normal "resting" state of the protein in the bacterial membrane absorbs light strongly in the green part of the visible spectrum. But decay of the first photoexcited state (the primary photoproduct) back to the resting state is a far from simple matter, occurring in a complex sequence of steps involving many intermediate states, not all of which are fully understood or characterized. Each of these intermediate states has its own absorption spectrum—each absorbs light optimally at a different wavelength— and an associated characteristic lifetime before it decays to the next intermediate state (fig. 2.17). In effect, the protein changes color each time it makes a step from one state to the other.

The rate at which the primary photoproduct, reached by absorption of green light, decays through the cascade of the photocycle back to the resting state depends on a number of factors, and in particular on the temperature of the protein's

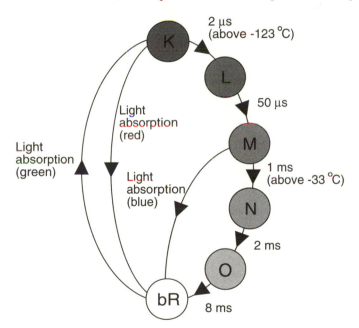

FIGURE 2.17 When it absorbs green light, the bacteriorhodopsin molecule is excited to the so-called **K** state (the primary photoproduct), which then loses its energy and returns to the resting state (**bR**) in a series of steps known as the *photocycle*. Each of the intermediate states in the photocycle has a characteristic lifetime, which depends on temperature. (Here ms is milliseconds and μs is microseconds.) Decay of the **K** state can be arrested by cooling the protein to below minus 195 degrees Celsius in liquid nitrogen. The photocycle can be frozen at the **M** state, meanwhile, by cooling to just minus 30 degrees. Memory devices have been constructed from cooled thin films of bacteriorhodopsin in which selected regions are switched between the resting state and the **K** or **M** states.

environment. At room temperature this decay occurs in a fraction of a second, but at lower temperatures it can be arrested at various intermediate stages, which can remain stable for long periods. This means that the optical properties of the protein can be switched essentially irreversibly by irradiation at low temperatures. In other words, irradiating specific regions of a cooled film of bacteriorhodopsin molecules makes the film transparent to certain wavelengths of light but opaque to others, a property that can be exploited in memory devices when a second beam is used to probe the film.

It is necessary to cool bacteriorhodopsin in liquid nitrogen to minus 195 degrees Celsius in order to trap the protein as the primary photoproduct (called the **K** state); otherwise the photocycle proceeds to subsequent stages. Trapping in the **K** state has been used in some optical devices, which have the advantage that the switching speed between the ground and **K** states is about a million times faster than that of conventional semiconductor devices—but liquid-nitrogen cooling is not an option for practical optical memories. Only a moderate degree of cooling (to around minus 30 degrees Celsius) is necessary, however, to trap the excited molecules in a later intermediate state of the photocycle called the **M** state. This absorbs light almost entirely in the spectral region where the resting state does not (that is, in the blue/violet part of the spectrum), a happy coincidence that makes the **M** state and the resting state virtually transparent in the regions where the other absorbs strongly. In addition, the refractive index of the **M** state is very different from that of the resting state, enabling films of bacteriorhodopsin to act as holographic storage media. Illuminating a film with an interference pattern in green laser light will switch the brightly illuminated regions to the **M** state; this sets up a modulation of refractive index in the film which acts as a diffraction grating to re-create the hologram when exposed to a readout beam. Diffraction efficiencies of around 3 to 8 percent can be obtained, which are low compared with inorganic photorefractive materials but serviceable nonetheless. Moreover, the protein can withstand over a million write/read cycles without a degradation in performance.

Norbert Hampp and colleagues in Munich, Germany, have used this behavior in holographic memories based on bacteriorhodopsin. Hampp's group has demonstrated that these devices can be used for pattern recognition—for matching a fragment of an image to the respective complete image among a galley of others, or for locating particular details of a picture within the whole. For example, the researchers have used bacteriorhodopsin in a holographic device called an image correlator, which compares two complex optical signals and identifies the location of coincidences between them (fig. 2.18). A memory device can be constructed on the same principle, which superimposes an input image over a whole series of other images stored holographically and searches for the maximum degree of correlation. This kind of memory, called an associative memory, possesses a facility for recognition akin to that of the human brain.

A new kind of bacteriorhodopsin memory architecture, which operates without the need for *any* cooling below room temperature, was made possible by a discovery made in 1994 by Robert Birge and coworkers at Syracuse University in New York. They found that the cascade of excited states in the normal photocycle of

FIGURE 2.18 A holographic image correlator compares two optical input patterns and searches for correlations between them. Norbert Hampp and co-workers have made an image correlator from bacteriorhodopsin, which they demonstrated using this drawing by Wilhelm Busch. The complete drawing comprised one input signal, and the other input consisted of fragments of the image: the child and the old man's right foot. The holographic device compared the images and identified a clear correlation between the fragments and one part of the full drawing. The output signal was a kind of spatial map of the correlations, in which a bright dot appeared in the region in which the correlation was identified in the full drawing. This kind of device would enable rapid searches of data sets to be made for a particular pattern—for example, to execute a word search in a text. (Images courtesy of Norbert Hampp, University of Münich, Germany.)

the molecule can be diverted to a new branch by kicking the molecules with another beam of light at the right moment. The protein molecules will sit in the new branch virtually indefinitely (for several years, at least) unless given a further optical kick back to the ground state (fig. 2.19). The point of diversion from the

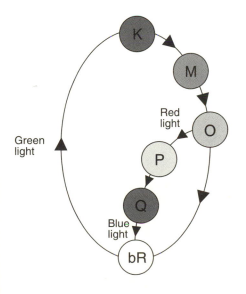

FIGURE 2.19 A room-temperature bacteriorhodopsin memory can be made by inducing a branching reaction on the photocycle when it reaches the **O** state, which happens a few milliseconds after the initial excitation with green light. If the **O** state is irradiated with red light, it can be switched to the **P** state along a new branch of the cycle; this decays rapidly to the **Q** state, which is stable for years. Blue light will convert the **Q** state back to the resting state (**bR**). So the switch from **bR** to **Q** can be effected by consecutive pulses of green and red light, with an interval of a few milliseconds between them; and the switch from **Q** back to **bR** can be effected with a pulse of blue light. These switching processes can be used in a three-dimensional memory.

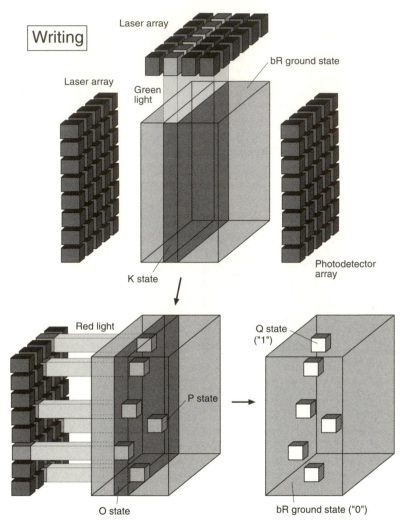

normal photocycle comes after the excited molecules have decayed from the **K** state down to the **M** state and thence to the so-called **N** and **O** states. This decay process takes around two thousandths of a second. The **O** state absorbs red light strongly, and this absorption takes it onto a new cascade of states, first to a **P** state and then to a **Q** state, where it stops.

So a memory device in which bacteriorhodopsin molecules are switched between the stable resting state and the long-lived **Q** state can be operated by first illuminating the films of the protein with green laser light, and then two milliseconds later delivering a pulse of red light to divert the molecules to the **Q** state. This switching process allowed Birge to develop a "paging" process to write and read information in bacteriorhodopsin molecules dispersed in a block of a polymer (fig. 2.20). His approach uses two crossed laser beams—one green, one red—to write data into the material. As switching occurs only where the two

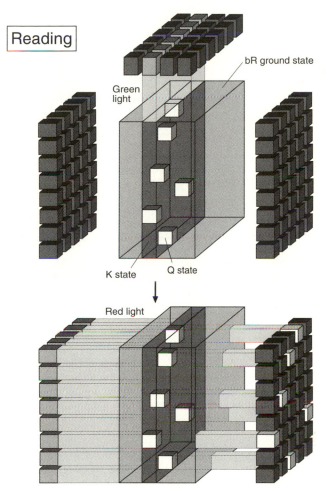

FIGURE 2.20 In this three-dimensional memory unit, binary data are stored in pixel cubes of bacteriorhodopsin, which is either in the resting state (**bR**), representing a "0," or in the **Q** state ("1"). Data are written in by a laser array one page at a time (*left*). First, a "page" of the memory medium is activated by green light, converting this slab of **bR** to the **K** state. When enough time has elapsed for this to have decayed to the **O** state, a red-light laser array writes the data into the activated page. The resting-state molecules are transparent to red light, but where the light encounters the **O** state it is absorbed and the bacteriorhodopsin molecules are converted to the **P** state. This decays rapidly to the **Q** state, which remains stable while the rest of the activated page completes the photocycle and returns to the ground state. This leaves the page with an imprinted pattern that has different optical properties to the rest of the storage medium.

To read the data back out (*right*), the page is first reactivated by the green-light laser array. All of those memory elements in this page that are *not* in the **Q** state absorb the green light and are switched to the **K** state. When these elements have reached the **O** state, the entire red-light array lights up. The **O**-state elements absorb the red light, but the **Q**-state elements are transparent to it—so the light passes through those elements and "lights up" the corresponding photodetectors on the other side, reconstructing the original data page in the photodetector array.

beams cross, specific small volume elements can be addressed within the three-dimensional memory. Although the memory is not holographic, it allows entire pages to be read out in a single operation, making the speed of the write and read processes comparable to that of conventional semiconductor memories. More-over, the speed increases as the memory gets larger in size, because that means that more data can be held, and simultaneously read out, in a single page. This way of handling data is called parallel processing. And because these memories are three-dimensional, their storage capacity is potentially much greater than that of conventional 2D devices—a 300-fold increase in capacity should be achiev-able. Birge hopes to develop these prototype memories into a miniaturized form compatible with existing computers in just a few years. He envisages a future generation of hybrid computers—made partly of inorganic semiconductors and partly of protein-based memory chips, which will be able to store more data in less space and to carry out a wider range of operations, like pattern recognition and associative memory processing. This kind of flexibility will be essential for the development of machines that show artificial intelligence.

But can these organic memories, made of even a relatively robust protein, truly compete in terms of economics, stability, and resilience with the magnetic media of conventional erasable memories? Making the bacteriorhodopsin itself is straightforward enough—one can grow cultures of the bacteria in fermentation vats and then extract the protein from the bacterial membranes by using centrifu-gal techniques. The protein is extracted in the form of a purple membrane—essentially a close-packed film of bacteriorhodopsin molecules. It can be stored in sucrose solution for many months without degrading. Birge's team has been able to accelerate the production rate of this membrane by growing genetically altered bacteria that produce ten times as much of the protein as those growing in the wild. (When applied to animals or plants grown for consumption, this kind of genetic modification carries a whole host of ethical and environmental questions; but it is less clear that the same issues apply to bacterial fermentation.)

The purple membranes can be deposited as thin films on substrates, or they can be incorporated into an inert polymer matrix for making three-dimensional mem-ory units. Such polymer-encapsulated proteins should be cheap and stable, and it should be possible, says Birge, to fabricate memory cards no thicker than optical disks from the material. He estimates that the cost per bit of memory could be around four or five times lower than that of current disk-based data storage units. So perhaps protein will turn out to be healthy for computers too.

Clever Stuff

SMART MATERIALS

> Machines do not grow, change, or metabolize of their own accord.
>
> —Kisho Kurokawa, *The Philosophy of Symbiosis*

Materials can replace machines. Substances that change their shape or properties in response to various stimuli—electrical signals, light, sound waves—can be used as switches and valves with no mechanically moving parts. These smart materials might find uses ranging from vibration dampers to artificial muscles. Coupling them to sensors and microprocessors will lead to intelligent systems that adapt their properties to their environment.

MOST OF US who saw *Terminator II* probably had the feeling that Arnold Schwarzenegger was outclassed; but it would take a materials scientist to tell you why. The distinction between Schwarzenegger's bioelectronic Terminator and his liquid-metal opponent is that between the old style of engineering and the way of the future promised by a new breed of materials called "smart" materials. True enough, the Terminator was far and away a more sophisticated piece of engineering than we could manage today, with its apparent amalgam of flesh and blood over a metal robotic superstructure; but the metal villain was something else again. It seemed to have no mechanical parts at all; rather, it was a seamless blob of metal that could heal itself, change its color and shape at will, and mimic any object. This was a material that was more than smart—it was diabolic.

That movie demonstrated more graphically than can any words why smart materials are revolutionary. If Schwarzenegger lost a hand, he couldn't simply grow another. His shape was fixed. He had moving parts, which could seize up or break down. There was a lot of scope for malfunction. His opponent seemed, in contrast, capable of almost anything.

Smart materials can be thought of as materials that replace machines—in other words, materials that carry out tasks not as a consequence of signals or impulses passed from one component to another, as the transmission of a car passes the power of an engine to the wheels, but as a result of their intrinsic properties. Consider a smart valve which shuts off when the flow of a fluid through it exceeds a certain rate. An old-style valve might include a flow meter that measures the flow rate, connected to a mechanical device that moves a barrier across the channel once the meter reads a certain value. The smart valve is no more than a tube through which the fluid passes; when the flow exceeds the critical value, the material of the tube expands until it pinches off the flow.

"Smart" is a loaded term, however. The *Terminator II* villain was smart in a way that smart materials are not (and, I suspect, never will be). He was fully able to process input data, to make decisions—he was intelligent, and apparently conscious. I've no idea how a liquid metal was able to acquire such acuity, and I'm sure that the movie makers had no idea either. We should not imagine that a smart material will engage us in dialogue—at least, not yet. There is no thinking involved in the behavior of a smart material, not in the normal sense of the word. A computer is smarter than the best smart material, and we remain, so far, smarter still.

But this is not to say that "smart" is a bad or an inappropriate term. It requires a word as dramatic as this to convey the striking difference between the old-style dumb materials and the new. Smart materials do not stand to the mercy of the elements in the way that, say, a concrete building or a Ford Mustang does, crumbling or rusting as the years pass. They respond to their changing environment.

Smart materials represent the epitome of the new paradigm of materials science, whereby structural materials are being superseded by functional ones. In the past, a change in a material's properties (its elasticity, say, or its volume) in response to a change in the environment was generally seen as a potential problem, as a thing to be avoided. A piston that expanded as it got hotter could seize up. The engineer's ideal was that a material should stolidly maintain its properties come what may—a steel girder should remain rigid, bricks should not fracture or crumble. *Functions* like switching or relaying a signal were things performed by *devices* made up of many components, each comprised of a material that displayed reliable, unchanging characteristics.

There will always be a demand for "dumb" materials of this sort. But smart materials have the potential to simplify engineering considerably. Moving parts have a tendency to break down (or break up), whereas smart devices in which the materials themselves do the job of levers, gears, and even of electronic circuitry, will contain less potential for malfunction. Even in applications where one might imagine that a dumb structural material would suffice, a degree of smartness may prove tremendously useful. What about an aircraft wing made of a substance that changes color when and where microscopic cracks develop? Or even one that automatically heals those flaws? A windshield that darkens when the light passing through it becomes more intense, and lightens again when a cloud passes over? A house built of bricks that change their thermal insulating properties depending on the outside temperature, so as to maximize energy efficiency? In the years to come, it will pay to be smart.

But some materials scientists hope for even more than this. They make a distinction in materials, just as we do (with questionable success) in people, between *degrees* of smartness. Some materials may simply be smart enough to respond each time a particular characteristic of their surroundings (such as temperature or pressure) changes. But others can be envisioned that get wise to such changes, that maintain a memory of what has transpired before and that learn from these previous experiences. Such a material gets smarter as it gets older.

Beyond this are material *systems* that can be claimed to possess a degree of real intelligence. What I have in mind here are materials that are hooked up directly to microprocessors. In response to some change in the surroundings, the material creates a signal and conveys it to the microprocessor; the latter figures out what properties the material needs to optimize its performance under the prevailing conditions, and feeds back a signal that will establish those properties in the material. Such a system would be very different from, say, a conventional photo-detector (light meter), where the photosensitive material feeds a signal to the processing circuitry in a strictly one-way exchange. In a corresponding intelligent system, the circuitry would be able in turn to alter, for example, the light sensitivity, the wavelength response, or the detector area of the photosensitive medium, depending on what the medium tells it about the light that is falling on it. We can draw a tentative analogy with the mechanism of the eye, in which the area of the pupil is adjusted automatically depending on the intensity of the light it receives.

INTELLIGENCE TESTS

At the simplest level, one could say that a smart material is one that responds to its environment. But that does not help us too much, because every material does this. In particular, when materials get hotter, they (almost invariably) expand. This thermal expansion is a matter of simple physics: hotter atoms vibrate more vigorously, and like gesticulating passengers on the Metro they take up more space, so that everyone else has to move away slightly. You could say that thermal expansion is a smart response, but you would have a hard time convincing engineers wrestling with jammed pistons that it is a type of behavior that is particularly novel or desirable.

All the same, humble thermal expansion can be, and has long been, used to develop materials structures that exhibit some of the characteristics that today are commonly seen as "smart." Most thermostats in domestic heating systems are based on a strip of two different metals pasted back to back. Different metals expand to slightly different degrees when heated; if the metal strips are stuck together, this means that the metal that expands most can reach its equilibrium volume in response to a temperature rise only by bending of the strip; that way, the strips stay back to back with their ends married, but the one on the outside of the curve occupies more volume than the one on the inside. This differential thermal expansion can be used to break a circuit contact and switch off the heating. Much of today's research on smart materials is geared toward producing switches (often called actuators) that, like the thermostat metal bilayer, are operated by changes in some environmental factor.

Then there are thermistors—electrical resistors that change their resistance as temperature changes. These devices can be used in sensor circuits in which the size of an electrical signal delivered to some device varies as the temperature changes. But again, changes in conductivity with changes in temperature are characteristic of just about any material that conducts at all. A decrease in conductivity with rising temperature is in fact the defining characteristic of a metal, and conducting materials that *don't* change their conductivity significantly as temperature is altered are rather special, carefully engineered materials in which a degree of smartness compensates for the usual thermal effect.

In examples such as these, smartness tends to be a matter of degree. Changes in properties induced by changes in environmental conditions are, in most ordinary materials, slight and gradual. Materials that are regarded as genuinely smart, meanwhile, might be designed so that such changes are abrupt or pronounced: in thermistors (which can be regarded as early examples of smart materials), a rather sharp drop in conductivity is required as temperature rises in order that the device be useful. And we will encounter below examples of materials in which swelling or shrinking in response to temperature changes is sudden and alters the volume severalfold, rather than by the tiny fraction induced by conventional thermal expansion. In general, the *kind* of behavior exhibited by this type of smart material might be analogous to the behavior shown by conventional materials, but the *degree* is much more marked, because the causes are different.

A fundamental distinction that can be made for any smart system is whether it is passive or active. These terms mean much the same as they do in microelectronics. A passive system responds to some external change without further assistance: a given stimulus brings about some change in the system that gives it different properties. Thermal expansion is an example of this: in response to a temperature increase, a material expands by a well-defined amount. In electronics, an example of a passive component of a circuit is a resistor: a given change in voltage across the resistor induces a change in current through it. An active system, meanwhile, is controlled not just by the stimulus received from outside but by some internal signal, which might or might not be triggered by the external stimulus. In electronics, a transistor is an active component: a separate voltage to the gate terminal of the device controls the relationship between the external applied voltage (the stimulus) and the output current (the response). In smart systems, an active response usually involves some kind of feedback loop, whereby a change in the system induced by an external stimulus triggers a feedback signal that "tunes" the system's response. Such systems can often adapt themselves to a changing environment rather than simply being passively driven by those changes. A typical example is a vibration-damping smart system: mechanical agitation triggers the feedback loop into providing a stimulus that counteracts the agitation and stabilizes the system. As the frequency or amplitude of the driving vibration changes, so the feedback loop modifies its effect to compensate.

These two kinds of behavior allow us to see another distinction: that between a smart material and a smart structure. Whereas a passive smart response can often be provided by a single-component system (say, a slab of an environmentally responsive material), an active response generally requires several components—it will be a composite system rather than a monolithic material. For example, a block of a smart material might alone be able to act as a valve: a certain stimulus, such as a change in temperature or an applied electric field, might induce a change in shape or volume that blocks off some fluid-bearing channel. But a vibration damper will typically combine a smart material with other components that tune the behavior of that material. It might, for instance, comprise a material whose stiffness can be controlled by an electric field, coupled to an electrical feedback circuit that controls the field applied to the material and a sensor that detects mechanical movement of the material. Vibrations picked up by the mechanical sensor trigger the feedback loop to alter the field applied to the material, changing its stiffness to a value that damps out the vibration and stops it from moving under the eye of the sensor. Such a system is a smart *structure*, comprising a smart material and several control elements (fig. 3.1). Perhaps the simplest way to envision this distinction between a smart material and a smart structure is to ask what happens if we cut them in half. If the smart function is retained, we have the former (it need not be homogeneous, however—some smart materials are composites); if the function is lost, the latter.

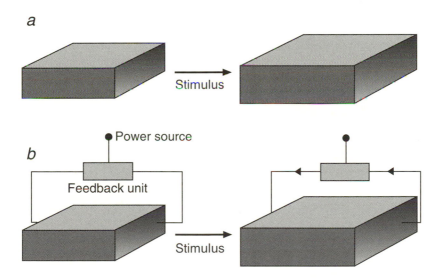

FIGURE 3.1 A smart material responds to a stimulus from the environment, such as an electrical impulse, a change in temperature, or a sound wave, by changing its properties—for example, it might become stiffer or might expand (*a*). This is commonly a passive response; it depends only on the nature of the stimulus. A smart *structure* is a multicomponent system, in which a smart material is coupled to other devices. Commonly these will include a feedback unit, which might be driven by an external power source (*b*), making the system an active one. The stimulus-induced change in the material then sends a signal to the feedback unit, which in turn produces an impulse that "tunes" the properties of the material. This raises the possibility of altering the material to counteract the stimulus—for example, tuning its stiffness in such a way as to damp out a vibration.

These examples also help us to understand another common categorization of smart systems: the division into actuators and sensors. Actuators can be regarded as switches. More generally, they are control devices: they make something happen, whether it is the closing or breaking of an electrical circuit, the closing or opening of a pipe, or the damping of a vibration. Sensors are detection devices: they detect changes in the environment, and generate some kind of signal to warn of that change. Most smart materials can be considered to fall into one or the other of these two categories; some can perform both functions, detecting some environmental change and carrying out some task in response to it. Smart structures often incorporate both sensors and actuators. A sophisticated composite smart structure will also commonly include a data-processing unit, which can greatly enhance the subtlety of the system's response. Often this means that the smart structure is hooked up to a computer or a microprocessing unit, which "thinks" about how to respond appropriately. If this thinking unit can learn by experience—for example, if it is a neural network system, which can be trained to recognize certain stimuli—then the overall smart structure starts to possess a real degree of intelligence. At least we would like to think so, because we are ourselves structures of this sort. Right now, my sensor units (eyes) send signals to my processing unit (brain), which controls my actuators (muscles), with appropriate feedback to fine-tune the positioning, to pick up my cup of coffee—a smart maneuver if ever there was one.

THE SHAPE SHIFTERS

In many of today's radio transmitters throbs a heart made from one of the earliest smart materials to be discovered: crystalline quartz. This natural mineral, a compound of silicon and oxygen, exhibits the property of *piezoelectricity*: when it is squeezed, it produces an electrical voltage. In transmitters, quartz crystals provide the oscillators that generate the radio signal, an electromagnetic wave vibrating at frequencies of around a million cycles per second. The quartz is connected to a driver which sets it vibrating, and these vibrations produce a corresponding electric field oscillating at the same frequency.

The piezoelectric properties of minerals were discovered in 1880 by the brothers Pierre and Jacques Curie in France, who observed the effect in single crystals of quartz and several other minerals. The appearance of an electrical voltage across a squeezed piezoelectric material is called the direct piezoelectric effect, and can be used to convert mechanical energy into electrical. It is exploited in microphones, where the periodic oscillations in air pressure that constitute sound waves impinge on a piezoelectric material and are transformed into periodic electrical signals. The effect also works in reverse—the converse piezoelectric effect—whereby applying an electrical field to the material induces a mechanical distortion, generally a contraction or expansion in the direction of the applied field. This can be used to transform electrical to acoustic energy: an oscillating electrical signal makes a sheet of piezoelectric material vibrate at the same frequency, and these vibrations are transmitted to the surrounding air as sound waves. So loudspeakers, such as those in telephone earsets, can also be made from piezoelectric materials. Quartz resonators are also used as timekeepers in computers and electronic watches.

The piezoelectric effect is a result of an asymmetric arrangement of atoms or ions in the material's crystal structure. In quartz, the silicon and oxygen atoms form tetrahedral units with an oxygen at each corner and a silicon in the middle. The tetrahedra are linked into helical chains, which makes one corner of each tetrahedron distinct from the other three—this corner points in a "special" direction. When the crystals are squeezed in this direction, the tetrahedra are deformed such that each acquires an electric dipole, an imbalance of electric charge along the squeezing direction (fig. 3.2). These dipoles add up to give the crystal as a whole opposite charges on opposite faces; this charge imbalance sets up an electric field across the material, which is then said to be electrically polarized. In general, all piezoelectric materials possess a nonsymmetrical arrangement of charge—an electric dipole—within each of the crystal repeat units (unit cells) when the material is squeezed.

Among the piezoelectric materials discovered by the Curie brothers was one called Rochelle salt, after the town of La Rochelle in France where it was synthe-

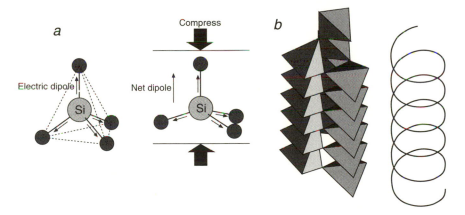

FIGURE 3.2 Quartz is a form of silica (SiO_2). Like all silica minerals, it consists of tetrahedral units in which one silicon and four oxygen atoms are linked together (*a*). There is an imbalance of charge between the silicon and the oxygen atoms (the latter have a slight negative charge relative to the silicon), and so there is an electric field—a *dipole*—along each silicon-oxygen bond. But these four dipoles point in different directions, and their net effect is to cancel each other out—there is no net dipole in the material, and so no spontaneous electrical polarization. When they are squeezed along a certain direction, the tetrahedra are deformed and the individual dipoles no longer cancel out—each tetrahedron has a net dipole. In most forms of silica, for each tetrahedron deformed in one direction there will be another deformed in the opposite sense, and so the net effect is again complete cancelation of dipoles in the material as a whole. But in quartz crystals the tetrahedra are joined in helical chains (*b*), which does not permit this kind of compensation to take place for squeezing along certain directions. Instead, the dipoles in each deformed tetrahedron add up, and the two opposite faces of the crystal develop opposite charges.

sized for use as a laxative in the seventeenth century. The salt is a compound of sodium, potassium, and tartrate ions (the latter being the ionized form of tartaric acid) and water. Unlike quartz, single crystals of this material have an electrical polarization even without squeezing, although squeezing the crystals makes the polarization larger. In 1920, J. Valasek showed that the direction of polarization could be reversed by placing a polarized crystal of Rochelle salt in an electric field of the opposite polarization. This effect is akin to the switching of the direction of *magnetic* polarization in a ferromagnetic material such as iron (see chapter 2), and it came to be ascribed the name *ferroelectricity* (somewhat misleadingly perhaps, as the ferro- prefix has nothing to do with iron in this case).

In crystals of Rochelle salt the unit cells have a spontaneous asymmetric charge distribution: the unit cell is neutral overall, but along the axis of polarization there is an excess of positive charge on one side of the cell and of negative charge on the other. The direction of polarization is switched by applying a field strong enough to drag the unsymmetrically placed ions into equivalent positions in the opposite direction: in effect, the unit cells are converted to their mirror image (fig. 3.3). Once the polarization is flipped in this way, it will stay like that unless flipped back again by another applied field.

In 1935 a second ferroelectric material was discovered: a salt called potassium dihydrogen phosphate (KDP), which proved to be just one of a whole family of phosphate and arsenate salts with ferroelectric properties. During the Second World War these materials were used in sonar devices, in which their piezoelectric properties were exploited for the detection of sonar waves. It became clear that in both KDP and Rochelle salt the origin of the ferroelectric behavior lay with the hydrogen atoms in the crystals. In KDP, each hydrogen atom is chemically bound to the oxygen of a phosphate (PO_4^{2-}) ion, but also forms a weaker bond, called a hydrogen bond, with the oxygen of a neighboring ion. This means that the hydrogens lie on an axis between two oxygens, but nearer one than the other; the asymmetric arrangement gives the unit cell a spontaneous polarization (fig. 3.4). An applied electric field can pull the hydrogen atom (which has a slight positive charge) to an equivalent position along this axis in which it sits closer to the other oxygen, flipping the direction of polarization. Above the Curie temperature (see fig. 3.3 caption) of about minus 150 degrees Celsius, the hydrogen atoms are distributed at random between these two positions in different unit cells, and so there is no net spontaneous polarization; below this temperature, the crystal becomes more stable if all of the hydrogens lie on the same side, and the material then acquires an overall ferroelectric polarization.

It was believed for some time that hydrogen bonds were essential for ferroelectricity, but in the late 1940s a material was discovered that contained no hydrogen. This material, barium titanate, is a ceramic, and it paved the way for today's generation of ceramic ferroelectric and piezoelectric materials, which form one of the main focuses of research in smart materials and structures. Barium titanate has a so-called perovskite structure, in which the barium ion sits at the center of a cube between eight octahedral titanate units (fig. 3.5). The ferroelectric behavior arises because, below the Curie temperature of 120 degrees Celsius, the ideal

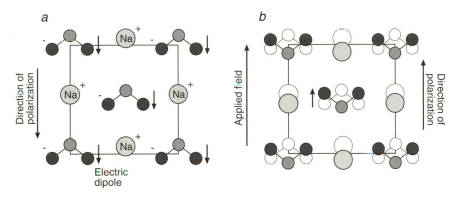

FIGURE 3.3 In ferroelectric materials such as Rochelle salt, the repeat units (unit cells) of the crystal structure have an imbalance of charge even without being squeezed (that is, a *spontaneous* polarization). Here I show this phenomenon in a simpler ferroelectric crystal, sodium nitrite. Each nitrite ion has an electric dipole as a result of its bent shape, and below sodium nitrite's so-called Curie temperature (164 degrees Celsius) all the ions may sit with their dipoles all pointing in the same direction, giving the crystal a spontaneous polarization (*a*). The direction of polarization can be flipped by applying a strong electric field in the opposite sense (*b*). Above the Curie temperature, any spontaneous polarization is lost because thermal vibrations randomize the direction in which the nitrite dipoles point, so that on average they cancel each other out. The transition from a nonferroelectric to a ferroelectric state is entirely analogous to the magnetic ordering transition in a ferromagnet below its Curie temperature.

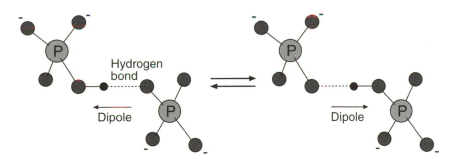

FIGURE 3.4 Potassium dihydrogen phosphate is a ferroelectric material by virtue of the hydrogen atoms in the crystal. Each of these sits asymmetrically between two oxygen atoms on different phosphate ions; the asymmetry arises because the hydrogen atom is covalently bound to one oxygen atom, but bound to the other only by the weaker bonds called hydrogen bonds. This creates an electric dipole, and the crystal as a whole has a spontaneous polarization. The direction of polarization can be switched by an electric field, which can pull the hydrogen atom to the equivalent position on the other side of the midpoint of the oxygen–oxygen axis.

perovskite structure is slightly distorted—the titanium ions lie off the center of the octahedra, slightly closer to one of the oxygen ions than the others. This asymmetry gives the octahedra a net dipole. Crucially, the direction of the titanium shift is the same in all the octahedra, so the material has a net polarization.

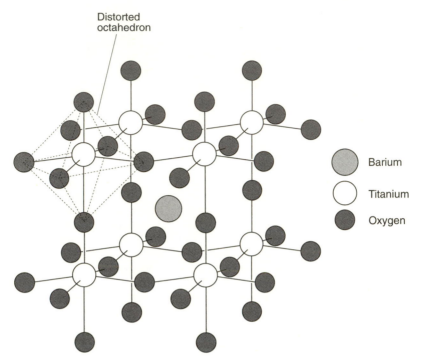

Distorted
octahedron

Barium

Titanium

Oxygen

FIGURE 3.5 In barium titanate, each titanium ion is surrounded by an octahedral arrangement of oxygen ions. But because the titanium is slightly displaced from the very center of the octahedron (and because this displacement occurs in the same direction throughout the crystal), each distorted octahedron possesses an electric dipole and the material is polarized.

Above the Curie temperature these shifts are randomized, so that on average the material appears to have a perfect perovskite structure with no asymmetry and no polarization.

Barium titanate has been long used in piezoelectric sonar devices, but in most applications it has now been supplanted by other, superior materials, most importantly lead zirconate–titanate (PZT). (These abbreviations, incidentally, seem perverse unless one writes out the chemical symbols of the elements involved—K for potassium, Pb for lead.) PZT is a mixture of two compounds, lead titanate and lead zirconate. The two mix intimately, to the extent that it is a matter of pure chance whether titanium or zirconium ions appear in equivalent positions from one unit cell to the next. Like barium titanate, PZT has a distorted perovskite crystal structure; but as well as the off-center position of the zirconium and titanium ions in their respective octahedra, the lead ion too is significantly displaced from its central location, and the material has a large spontaneous polarization.

But just as ordinary iron does not have a spontaneous magnetization at room temperature (even though this is way below iron's magnetic Curie temperature), so ferroelectric materials do not show spontaneous electric polarization below their Curie temperatures. In both cases the ordering occurs in microscopic domains, with the polarization in different domains pointing in different directions (see fig. 2.4). And just as iron can be magnetized by a strong magnetic field, which realigns all the domain magnetization directions with the applied field, so PZT can be made into a permanent ferroelectric by placing it in a strong electric field. This field-induced alignment of the ferroelectric domains is called poling.

The Right Moves

PZT and related ferroelectric materials are widely used as sensors of acoustic waves and mechanical motion and vibration. In sonar devices, they respond to ultrasonic waves—a passive application. Sensors have been developed in which the impact of a mere raindrop is sufficient to trigger a piezoelectric signal. These sensors could be used to control the speed of automobile windscreen wipers, which would be adjusted by a processing circuit to give the optimum response to the amount of rain that the sensor detects.

An active piezoelectric device for vibration damping has been developed by Robert Newnham and colleagues at Pennsylvania State University. This device acts as a material whose stiffness can be varied continuously, a property that allows one to damp out vibrations over a wide frequency range. The trouble with most passive vibration dampers, such as automobile suspension springs, is that they have a certain resonant frequency—when exposed to a vibration of that frequency, the spring vibrates in sympathy, like a guitar string resonating to a sound wave, and the amplitude of vibration becomes very large. Anyone who has sat at the back of London's double-decker buses will be familiar with this effect: when the bus goes up and down over bumps in the road at a frequency that matches the suspension's resonant frequency, you have to hold on to your seat.

Newnham's smart structure avoids this problem of resonance by tuning its stiffness such that its resonant frequency is always far from the frequency of the vibrations to which it is subjected. Structures like this could also be useful for shock absorbers: they could be made stiff to resist stresses that are applied gradually, but soft and compliant to absorb the energy of a sudden impact.

The device works by coupling together two stacks of piezoelectric disks, which can be made to vibrate by applying oscillating electric fields. The stacks are linked mechanically via a rubber gasket, which transmits the vibrations of one stack to the other (fig. 3.6). One of the stacks is set vibrating by an applied field (stacks of piezoelectric disks are used rather than single disks because they require a much lower operating voltage to obtain the same effect). The vibrations are transmitted through the rubber gasket to the top of the other stack. Here there is a piezoelectric pressure sensor, which generates an oscillating electrical signal in step with the vibrations of the first stack. The signal is fed into an external feedback circuit, which produces another electrical signal to drive the second stack into vibration. If the feedback circuit produces a driving signal in step with that generated in the piezoelectric sensor by the motions of the top stack, then both stacks vibrate in step. This leads to resonance between the stacks and allows

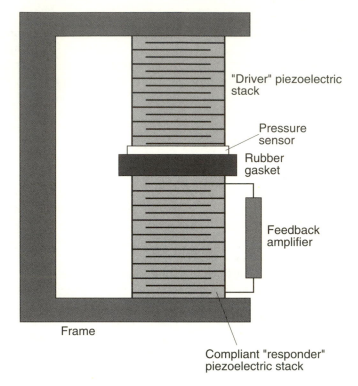

"Driver" piezoelectric stack

Pressure sensor

Rubber gasket

Feedback amplifier

Frame

Compliant "responder" piezoelectric stack

FIGURE 3.6 Vibrations in a material can be damped if its stiffness (compliance) can be adjusted to avoid resonant vibration at all frequencies. Such a "variable-compliance" device can be constructed from two stacks of piezoelectric layers, coupled through a rubber gasket which transmits vibrations from the upper to the lower stack. A feedback circuit registers the vibrations in the upper stack and sends an electrical signal to the lower stack to excite it into vibration. If the vibrations in the upper and lower stacks are in step, the material is very compliant; if they are out of step, it is stiff. This is a smart structure of the type shown in figure 3.1*b*.

a small driving force on the first stack to excite large deformations of the two stacks together—the device acts like a very compliant material. If, however, the feedback circuit ensures that the second stack is driven exactly out of step with the first, the oscillations in each stack "collide" and the device acts like a stiff material. The effect is rather like that of two people jumping on a single trampoline: if they jump together, they can reach greater heights than each alone, but if one lands while the other is taking off, the trampoline seems to lose its "spring."

To turn the device into a vibration damper, it can be coupled to another feedback circuit that varies the stiffness of the structure until the amplitude of vibration is minimized. Active vibration dampers of this sort are being explored for use on space structures such as satellites, where precise positional control is crucial and there is no gravity to help damp out vibrations, and the Japanese motor company Toyota has developed a smart automobile suspension system that uses such controlled-compliance devices.

To make useful actuators and switches from ferroelectric and piezoelectric materials, one generally wants the material to undergo large distortions and to generate large forces when only a modest electric field is applied. Commonly a fast response time is desirable too: typically, the maximum displacement must be achieved in about ten millionths of a second. To obtain these characteristics, it is usually necessary to use composite designs like Newnham's piezoelectric stacks. The Japanese company NEC uses multilayer stacks in dot-matrix printers. A stack of around a hundred thin piezoelectric layers drives a wire arm onto an ink ribbon to leave a dot imprinted on the paper below (fig. 3.7). The multilayer stacks themselves generate only a few micrometers of displacement, which is not enough to be useful on its own—the print head would have to be kept too close to the paper. So the motion is amplified by a hinge unit, which brings about displacements of up to half a millimeter. These devices can be much faster than conventional printers (in which the print head is driven by electromagnetic forces) while using less power.

In some applications of piezoelectric materials the prime objective is to obtain a large amplitude of motion. The mechanical distortion of a piezoelectric material driven by an electric field can be enhanced by a design akin to the bilayer metal strip of thermostats. When a strip of a piezoelectric material is glued back to back with a strip of a nonresponsive but elastic material, with the polarization of the piezoelectric directed perpendicular to the strip, the contraction of the piezo-strip in response to an applied field will be hindered by the other strip and the bilayer composite will bend (fig. 3.8). In this way the end of the strip can be displaced by as much as several hundred micrometers—but the response is rather slow, and the force generated is small.

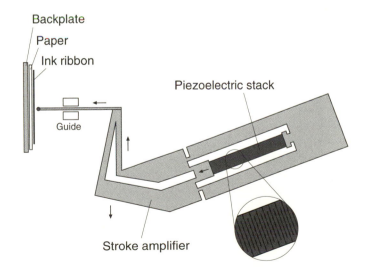

Backplate

Paper

Ink ribbon

Guide

Piezoelectric stack

Stroke amplifier

FIGURE 3.7 Piezoelectric stacks are used to drive the printing arm in some dot-matrix printers. The small displacement generated by the stack is amplified by the design of the arm structure.

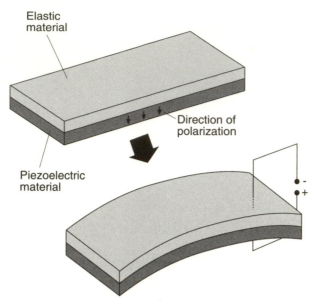

FIGURE 3.8 To generate large displacements in piezoelectric devices, a bilayer structure is commonly used in which a piezoelectric slab, with its direction of polarization perpendicular to the slab face, is glued to an elastic backing strip. When an electric field is applied, contraction of the piezoelectric strip is hindered by the backing strip, and the structure bends. The motion of the ends of the strip is much greater than the deformation of the piezo-strip on its own.

Robert Newnham and Kenji Uchino have developed a composite structure that combines some of the advantages of multilayers and bilayers. They constructed the usual multilayer structure, with the direction of polarization aligned perpendicular to the layers. When an electric field is applied across the layers, they expand in the direction of the field and contract laterally. But in Newnham and Uchino's device this expansion is then amplified by flexible metal plates fixed to the top and bottom of the stack, with half-moon-shaped cavities in the central region (fig. 3.9). The lateral contraction of the multilayer makes the central part of the metal plates bulge upward, adding to the expansion perpendicular to the layers. These devices, which Newnham calls moonies, can generate around eight times as much displacement as a multilayer stack by itself, but without compromising the quick response or strong generated force. Devices like this are now being used to control the pointing direction of a laser beam in miniaturized laser scanners.

While piezoelectricity is a characteristic of rather special materials with asymmetric crystal structures, all nonconducting materials show a related, but usually smaller, mechanical response to an applied electric field called *electrostriction*. This property relies on the fact that even apparently symmetrical crystal structures are not perfectly so, because the ions vibrate in a slightly nonideal way. Vibrating atoms or ions in a crystal can be thought of as tiny balls on springs—as they are displaced from their equilibrium position (where they would sit if there were no vibrations), they experience a restoring force that pulls them back. When this restoring force increases in direct proportion to the displacement, the ions are said to exhibit harmonic motion, the "ideal" kind of vibrational behavior exhibited by springs and pendulums for small displacements. Harmonic oscillations are easy to describe mathematically, and were exhaustively studied by Isaac Newton and Robert Hooke in the seventeenth century. In real crystals, however, the vibrations are not harmonic; the restoring force varies with displacement in a more complex way, and in particular equal displacements in different directions may generate different restoring forces. The asymmetry that this introduces means that an applied electric field will actually generate a very small displacement, and a consequent small polarization, in any crystal.

In a few materials, however, electrostrictive effects are appreciable. In the ceramic material lead magnesium niobium oxide (PMN), for instance, electrostriction gives rise to a mechanical distortion comparable to that induced by the piezoelectric effect in PZT. These materials are therefore also useful in applications

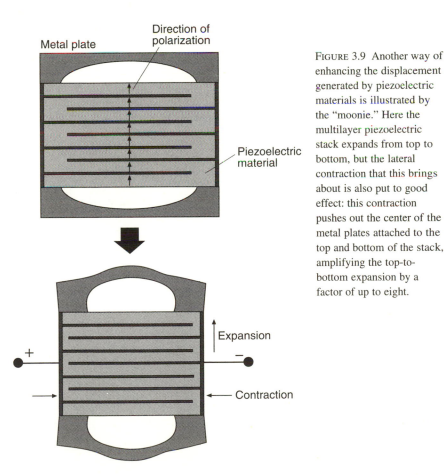

FIGURE 3.9 Another way of enhancing the displacement generated by piezoelectric materials is illustrated by the "moonie." Here the multilayer piezoelectric stack expands from top to bottom, but the lateral contraction that this brings about is also put to good effect: this contraction pushes out the center of the metal plates attached to the top and bottom of the stack, amplifying the top-to-bottom expansion by a factor of up to eight.

where one needs to couple electrical and mechanical energy. One such application is in the development of smart and adaptive mirrors for astronomical imaging. Kenji Uchino has fabricated mirrors in which silvered glass is bonded to a series of electrostrictive layers that enable one to make minute adjustments in the mirror's shape. In response to applied electric fields, each layer bends the glass (which is slightly elastic) in a different direction until the desired shape is attained. Mirrors whose shape can be continuously adjusted with great precision are much in demand by astronomers, who use them for focusing the light from distant stars and galaxies onto photographic plates and other image-recording devices. Adaptive mirrors could compensate for the image distortions induced by atmospheric turbulence, by quickly changing their shape to cancel out its effects.

Smart ceramic actuators that change shape in response to electrical stimuli are being considered by several aerospace companies as alternatives to mechanical moving parts on aircraft. 1995 saw the first flight of a test model aircraft with shape-changing wings. In this model, designed and built by researchers at Auburn University in Alabama and called the Mothra, all of the ailerons and tail flaps that are used to control the flight of conventional aircraft were replaced by wings and tail fins containing piezoelectric actuators that altered their shape. One advantage of smart wings is that they can be continually adapted to maximize aerodynamic (and thus fuel) efficiency in a way that is just not possible for today's aircraft. The wings of commercial aircraft can achieve maximum efficiency only at specific flight speeds (just as an automobile's engine achieves optimal fuel efficiency for a given speed); but smart wings could be molded to the most efficient shape for any speed. Aircraft engineers are also interested in using smart actuators to reduce cabin noise. Most of this noise comes from the vibration of an aircraft's fuselage, rather than noise produced directly by the engines. The present generation of noise suppression systems use audio speakers that broadcast "anti-sound": sound waves whose peaks coincide with the troughs of the sound waves that they are combating, thereby canceling out the waves. Smart piezoelectric sound suppressors would be more compact and responsive, and would fight the cabin noise at its source—not by canceling the noise produced by vibrations, but by canceling the vibrations themselves. They would be intelligent structures embedded in the fuselage body, in which several piezoelectric devices communicate with one another. One set of devices act as sensors to detect vibrations of the fuselage; they transmit this signal to piezoelectric actuators, which then vibrate so as to cancel the triggering vibrations—much as the piezoelectric stack depicted in figure 3.6 senses and damps out vibrations. The McDonnell Douglas aerospace company has already built a prototype passenger aircraft with a smart fuselage of this sort.

Notice that in this example we see piezoelectrics being used both as actuators and as sensors. I've talked here mostly about *actuator* applications—that is, about the conversion of electrical to mechanical energy. But sensing technologies, in which the reverse process is generally invoked, constitute an equally important branch of smart-materials research. The electric field that is set up when a piezoelectric material is squeezed provides a measure of the mechanical deformation it experiences, and this enables the use of such materials in pressure and strain sensors, as well as in the kind of vibration sensors that I have touched upon already. Some polymeric materials, most notably polyvinylidene fluoride, are piezoelectric, and they lend themselves particularly well to sensing applications as they can be produced in cheap, flexible sheets of large area. Piezoelectric polymers are used in some pressure-sensitive keyboards, in which an array of electrodes picks up the electric fields induced by local compression of a polymer sheet beneath the keys.

Squeezed Magnets

In 1847 the Scotsman James Joule found that when an iron rod is magnetized it contracts very slightly. This effect, called *magnetostriction*, is now recognized as a smart and potentially useful response—the magnetic equivalent of electrostriction.

Magnetostriction is a property of all ferromagnetic materials. But just as was the case for electrostriction, practical exploitation of the phenomenon had to await the identification of a material in which the magnetostrictive effect is large enough to be useful. That discovery was made in 1971, when an alloy of iron and the rare metal terbium was found to have a huge magnetostrictive effect at room temperature—about two hundred times larger than that of iron, and in the opposite sense (the material expands, rather than contracts, when magnetized). Soon afterward, a similar alloy of iron with samarium (another rare metal) proved to have a magnetostrictive effect of similar magnitude but, like iron, in the shrinking sense. At the low temperature of minus 196 degrees Celsius (the boiling point of liquid nitrogen), even larger size changes are seen in pure terbium, in the related metal dysprosium, and in compounds of the two: magnetized rods of these materials can undergo a length change of about one percent, about five times larger than the change at room temperature. For small applied magnetic fields the change in length increases almost directly in proportion to the strength of the field, but as the field is increased the length change "saturates," so that no further change can be brought about by making the field stronger.

Magnetostriction is the result of the reorientation of atomic magnetic moments in an applied magnetic field. As we saw in chapter 2, magnetic atoms are rather like tiny bar magnets whose poles will swing around like compass needles to line up with an applied magnetic field. In the simplest picture of a ferromagnet the positions of the magnetic atoms are independent of the direction in which the needles point—it is as if each compass is fixed firmly in place, but with a needle that is free to swing. But the real picture is not quite so simple. The direction in which the needles point influences the interactions between neighboring atoms, because both the magnetic moment and the interatomic attractions are mediated by the same entities: the outermost electrons in the atoms' electronic shells. This means that the interactions of an atom with its neighbors are not equivalent in all directions, but differ slightly in the direction of magnetization relative to other directions, and the crystal structure is distorted slightly as a result.

Within a nonmagnetized ferromagnetic material there exist many domains of different magnetization, so that the overall distortions cancel out. But when the material is magnetized by an applied magnetic field, the distortions occur all in the same sense, and the bulk material undergoes a change in shape. The nature of the distortion depends on the relative orientation of the applied field with respect to the crystal axes of the domains—a magnetic field applied along one crystal axis will generally produce a different degree (or even a different sense) of distortion relative to some other axis. In polycrystalline materials, which are patchworks of small crystallites with their symmetry axes oriented at random, the absolute orientation of the applied field makes no difference to the degree of magnetostriction obtained; but for single-crystal materials, in which the crystal axes remain in the same orientation throughout, the response to a magnetic field will be anisotropic—different in some directions to others. In particular, there will be a direction in which the magnetostrictive distortion is maximized. To make magnetostrictive devices, therefore, the ideal would be to use materials either that are polycrystalline yet with an overall preferential alignment of crystallite orientation, or that are single crystals.

A typical magnetostrictive actuator consists of a long rod surrounded by an electromagnetic coil, which generates a magnetic field when a current is passed through it (fig. 3.10). The maximum change in length of the rod would be obtained by making the rod from a single crystal with its most responsive axis lying along the rod's axis. But it is difficult to grow materials to this specification—crystals of the most commonly used magnetostrictive material, an alloy of ter-

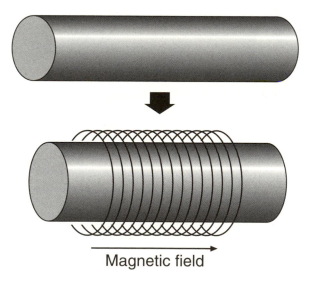

Magnetic field

FIGURE 3.10 A magnetostrictive material changes shape when placed in a magnetic field. Here a rod contracts in the magnetic field produced by electromagnetic induction when an electric current is passed through the wire coil.

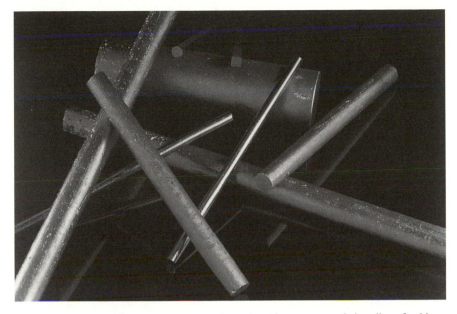

FIGURE 3.11 Terfenol-D, rods of which are shown here, is a magnetostrictive alloy of terbium, dysprosium, and iron developed at the Naval Ordnance Laboratories in Washington, D.C. It is used in commercial magnetostrictive actuators. (Photo courtesy of Etrema Products Inc., Ames, Iowa.)

bium, dysprosium, and iron called Terfenol-D (fig. 3.11), grow most readily (and most inconveniently) along a direction perpendicular to the most responsive axis. Nonetheless, the degree of magnetostriction that can be obtained in this direction is still sufficient for many applications.

Because magnetostriction can be electrically induced by an electromagnetic induction coil, magnetostrictive devices can be used like piezoelectrics to convert electrical to mechanical energy and vice versa. They therefore find similar applications—as sonar sensors, for example, and vibration dampers and tunable-compliance materials. But magnetostrictive materials have the additional advantage that they are very robust—they are metal alloys, not brittle ceramics. In addition, piezoelectric and ferroelectric materials can fail when driven by high electric fields, because these can cause dielectric breakdown, essentially the same process as that in which sparks leap across two electrodes through air. But there is no analogous breakdown process for magnetic materials.

To obtain large displacements, magnetostrictive ribbons and bilayer strips are used. Ribbons are also used in sensor devices that are highly sensitive to small mechanical disturbances. These instruments employ amorphous metal alloys as the magnetostrictive component rather than the crystalline alloys discussed so far. The most common of these materials is Metglas 2605SC, an alloy of iron with boron, silicon, and carbon. The atoms in this alloy are arranged irregularly rather than being stacked with crystalline order: this means that there exist a great many possible atomic arrangements, each more or less as stable as the others. Changing the orientation of magnetic moments of the iron atoms in the alloy by applying a magnetic field can change the interactions between these and their neighboring atoms just enough to tip the balance and induce a rearrangement to some other structure, which can bring about a change in shape. Amorphous metal alloys show a smaller magnetostrictive distortion at the saturation point than do the crystalline iron–terbium alloys, and they generate smaller forces, but they provide a larger response for small applied fields. Conversely, small mechanical distortions of these materials create a larger change in magnetic properties. They are used in sensitive strain gauges and in devices that measure acceleration forces.

Light-Footed

Kenji Uchino has used a modified version of PZT to create a device that couples optical and mechanical energy—converting light to motion. Flashes of light cause the material to expand, creating an acoustic wave in the surrounding air with the same frequency as that with which the light flashes on and off. In this way, Uchino has created "photophones," light-driven loudspeakers. These might one day be useful for converting the light pulses of a telecommunications optical-fiber cable directly into a voice in the earpiece of a telephone at the receiving end, without the need for any intermediate electronic processing.

Uchino's material is in one sense a conventional piezoelectric material: it is based on a variant of PZT in which some of the lead ions are replaced with lanthanum, giving lead lanthanum zirconate–titanate (PLZT). This material is used in several piezoelectric actuators, and has piezoelectric properties that are slightly better suited to some applications than those of PZT itself. But Uchino has modified this ceramic material further so that it produces its own electric field in response to light—it exhibits the photovoltaic effect, the conversion of light into electricity (chapter 6). Looked at this way, the photophones *do* use electricity as an intermediary between light and sound, but the "electronics" are all intrinsic to the material itself.

Uchino found that when PLZT is doped with small amounts of tungsten oxide it acquires a new property: *photostriction*, whereby irradiating the material with (violet) light causes a change in shape. The origin of this effect is still not completely clear, but Uchino believes that the lanthanum and tungsten ions absorb the light, which causes them to eject an electron. Because of the asymmetry of the crystal structure and the consequent spontaneous electrical polarization of the material, the electrons drift in the direction of this asymmetry, enhancing the ferroelectric field that already exists. The effect is equivalent to applying an external electric field to induce a piezoelectric response.

Uchino used this doped PLZT material to make a light-driven actuator. He constructed a bilayer strip of back-to-back ferroelectric PLZT layers with opposite polarization; but unlike the bilayers mentioned earlier, these had their polarization directions *parallel* to the strip faces. When this double strip is illuminated on one side, motion of the photo-excited electrons to one end of the illuminated strip sets up a voltage of 10,000 volts between the top and bottom of the strip, causing it to expand. Because the polarization axis of the other strip lies in the opposite direction, this same large electric field induces a contraction. Thus, expansion of one strip and contraction of the other causes the bilayer to bend appreciably. This bending breaks a circuit connection between an electrical contact attached to one end of the bilayer and a contact attached to the immobile support. In principle, the bilayer strip will straighten itself when the light is switched off, but in practice this process is slow because the photo-excited electrons take some time to return to their parent ions in the dark. Reconnecting the circuit can be effected more rapidly by irradiating the second strip, whereupon the bilayer bends back in the other direction.

Uchino has used this bilayer structure to create a device that walks in response to light pulses. The walker has two legs, each of which is a bilayer strip of doped PLZT glued at one end to a plastic block. It advances like an inchworm, with alternate bending of first one leg and then the other in response to a sequence of light signals (fig. 3.12). Uchino's walker can advance across a surface at the leisurely pace of about a millimeter per minute.

That, of course, is hardly going to win any races, but it does mean that the position of a mechanical moving device of this sort can be controlled with great precision. The flexing of the walker's legs happens very quickly, but the tricky

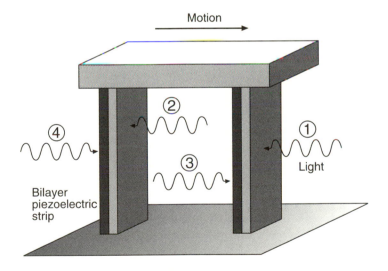

FIGURE 3.12 When the piezoelectric material PLZT is doped with tungsten oxide, it becomes photostrictive: it changes shape when illuminated with light. Kenji Uchino has used this material to make a light-activated walking device, in which bilayer legs of PLZT flex one way and another in response to a sequence of light pulses (1–4). Light shining from the front of the walker causes the forward leg to bend; shortly after, a second pulse is directed at the leading face of the back leg. These consecutive bending movements cause forward motion. Then a pulse is directed onto the rear face of the forward leg, bending it in the opposite direction, and finally the back leg is bent likewise.

part is getting that flexing to be large enough to generate significant forward motion, as well as preventing slippage of the "feet" on the surface over which they walk. This fast flexing is what Uchino has put to work in his photophones. Rapidly alternating the illumination of a single bilayer strip on one side and then the other causes the strip to vibrate, setting up a corresponding sound wave in the air. Uchino has been able to induce vibrations of up to eighty cycles per second—an acoustic wave of this frequency is just within the audible range of humans. To create more easily audible sounds, he needs to be able to increase the flexing rate by two or three times. Then it will truly become possible to hear light sing.

You Must Remember This

I have often wondered why there is not, to my knowledge, a device on the market for straightening tent pegs. The sight of a pile of bent pegs, some deformed into a series of ripples where they have been crudely and repeatedly twisted back to something approximating their original shape, always fills me with dismay. Surely there is a commercial opportunity here, if only someone more inventive and enterprising than I would devise a solution.

But they had better be quick, because there is already a class of smart materials that promises to provide the remedy. Deformed tent pegs made of these materials could be gently heated over the Calor gas camping stove, whereupon they would, as if by magic, stretch out into their pristine form. Such materials—alloys of metals such as nickel and titanium—seem to possess a memory of their original shape, to which they will revert when deformed and then heated (fig. 3.13). They are called *shape-memory alloys* (SMAS).

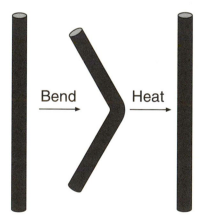

Bend   Heat

FIGURE 3.13 When deformed and then heated, shape-memory alloys return to their original shape.

The uses envisioned for these materials are, I must admit, a little more profound than making campers' lives easier. Shrink-fit metal tubing for hydraulic lines on aircraft and plumbing on ships and submarines, and for joins in undersea pipelines, are one such. The materials are generally prepared in a shape-memory state well below room temperature, stretched at this same temperature and fitted

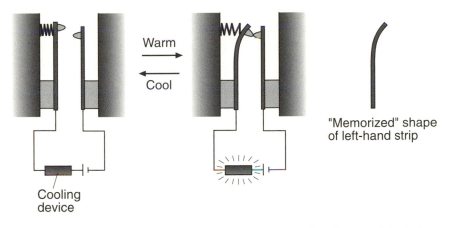

Warm

Cool

"Memorized" shape
of left-hand strip

Cooling
device

FIGURE 3.14 A thermostat device made from a shape-memory alloy. The arm of the device is prepared in a bent state—it is this shape that the arm "remembers." It is held straight by a retaining spring, but when the temperature rises so that the memory returns, the arm regains its bent shape and a circuit contact can be actuated (as here) or broken.

over the junction of two pipes, and then warmed to room temperature. As they warm up, the tubes shrink to their original state, giving a tight seal. Tubes six inches in diameter made from SMAS have been installed in undersea pipelines at depths of up to 100 meters, by taking the still-cold tubes down in a diving chamber and allowing them to warm at depths to the temperature of the surrounding water. Making joints this way is much easier than trying to perform welding operations under the sea.

Another obvious application is in thermostats, where the alloys replace the bimetallic strips in conventional use (fig. 3.14). Thermostats of this kind based on SMAS are already being used for controlling a range of heating and cooling devices, including clutch fans in car engines, and have been incorporated into mechanisms for operating greenhouse windows or ventilators.

In principle, these materials might find their way into just about any device in which one wants to convert heat into a mechanical response. A simple switching response of the kind used in the devices above is easy to arrange; but one can also imagine using SMAS in heat engines, in which heat is used to drive some kind of mechanical work. For example, the mechanical motion induced by exposure to heat might be used to lift a load or to pump a fluid. Prototype heat engines have been built from SMAS, but they are not particularly efficient. Nonetheless, they might be useful for extracting energy from so-called low-grade heat sources, which do not have much energy content relative to their surroundings—for example, industrial coolant water and geothermal sources such as hot springs.

When incorporated into composite materials, shape-memory alloys can provide a highly temperature-sensitive stiffness. This stiffness can be controlled by, for instance, passing an electrical current through the material to heat it. For example, imagine a composite material containing SMA fibers that bend in opposite directions when heated, held within a surrounding matrix. Warming the material will induce no overall deformation, since the stresses imposed by the fibers will balance each other out; nevertheless, the composite will then "hold on to" this frustrated strain, which will alter its stiffness. Materials whose stiffness can be subjected to this kind of active control might be very useful as vibration dampers, as we saw earlier. And systems that incorporate both controllable bending, as offered by pure SMAs, and controllable stiffness, offered by SMA composites, would be able to act rather like mechanical muscles.

Bend It—Just a Little Bit

The origin of the shape-memory effect lies in the fact that the crystal structure of the shape-memory alloy changes when it is heated. There is nothing particularly unusual about this in itself; many crystalline materials undergo substantial changes in structure (that is, in the way the atoms are packed together) as their temperature is changed. These changes are generally termed solid-state phase transformations. They are driven by the fact that the energies of atomic packing arrangements are temperature-dependent, so that one particular arrangement might have a lower energy than another at one temperature but a higher energy at a different temperature. The change in structure involves the shuffling of atoms into their new positions.

But in the kind of transformation that lies at the heart of the shape-memory effect, called a martensitic transformation, the change in structure is *not* effected by independent diffusion of individual atoms. Rather, it involves a concerted deformation of the crystal structure, as if it were skewed by a shear force. During a martensitic transition, the crystal structure acts rather like a hinged network of struts joining the atoms: the struts are not broken and rearranged, as in a diffusive transition, but rather the entire network becomes skewed as the atoms are displaced en masse. The resulting skewed phase is called a *martensite*, and the unskewed "parent" phase is an *austenite* (fig. 3.15). The transformation from an austenite to a martensite takes place in several metal alloys when they are cooled. Steel, for instance, undergoes a martensitic transformation, and the resulting change in atomic structure gives the material its strength. Steel is not a shape-memory alloy, however, because the transformation does not satisfy the prime criterion for the shape-memory effect: reversibility. In SMAs the austenite phase can be regained by heating the material back up again (martensitic transformations of this sort are said to be thermoelastic). The reverse transformation does not generally take place at the same temperature as the initial austenite-to-martensite transformation; instead it occurs at a slightly higher temperature (fig. 3.15). This behavior—reversibility but at different temperatures—is known as hysteresis.

The tilting of the atomic lattice in a martensite can generally slant in one of several equivalent directions. If we think of skewing a square arrangement of atoms into a rhombus, for instance, we can do so in four equivalent ways (fig. 3.16). If this tilting occurred in the same sense throughout the material, it would undergo a spontaneous deformation as it was cooled through the martensitic transformation. But what tends to happen in practice is that some atomic planes

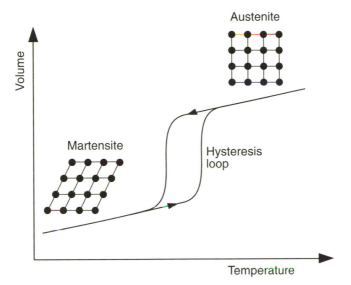

FIGURE 3.15 The shape-regaining behavior of SMAs involves a transformation between two states of the material with different crystal structures: a high-temperature "parent" state called an austenite and a low-temperature "skewed" state called a martensite. The transformation from the former to the latter is called a martensitic transformation, and takes place as the temperature is lowered. This transformation is accompanied by a change in several characteristics of the material, such as its length, its volume, and perhaps its electrical conductivity. The reverse transformation, from a martensite to an austenite, occurs when the temperature is raised, but the onset takes place at a higher temperature than the martensitic transformation. This difference in transition temperature when the transformation is approached from one side or the other is called hysteresis.

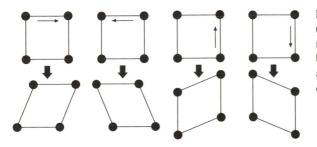

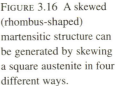

FIGURE 3.16 A skewed (rhombus-shaped) martensitic structure can be generated by skewing a square austenite in four different ways.

slew one way and some slew the other. Then the material becomes divided into microscopic domains of alternating tilt. The opposing tilts cancel each other out, and the shape of the material as a whole is more or less unaltered. It is then said to be in a self-accommodating state.

Because the austenite and martensite phases are related to each other by a simple tilt of the crystal structure, the original austenite lattice can be regained by heating the martensite to induce a "backwards shear." When the martensite is divided into several self-accommodating regions, this backwards shear occurs in the opposite sense in each region, just as if the material were ironing out its kinks. But what if the material is first deformed while in the martensite phase?

A normal metal can respond elastically to a small amount of bending—it takes up the strain by elastic deformation of the crystal lattice, and springs back once the stress is released. But more pronounced bending (deformations of typically just a few percent from the original shape) introduces defects such as misalignments of the rows of atoms (called dislocations) into the crystal structure, which cannot then easily be rectified. The dislocations travel through the lattice like cracks, leaving it permanently defective. This is called plastic deformation. The dislocations can be removed again only by annealing the metal to get the atoms back into a regular packing arrangement—this is what blacksmiths are up to when they hammer away at hot steel. Of course, we could simply bend the deformed metal plastically back to an approximation of its original shape, but this does not remove the dislocations; indeed, it introduces more, which is why repeated plastic deformation back and forth weakens a metal until it breaks.

Deforming a self-accommodating martensite is another matter, however. The deformation appears to be plastic, because the material does not (once the normal elastic limit is exceeded) simply spring back into place. But the strain is actually taken up *without* any permanent damage to the material's microstructure: the crystal lattice simply skews under the stress (fig. 3.17). Heating the material up then causes a reverse skewing that regenerates the original austenite structure. As the various domains of the bent martensite alloy revert to the austenite, the material is pulled back into its original shape. So you could describe the overall process as follows. When transformed from an austenite to a (self-accommodating) martensite, the SMA develops wrinkles in all possible directions. But a subsequent deformation of the martensite pushes all the wrinkles in the *same* direction. Then, when the wrinkles are smoothed out by the reverse transformation to an austenite, the material as a whole regains its shape.

The first SMA discovered, by researchers at the U.S. Naval Ordnance Laboratory in 1965, was an alloy of nickel and titanium, patented under the name Nitinol. When prepared in the austenite phase, cooled to the martensite, deformed, and reheated, Nitinol regains its original shape for deformations of up to 8 percent. When bent by more than this, recovery is not complete. This recovery range remains much the same for more recent SMAS: the record is currently about 10 percent. It is unlikely that there will be much further improvement on this, because it is close to the fundamental limit set by the extent to which the atomic planes become displaced in the martensitic transformation itself. Many of the new materials are alloys of copper and zinc (that is, essentially brass) mixed with other metals such as aluminum, tin, and gallium. All of these alloys are crystalline: the different metal atoms do not simply occupy lattice sites at random, but form a regularly repeating unit.

Nitinol wires are now being used as the muscle fibers of robotic devices. Japanese researchers have constructed robot hands of this type that are controllable

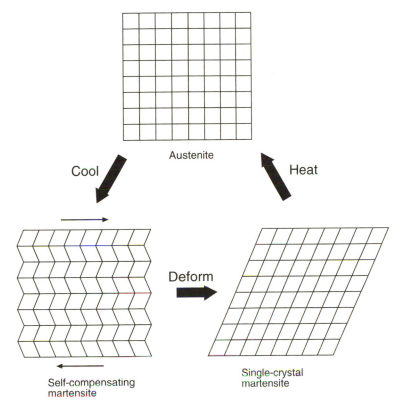

FIGURE 3.17 A self-accommodating martensite has domains that tilt in different directions, such that there is no net change in shape of the material in passing through the martensitic transformation. When such a material is deformed, it can accommodate the strain without disruption of the regular crystal structure, simply by altering the tilt direction in some domains. (In a normal metal, in contrast, deformation introduces defects in the crystal.) When the deformed material is warmed back to the austenite phase, there is only one way that the skewed structure can revert to the square structure: by pulling the whole solid back to its original, undeformed shape.

enough to pick up paper cups of water. Composite materials threaded with Nitinol wires have also been developed for use as vibration dampers, in which shape changes in the Nitinol wires alter the rigidity or internal stresses of the composite, and as stress reducers, in which shape changes set up an internal stress that opposes an external one.

An intriguing characteristic of some shape-memory alloys is that they seem to be able to learn a new shape. If they are repeatedly taken through the cycle of cooling to the martensite, deforming to a new shape, and then heating to regain the austenite phase *but under a constraint that holds the deformed material in place*, then after many cycles the deformed shape is taken up automatically when the material is cooled. The material can then adopt a different shape reversibly by

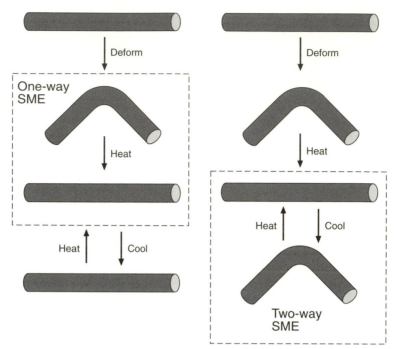

FIGURE 3.18 Shape-memory alloys can learn a new shape. If repeatedly deformed in the martensitic state and then heated to the austenitic state while constrained so that the deformation is maintained, the material learns to adopt the deformed shape automatically when cooled to the martensite. The shape can then be altered reversibly simply by heating and cooling. This is called a two-way shape-memory effect.

merely heating and cooling through the phase transformation, without applying a stress. This is called a two-way shape-memory effect (fig. 3.18). It is the result of stresses that accumulate in the austenite parent phase when the material is constrained in the deformed state before being heated. These stresses bias the transformation back to the martensite in such a way as to form preferentially just one of the tilted forms, rather than the self-accommodating mixture. This kind of two-way training can generate, for example, a spring that expands and contracts reversibly in response to temperature changes (which could be triggered by passing an electrical current through the material).

INSTANT FREEZING

Freezing of liquids inside a car engine is generally a thing to be avoided at all costs, as drivers of diesel vehicles sometimes discover to their cost in the depths of winter. But a class of liquids that can be frozen and melted again at the push of a button promise to be a great boon for engine design—they would banish clanking, grinding clutch problems forever. Such a liquid could be used in a contact-free clutch, where the bringing together of two plates in relative motion would never be necessary. Engaging clutch disks is a horrible engineering problem, incurring considerable wear on the surfaces due to the energy that must be dissipated as they lock together. But imagine instead that the two disks are spinning inside a compartment containing a liquid that can be solidified and melted almost instantaneously. When the substance is in the liquid state, the two disks can move independently—the clutch is disengaged. But if the liquid is abruptly frozen, the two disks become locked together, as if both encased in a block of ice—yet they never need to meet (fig. 3.19).

Fluids like this, whose flow (rheological) properties can be controlled by some external agent such as an electric or a magnetic field, are now being developed for applications of this kind. *Electrorheological fluids* can be switched to a highly viscous or semisolid gel-like state in a matter of a few thousandths of a second by applying a strong electric field. The solidified state has a measurable stiffness and resistance to shear. These smart fluids can be made "intelligent" by coupling them to sensor devices which, for example, detect sudden movements such as those caused by an impact, and respond by applying an electric field to the fluid. These systems would then act as shock absorbers—as the fluid becomes increasingly viscous, more of the energy of the impact can be dissipated within it. If the extent of fluid thickening depends on the magnitude of the shock received, the fluid

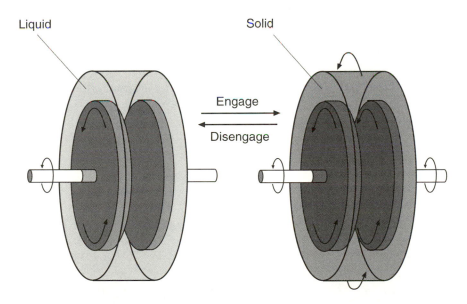

FIGURE 3.19 A clutch that requires no contact to be made between moving solid components can be devised by enclosing the two clutch plates in a compartment containing a liquid that can be reversibly solidified. When it is frozen, the liquid locks the plates together; when it returns to the liquid state, the clutch is disengaged. Such a device would suffer from far less wear on the plates, because they never have to come into contact.

acts as a responsive spring, adapting its resilience to suit the blow just as a boxer rolls with the punches. Electrorheological fluids might also find applications in valves with no moving parts, and in devices for damping or isolating mechanical vibrations.

In these instances we see again the replacement of a mechanical operation by a change in material property. The electrorheological effect—the change in flow properties of a liquid in response to an electric field—was first reported by the American scientist Willis Winslow in 1949. Clearly the commercial possibilities were not lost on him: in a style that is very much in tune with common practice in modern research, he had patented the discovery two years earlier. Winslow was studying the properties of suspensions, in particular of starch and of silica gel in mineral oil. The silica gel becomes dispersed in the liquid as tiny, near-spherical silica particles typically a few micrometers in size, which are visible in a conventional optical microscope. When the suspensions were placed between two metal plates and a strong electric field (of the order of a few thousand volts per millimeter) was set up across them, they "froze" into a solidlike state. If the plates were moved relative to one another with sufficient force, the resulting shear on the frozen fluid caused it to flow again, but like a highly viscous liquid. The shear force needed to induce this flow increased as the strength of the field increased— in other words, the frozen state was stronger in stronger electric fields. Winslow was able to see that in the frozen state, the disposition of the suspended silica particles was very different from that in the normal fluid: they were no longer distributed at random but became linked into chains parallel to the direction of the electric field (that is, perpendicular to the plates) (fig. 3.20). These chains, which form within microseconds of switching on the field, aggregate to give the material a fibrous, mechanically stiff structure.

This behavior is not hard to understand. The electric field pulls positive charges in the suspended particles onto the face of the particle that is directed toward the negative plate, and negative charges to the other face. This arrangement is sustained if the electrical properties of the particles differ from those of the surround-

FIGURE 3.20 Electrorheological fluids are suspensions of fine particles that form into chains in an electric field. These chains tangle into a fibrous structure, thickening the fluid. Shown here is the aggregation of polymer particles suspended in silicone oil. (Photograph courtesy of Keith Blackwood and Harry Block, Cranfield University, U.K.)

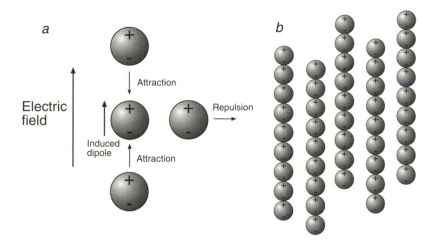

FIGURE 3.21 In an electric field, the particles in an electrorheological fluid acquire an induced electric dipole. The particles then attract one another along the axis of the dipole, but repel each other in the other directions (*a*). As a result, they aggregate into chains in which all the dipoles are aligned (*b*).

ing medium in which they are suspended: specifically, if the tendency to be electrically polarized in this way, quantified by the so-called dielectric constant, is different. In the case of Winslow's suspensions, a mismatch between the dielectric constants of the particles and the liquid medium was achieved as a result of the presence of small amounts of water on the surface of the particles—water has a different dielectric constant from oil.

So in the electric field the particles acquire an electric dipole—an asymmetric imbalance of charge, the electrical equivalent of the magnetic dipole of magnets. Just as the north pole of a magnet will be attracted to the south pole of another, so the opposite electrical poles of the particles attract each other, causing them to line up in chains with the dipoles, and thus the axis of the chains, parallel to the electric field (fig. 3.21). The chains initially formed in this way then aggregate into bundles that give the material its fibrous structure, a process called coarsening.

Despite his patent, Winslow never made much profit from his discovery because the materials that he used were not well suited to practical applications. For one thing, they were too abrasive, being in effect slurries of very fine sand. Furthermore, relying on the particles' water films to provide a large difference in dielectric constant is not a good idea, first because the water has a tendency to evaporate if the fluid gets warm and second because the water films provide a pathway for charge to flow along the particle chains. The latter is undesirable because the flow of charge consumes power, whereas low power consumption is a prerequisite for applications. The problem of abrasion was solved in the early 1980s when a group from Sheffield University in England developed electrorheological fluids in which the particles were made of soft polymers with ionic groups attached. These still relied on adsorbed water to acquire their electric dipoles, however; but more recently teams at Cranfield University in England and at the University of Michigan have prepared electrorheological fluids that require no water. They are based on suspensions of particles of polymers or ceramic materials in highly insulating liquids such as silicone oil. These materials remain active at high temperatures—up to 200 degrees Celsius, compared with a maximum operating temperature of around 80 degrees for water-activated silica suspensions. This means that the new materials have much more potential as shock and vibration absorbers, applications in which the dissipation of energy will heat up the fluid. The introduction of these water-free systems has led to pilot-scale initiatives by companies such as Bayer and Bridgestone to produce electrorheological fluids commerically, and real applications in engineering may not be too distant. Smart clutch fluids, however, will need to be able to dissipate considerably more heat than can be handled by the present generation of electrorheological fluids.

One problem in putting these materials to practical use is that the suspended particles have a tendency to settle out if the fluid is left to stand, just as silt will settle in a muddy pond. Particle settling can be reduced by the simple expedient of using smaller particles (which therefore experience a smaller gravitational force) and by adding surface-active compounds, rather like soap molecules, to prevent particles from clumping together into heavier aggregates.

While electrorheological fluids have not yet reached the commerical market, analogous smart fluids which take advantage of magnetic rather than electrical interactions are beginning to make their presence felt. These *magnetorheological fluids* work on much the same principle of chain formation by the sticking-together of tiny suspended particles, but this time the attraction is magnetic, and is switched on by applying an external magnetic field. Magnetorheological fluids are already used in the braking systems of some home-exercise machines, and there are plans for using them in motion-damping systems for vehicle seats and for buildings in earthquake-prone locations.

Because the chain structure breaks down above a certain applied shear stress, making the material fluidlike again, electrorheological fluids are like solids that can be fractured and reformed as many times as one likes. This property might be useful in devices such as release mechanisms or safety catches that signal the accumulation of a dangerous stress level: the breakdown of the solidlike structure and resumption of fluid flow when the stress gets too great might trigger some safety response in the device, but the broken catch can be subsequently reformed rather than having to be replaced. A material that can be smashed and reformed again and again . . . shades of *Terminator II* after all?

SOFT MACHINES

Most of the body's tissues can be regarded as smart materials: blood vessels contract in response to a fall in temperature, the cilia in our ears flex and pull open the gates of ion channels in response to sound, the heart is a self-synchronizing pump, the eye's iris contracts or dilates in response to light, and electrical impulses from the neuromuscular synapses induce changes in length of the muscle fibers. But all of these materials are soft, whereas the smart sensors and actuators described earlier are generally made of hard inorganic materials. In biomedicine, there is a great deal of interest in developing soft smart materials that resemble more closely the body's own. Many of the biomedical materials described in chapter 5 are soft polymers, but they are not, in general, too smart: when employed as valves, for example, they must be controlled by conventional mechanical drivers. Soft materials that flex, swell, or vibrate in response to changes in their environment might have enormous potential as artificial tissues and as components of biotechnological devices.

A wide range of such materials has now been devised. Most soft smart materials are polymers whose interaction with a surrounding solvent can be modified drastically by a change in the environmental conditions. Most commonly this modification induces an abrupt collapse of the polymer chain into a tangled ball, or conversely an abrupt unraveling into an extended strand. Such polymers are especially useful when formed into gels: networks in which the chainlike molecules are either crosslinked by chemical bonds or physically entangled into nets. Gels can accommodate solvents (water, in the case of so-called hydrogels) in the spaces between their strands. The amount of solvent held between the interlinked polymer chains depends on whether the chains are extended (so that the network is an open mesh that can take up plenty of solvent) or collapsed (which squeezes the solvent out). In many gels, the amount of absorbed solvent changes slowly as the ambient conditions (say, temperature or acidity) are altered. These changes induce swelling or shrinking of the gel, just as a dried fruit swells as it absorbs water. But in the 1970s certain polymer gels were discovered that undergo abrupt and pronounced swelling or shrinking in response to slight changes in the environmental conditions such as temperature. In these gels, a small stimulus can induce a large response. They can therefore potentially act as mechanical switches or valves that expand or contract when some critical control parameter is exceeded. Abrupt changes in a gel's volume have now been demonstrated in response to a whole range of stimuli, such as light or electric fields, and these polymer gels are finding applications as artificial muscles in robotics and for the controlled delivery of drugs into the body.

Collapse and Recovery

Environmentally responsive polymers alter the conformation of their chains in response to changes in the surrounding medium. One of the best studied of these is N-isopropylacrylamide, or NIPAAm. Isolated chains of NIPAAm are water soluble in cool solutions, but if the solution is warmed up the strands collapse and precipitate out. This precipitation behavior seems surprising at first, because we are more used to substances that become *more* soluble in warmer water: hot coffee will dissolve more sugar than cold. The key to the polymer's behavior lies with the nature of the chemical groups along the chain: like most polymers that show this behavior, NIPAAm contains both water-soluble (acrylamide) and water-insoluble (isopropyl) groups. In solution, the former favor an open, elongated chain in which the water-soluble groups can interact with surrounding water molecules, but the latter favor a collapsed state in which the water-insoluble groups can be shielded from water.

The elongated state is rendered stable at low temperatures by hydrogen-bonding interactions between the polymer's amide groups and water molecules (fig. 3.22). This binding is energetically favorable, but it comes at the cost of decreasing the mobility of water molecules in the vicinity of the chains, which decreases their *entropy*. Entropy is a measure of the disorder of a system and can be decreased only if compensated by a sufficiently large energetic gain—that is, only if the process liberates heat (by, for example, the formation of chemical bonds).

But the price of a decrease in entropy rises as the temperature rises, whereas the compensation provided by the liberation of heat does not increase proportionately. So there comes a point at which the heat given out as water molecules stick to the polymer chains is not enough to compensate for the loss of freedom of the bound molecules. At this temperature (called the lower critical solution temperature, LCST) the polymer chains throw off most of their bound water, and the presence of the water-insoluble isopropyl groups then leads them to collapse into bundles (fig. 3.22). This abrupt change in conformation of the polymer in response to a temperature change is a kind of smart behavior that can be exploited to good effect.

For instance, a polymer strand attached chemically to another molecule can act as a "precipitation switch" by means of which it can be pulled selectively out of a solution. This can be used for separating and purifying biomolecules. Allan Hoffman, Patrick Stayton, and colleagues at the University of Washington in Seattle have used a smart polymer "switch" to control the ability of a protein to bind a small molecule. They attached a NIPAAm chain to the protein streptavidin, which has a binding pocket that seizes very tightly onto the molecule biotin (called its affinity ligand). The researchers used genetic engineering to incorporate right next to the protein's binding site a chemical group to which the polymer could be linked (fig. 3.23).

When the polymer chain remained extended in solution, the modified streptavidin bound biotin in the usual way. But heating the solution to above 32 degrees Celsius, while having no significant effect on the protein molecule's structure, triggered the collapse of the polymer chain; and the collapsed mass then blocked the biotin binding pocket and prevented biotin molecules from docking (fig. 3.23). In this way the researchers were able to switch the binding ability of the protein on and off by a very slight change in temperature. They propose that this reversible on-off switching of a protein's function might have a wide range of uses. For instance, if the same trick were performed on an enzyme, which catalyzes a chemical reaction rather than just binding a molecule, one could control the reaction rate at will—something that could be potentially useful in the enzymatic fermentation processes now widely used by the food and pharmaceutical industries. They also speculated that the release (rather than the binding) of a

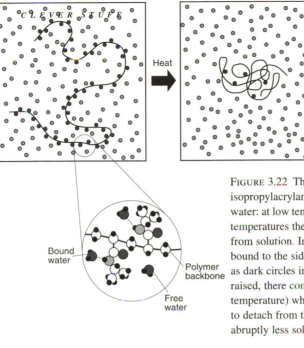

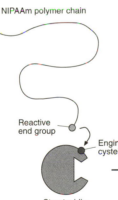

Bound water

Polymer backbone

Free water

FIGURE 3.22 The chainlike molecules of the polymer N-isopropylacrylamide (NIPAAM) are thermally responsive in water: at low temperatures they are extended, but at higher temperatures they collapse into a bundle and precipitate from solution. In the extended state, water molecules are bound to the side groups along the polymer chain (shown as dark circles in the top left frame). As the temperature is raised, there comes a point (at the lower critical solution temperature) where it is favorable for the water molecules to detach from the chains. The polymer then becomes abruptly less soluble, and the chains collapse.

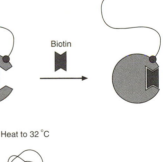

NIPAAm polymer chain

Reactive end group

Engineered cysteine unit

Biotin

Streptavidin

Heat to 32 °C

No binding

FIGURE 3.23 Binding between a protein and its affinity ligand (a molecule that fits the protein's binding site) can be turned on and off by attaching a collapsible polymer strand to the protein near the binding site. Allan Hoffman and coworkers have used this idea to control the binding between the protein streptavidin and its affinity ligand, biotin. They attached a strand of NIPAAM to the protein by genetically engineering an attachment site near the protein's binding site. When the NIPAAM chain was extended, binding of biotin could take place; but collapse of the chain above the LCST blocked the binding site.

ligand could be triggered in the same way—the collapse of a polymer appendage around a binding site could push the ligand out. If the ligand is a drug molecule, this would provide a means for triggering the release of a drug within the body, for example by warming the region in which the drug is to be delivered.

Gels That Swell

When chains of NIPAAm are cross-linked by chemical bonds they form a hydrogel, which can take up large amounts of water in the spaces between the interlinked chains. The consistency of the water-swollen network varies from jelly-like to rubbery, depending on the degree of cross-linking and the amount of water it has absorbed. If the temperature is raised above the polymer's LCST, the chains collapse and water is squeezed out from between them, leading to shrinkage of the gel (fig. 3.24a). The temperature at which total collapse occurs can be varied by varying the composition of the polymer—for example, by combining NIPAAm molecules with acrylamide molecules in the chains in different ratios. For some compositions, shrinkage of the gel is gradual as the temperature is raised, but for other compositions it can be very abrupt—the gel's volume can change by a factor of a hundred as the temperature changes by just a few degrees (fig. 3.24b). This sudden change in volume is called a volume transition. In practice, "sudden" refers to the fact that only a tiny change in temperature triggers the transition; it does not mean that the transition is quick. A gel may take hours, days, or even weeks to swell or shrink to its final equilibrium volume, depending on how long it takes the polymer network to take up or expel the water—this in turn depends on the size of the gel.

These NIPAAm-based gels are temperature responsive. A polymer gel made from cross-linked chains containing acrylic acid and acrylamide, meanwhile, undergoes volume transitions in response to changes in acidity (pH). Here the transition is driven by changes in the chemical state of the acrylic acid monomers. In alkaline solution, these lose hydrogen ions to become negatively charged acrylate groups, which repel each other and cause the gel to swell; in acidic solution, the acrylate groups regain a hydrogen ion and are neutralized, and the gel shrinks. Toyoichi Tanaka at the Massachusetts Institute of Technology has shown that these gels can also be made to swell or shrink in response to electric fields: when the ionized form is placed in an electric field, the hydrogen ions are attracted to the negative electrode. The effect is rather like removing the gas from an inflated balloon: it causes the gel to shrink. A polymer gel that changes shape in response to an electric field might be useful as a soft switchable valve or as an artificial muscle. Kenji Kajiwara from the Kyoto Institute of Technology in Japan has made a robot hand in which cylindrical polymer fingers flex in response to electric fields: because they are soft, these robot fingers are able to pick up delicate objects such as eggs.

One of the major uses envisioned for smart polymer hydrogels is as agents of drug delivery. The aim here is to release the drug in a controlled way—at a steady rate over a long period, perhaps, or suddenly in response to a change in environ-

a

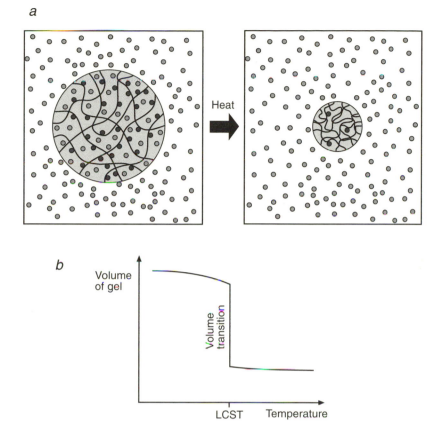

b

FIGURE 3.24 A cross-linked polymer gel shrinks above the LCST of the polymer, when the chains collapse and solvent is squeezed out from the polymer network (*a*). This collapse can be very abrupt, reading to a sharp volume transition in which the macroscopic volume of the gel changes discontinuously (*b*).

mental conditions such as acidity as the drug carrier passes from one organ to another. Hydrogels can keep drug molecules entrapped while in the collapsed state, but can release them if the gel swells and the polymer network is penetrated by water. The general strategy is to load tiny spheres of gel with drug molecules by letting them diffuse into the gel in the swollen state and then entrapping them there by shrinking the gel. The microspheres are then injected or ingested in the body, and they release their contents when they swell in response to some stimulus. Drugs could be released specifically in the gut, for example, when encapsulated in pH-sensitive, acidic gel microspheres that swell as they pass from the gastric system of the stomach (at a lower pH) to the enteric system of the gut (at higher pH).

The specificity of enzyme-catalyzed biochemical reactions—the fact that enzymes will transform one kind of molecule while ignoring all others—can be exploited to create gels that react to a specific chemical stimulus. In many bio-medical applications this might be the most useful kind of smartness of all, be-cause it allows one to develop materials that stay in tune with the body's bio-chemistry. This, indeed, is how many chemical processes in the body work: the generation of a particular molecule acts as a trigger to stimulate some other pro-cess. This sort of chemical response can enable the body to maintain constant conditions in the face of a changing environment—for example, if the concentra-tion of some compound in the cellular fluid exceeds a certain threshold, this trig-gers cellular production of an enzyme or some other biomolecule that acts to remove the trigger compound. That's how glucose is metabolized: rising levels of glucose in the bloodstream stimulate the pancreas to produce the hormone insu-lin, which mediates glucose metabolism.

A volume-changing polymer gel can be envisioned that would mimic this ac-tion of the pancreas in people with insulin-dependent diabetes (who cannot pro-duce insulin), not by synthesizing fresh insulin but by releasing preloaded insulin when glucose levels get too high. The key would be to attach to the gel an enzyme that reacts with glucose to bring about some kind of environmental change, such that the gel is induced to swell and release its charge of insulin. One such enzyme is glucose oxidase, which oxidizes glucose and generates a change in acidity in the process. If glucose oxidase were attached to a shrunken pH-sensitive gel loaded with insulin, a rising level of glucose would cause the pH in the vicinity of the gel to increase as the glucose is oxidized, eventually triggering swelling of the gel and release of the insulin, which would then allow the body to go about its normal insulin-mediated business of glucose metabolism. On page 240 I describe a smart polymer device of this sort.

Polymer gels are getting smarter. In 1995 Yoshihito Osada and Atsushi Matsuda, researchers at Hokkaido University in Japan, described a polymer gel with a memory, like the shape-memory metal alloys mentioned earlier in this chapter. A copolymer of acrylic acid and *n*-stearyl acrylate, their material showed a shape-memory effect when swollen with water. In this state, the copolymer behaved as a hard plastic below 25 degrees Celsius but was soft and elastic above 50 degrees. So the material could be molded or bent into different shapes above this temperature, which it would retain when cooled.

The researchers made a rod of the material by forming the copolymer in a glass tube and then swelling it with water. They heated the rod to 50 degrees, formed it into a coil, and cooled it to room temperature, so that the rod became set in the coiled shape. When warmed again to 50 degrees, the coil unwound and stretched out to its original rodlike shape within 15 seconds (fig. 3.25). Without this shape-memory effect, the coil would simply have become soft when warmed, but would have shown no tendency to straighten out. The researchers found that a transfor-mation of the gel's molecular structure took place at around 50 degrees in which the side chains of the stearyl acrylate units switched from a low-temperature, ordered packing arrangement to a high-temperature, disordered state. They sug-gested that this transformation plays much the same role as the martensite-to-austenite transition of shape memory alloys, so that as the transformation is

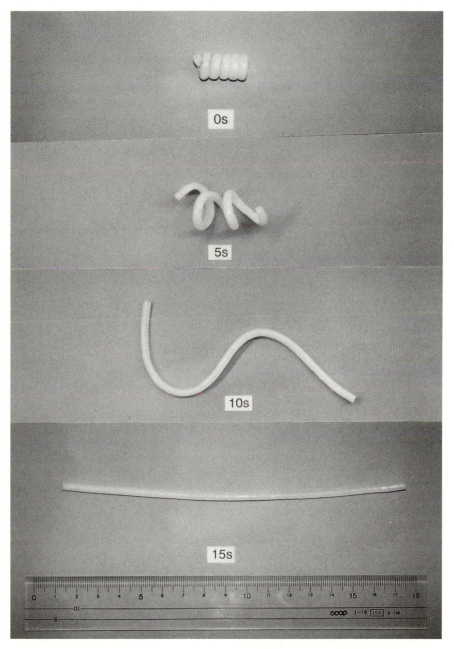

FIGURE 3.25 A shape-memory effect in a polymer gel. The gel is initially formed in a rodlike shape like that in the lowest frame. At room temperature the water-swollen gel is a hard plastic, but when heated to 50 degrees Celcius it becomes soft and malleable. At this higher temperature the rod is formed into a coil and then cooled to room temperature to "freeze in" the coiled shape. When heated again to 50 degrees, the gel remembers its original shape and uncoils spontaneously in 15 seconds. (Photograph courtesy of Yoshihito Osada, Hokkaido University, Japan.)

traversed the material regains its original, undeformed state. They believe that this memory effect in a soft material might find medical uses, for instance as the basis of a heat-activated diaphragm or surgical clamp.

SMARTER BY THE DAY

Materials that will repair their own flaws, or that will provide advance notice of impending failure, will be of tremendous value to the aeronautical and aerospace industries. Those that can be made to move and flex in response to external stimuli will surely find uses in medicine and robotics. And those that perform mechanical functions that are presently carried out by multicomponent devices with moving parts should find applications in all corners of engineering. Already, smart structures in which actuators and sensors are coupled to information-processing units are finding their way onto the civil engineer's palette. Bridges are being equipped with monitoring devices that provide a record of their state of health, warning of slippage, crumbling, or corrosion of the component parts. The Kingston Bridge in Glasgow, one of the busiest highways in Britain, has been kitted out with arrays of position and temperature sensors that supply a constant update on movements caused by subsidence and environmental factors. In earthquake-prone Japan, meanwhile, several new buildings in Tokyo and Osaka have been given vibration-damping and variable-stiffness devices to counteract the seismic hazard. Many of these devices are sophisticated pieces of engineering in their own right, but at their heart are smart materials. Some researchers envision smart structures laced with fiber-optic "nerves" that will tell us exactly what the structure is feeling at any instant—information that could be critical in civil engineering and aerospace applications, where catastrophic failure can be literally just that.

A report on smart materials prepared by the Japan Science and Technology Agency in 1989 sees their influence extending into all aspects of our lives. It proposes that "materials in daily life, including clothes, building materials and daily necessities" will be targets for smart engineering "for greater comfort and convenience, depending on the environmental conditions." Clothes that become better thermal insulators as the weather gets colder, food wrapping that changes color to warn of an approaching "use by" date: the implication behind such ideas is that high technology need not mean ever fancier and more complex electronic gimmickry, with more and more components crammed into an ever smaller space; rather, the future may hold an increasing simplicity, as materials replace machines, and as our environment becomes adaptive rather than more and more tightly controlled.

Only Natural

BIOMATERIALS

> . . . so, over that art
>
> Which you say adds to nature, is an art
>
> That nature makes.
>
> —William Shakespeare, *A Winter's Tale*,
>
> Act 4, Scene 4

In addition to supplying useful materials in their own right, the living world provides endless inspiration for the design of synthetic materials with sophisticated structures and functions. Nature achieves this sophistication through highly organized fabrication methods and hierarchies of structural features.

THE WORLD of advanced materials is a highly competitive commercial arena, but there is one organization that could never fail. Here are some of its products: polymer fibers stronger than steel; photodetectors more sensitive than the best that mainstream photonic technology has to offer; tough ceramic composites fashioned into elaborate patterns visible only under the microscope; data storage media that carry one bit of information for every three molecular units. Yet it is no wonder that this organization excels so far beyond its rivals, for it has been around for rather longer. For more than three and a half billion years longer, as far as we can tell. The company is Life, and its materials are the most advanced on the planet.

Humankind has long benefited from these products. Traditionally we have measured humanity's progress in materials exploitation according to our ability to fashion the products of the inorganic rather than the organic world: stone, iron, metal alloys, silicon. But it is likely that humans used wood as a tool at the same time as they developed a facility for shaping stone. (As wooden artifacts do not survive for long, however, the Wood Age would be a tricky thing to pin down.) Today, we still do not have a material that rivals wood in its subtlety of structure and property. Bone, too, no doubt found early use as a tool (though perhaps not so early as Stanley Kubrick would have us believe in *2001: A Space Odyssey*). We are only now on the threshold of being able to produce synthetic substitutes that come close to fulfilling the same role. And their combination of extreme toughness, insulation, and water resistance makes animal pelts and hides as sophisticated a material as Neolithic peoples, or we today, could wish for.

Perhaps inorganic materials are considered to delineate our technological progress because we take natural biological materials for granted: they are readily available, and they do not require complicated feats of chemistry and engineering to turn them into useful forms and give them useful properties, because billions of years of evolution have already refined their characteristics to an astonishing degree. While we can improve on iron by turning it into steel, we haven't made much significant improvement to wood or to leather, for there is no need.

Today we are seeing a return to fashion of natural materials, after the celebration of synthetic plastics in the 1950s and 1960s. ("I want to *be* plastic," said Andy Warhol.) It is common to regard these plastics as representative of all that is "unnatural" about modern materials. We think perhaps of nylon, whose use in textiles was once lauded but is now very much passé. We think of polyethylene (often called polythene), accursed for its overenthusiastic use in packaging, and of polystyrene, which in its expanded, porous form provides ubiquitous protection for fragile electronic goods. These materials are tough, durable, relatively cheap, and easily molded, and it is hard to deny their utility. But they are often now regarded as aesthetically displeasing, the materials of a throwaway culture generated by the irreversible consumption of ever more precious oil reserves. When we are finished with them, they sit around on waste dumps or on street corners, looking ugly but resisting degradation. If we burn them, we release toxic and evil-smelling fumes.

It is easy to forget, then, that the plastics industry has its origin in natural materials, which we learned to harvest and to modify to our needs long before we knew how to make comparable materials of our own design. And like synthetic plastics, just about all of nature's structural materials are polymers: large molecules built up by the assembly of many identical small units (the word *polymer* is an adaptation of the Greek for "many parts"). Most natural polymers have a chainlike structure, with the constituent units (called monomers, or "single parts") linked together via chemical bonds in linear fashion.

Perhaps the greatest value of biological materials does not lie with their "naturalness" in itself but in their potential to serve as models for the advanced materials of the future. They provide endless inspiration, and the way that they fulfill their roles displays boundless ingenuity. In recent years, as the potential for controlling materials structure and properties has blossomed, it has become ever more apparent how natural materials can serve as a guide for the design of synthetic ones. This has led to an ever increasing desire to understand how these natural materials are themselves forged, something that requires the collaboration of diverse disciplines ranging from engineering to genetics.

AT THE HEART OF LIFE

What is life? That question has been asked again and again. The physicist Erwin Schrödinger and the biologist J.B.S. Haldane, both giants of modern science, each wrote a book of that title in the 1940s. (Only Haldane confessed at the outset, however, that he was not going to answer the question.) It is only a refinement of the same question to ask, "What is an elephant?" And the only real answer can be a paraphrase of the response of the proverbial blind Indian sages: it depends on which part you're looking at. So: life can be a series of chemical reactions, it can be an interdependent ecosystem, it can be a network of information exchange, it can even be a God-given force if you so wish.

That makes me feel a little better about saying that life is also a materials production line. Lacking in romance as this description is, it stands to reason. We were built from a fertilized egg almost too small for the eye to see. All this fabric—this skin and bone and blood and muscle—was synthesized, and continues to be synthesized as you read. We have been assembled from a single cell.

As I want to talk in this chapter about the products of that assembly line, it makes sense to start with that single cell, which houses all the blueprints. It is a repository for a biomaterials tool kit—for molecules that between them can synthesize all of the materials an organism needs, including all the components of another, identical cell. The cell is a materials factory that can copy itself.

So let's look first at this tool kit. It consists primarily of the two most important kinds of polymer in biology: proteins and nucleic acids. The study of these tools and how they operate occupies more scientists than any other discipline today: it is the field of molecular biology. Proteins and nucleic acids are life's *functional* polymers, and together they orchestrate the vast majority of the body's biochemical processes. They are the engineers of our lives from conception to demise. From what are these tiny molecular devices made?

PROTEINS—NATURE'S MACHINERY

The Building Blocks of Life

Industrial polymer chemistry is an eclectic science—in its dazzling array of products you will find a diverse collection of molecular building blocks, linked together in all manner of architectures. Nature is more conservative in her building materials, but far more subtle in the way she puts them together. Just a handful of monomer units suffices to provide all of nature's complex machinery and materials. Proteins, surely the most versatile of natural polymers, are made up of just twenty varieties of small molecular units called amino acids, linked together in chains (fig. 4.1). Amino acids contain both a carboxylic acid group (like that in acetic acid, or vinegar) and an amine group, a so-called base. Acids (or to be more

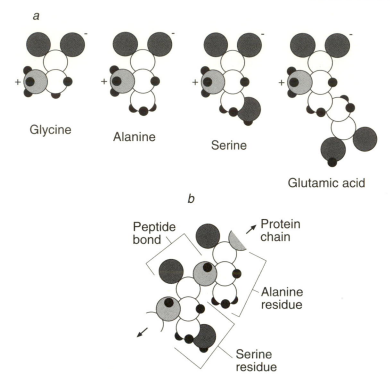

FIGURE 4.1 Amino acids are the building blocks of proteins, of which twenty varieties are found in nature. Here I show four examples (*a*). Each amino acid has an amine group (-NH₂; nitrogen atoms shown as light gray) and a carboxylic acid group (-COOH; oxygen atoms shown as dark gray). In neutral solution, both groups are ionized: the carboxylic acid loses a positively charged hydrogen ion, and the amine group gains one. In proteins, amino acids are linked into long chains via peptide bonds between the amine group of one amino acid and the carboxylic acid group of another (*b*). The remnant of each amino acid group in the chain is called a residue.

precise, the common class of acids called Brønsted acids) are characterized by their propensity to lose a positively charged hydrogen ion (a proton). Bases have the opposite tendency to take up hydrogen ions, forming a positively charged (protonated) chemical group.

Amino acids are (with the exception of glycine) *chiral* molecules, which means that they can each exist in two forms (enantiomers) that are identical except that one is the mirror image of the other—like our left and right hands (fig. 4.2). There is a convention in chemistry for labeling these two isomers "left-handed" and "right-handed"; all amino acids in nature are of the left-handed variety, although right-handed versions can be synthesized in the laboratory. The acid and base groups of two amino acids will readily react with each other in a manner that forges a chemical link between them and kicks out a molecule of water in the process. The link is called a *peptide bond*, and it represents the glue that holds

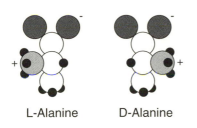

L-Alanine   D-Alanine

FIGURE 4.2 All amino acids with the exception of glycine contain four different chemical groups around a central carbon atom (glycine contains two identical groups—two hydrogen atoms). Because these are arranged in a tetrahedral manner, there are two mirror-image forms of the molecules, denoted by the prefixes D and L. The D-form rotates plane-polarized light to the right, and the L-form rotates the plane to the left. All amino acids found in nature are of the L ("left-handed") variety.

together a chain of amino acids in a protein. Polymers of amino acids are called *polypeptides*; the word "protein" is traditionally reserved for a functional polypeptide produced by cells (as opposed to, say, a random assembly of amino acids), although we will see later in this chapter and in chapter 9 that researchers are now able to produce peptide polymers that deserve the label "synthetic proteins." The seemingly endless repertoire of properties of proteins—from catalytic molecules called enzymes to intercell messengers, guardians of the immune system, and structural materials like silk—stems largely from the differences in the sequence of amino acids along the chain.

This sequence determines the shape—the conformation—that the polypeptide chain adopts. Every different protein has a different shape, but one can distinguish structural motifs that appear again and again in different proteins with entirely different functions. One such is the α-helix, whose structure was first deduced by Linus Pauling in the 1950s (fig. 4.3). Here the polypeptide chain curls up into a coil as a result of weak chemical bonds, called *hydrogen bonds*, between the carbonyl (C=O) and amino (N–H) groups of the peptide bonds. The spiral twist allows every carbonyl group to be hydrogen-bonded to the hydrogen of the amino group in the amino acid three monomer units away. It turns out that for polypeptide composed of left-handed (that is, natural) amino acids, an anticlockwise (right-handed) coil is slightly less crowded, and so more stable, than a left-handed helix—so the α-helix is a right-handed spiral.

These and other motifs make up the so-called *secondary structure* of proteins: the way in which the polypeptide backbone is folded up. The *primary structure* is the molecular structure of the backbone itself: the sequence of the amino acid constituents (and also the location of occasional chemical bonds between different parts of the chain). But functional proteins generally derive their activity from still higher structural levels: the *tertiary structure*, which describes how secondary structural motifs are packed together to give the protein its three-dimensional shape; and the *quaternary structure*, which refers to the grouping together of individual proteins into multiunit clusters that collectively carry out a task (fig. 4.4). We can thus see a hierarchy of structural organization in these biomolecular substances. Functional proteins are *big* molecules (or aggregates of several molecules), often visible individually under an electron microscope.

Functional proteins generally have a compact, globular shape. Most of these

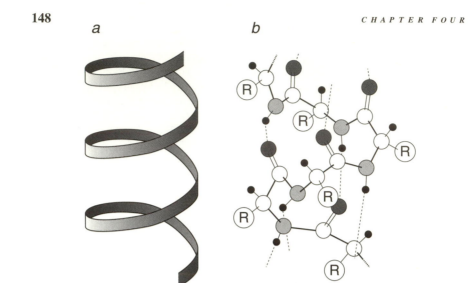

a   *b*

FIGURE 4.3 In the α-helix of proteins, the polypeptide chain is coiled into a spiral (*a*) with a right-handed "thread," which is held together by hydrogen bonds (*b*; dashed lines) between the amino acid constituents. This is one of the fundamental motifs of protein secondary structure.

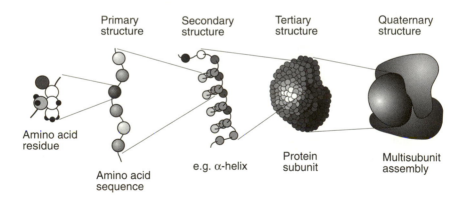

Primary   Secondary   Tertiary   Quaternary
structure   structure   structure   structure

Amino acid
residue

Amino acid   e.g. α-helix   Protein   Multisubunit
sequence   subunit   assembly

FIGURE 4.4 Several levels of structure can be distinguished in proteins. The primary structure is the sequence of amino acid groups (residues) along the polypeptide chain. These chains are folded into various secondary structural motifs, such as coils, sheets, and ribbons. The motifs are packed together to give the overall molecule a distinctive shape, the tertiary structure. Some functional proteins contain several distinct polypeptide chains, called subunits; this association of several subunits corresponds to the protein's quaternary structure.

are enzymes, biological catalysts that assist the biochemical processes of the body. Not only do these biochemical reactions often involve a high degree of selectivity and precision—for example, the enzyme must select precisely the right initial molecule from a seething horde of others, and must then make a chemical modification at exactly the right place on this reactant molecule—but they must be performed under the extremely mild conditions of the body, using moderate temperatures and pressures and in solutions that are usually only modestly acidic or alkaline. The catalytic role of an enzyme has two aspects: the protein must reduce the energy required to initiate the reaction (for example, the energy barrier to breaking of a chemical bond), and it must do so for just the particular reaction pathway required and not for other, competing pathways that would generate different products. Enzymes perform this function by providing a kind of micro-environment that is specifically designed to facilitate a given reaction: they provide a comfortable housing for the target molecule (into which other molecules will not fit), a cushioning environment that reduces the "pain" of chemical modi-fication (more precisely, that reduces the energy of the intermediate states through which the target molecule must progress on its way to becoming the product), and an array of operating units that perform the required molecular surgery. The latter are often so-called prosthetic groups, which are nonpeptide components added to the protein after the polypeptide chain has been synthesized in the cell. These groups commonly contain metal ions, which are adept at catalysis and which also feature in much industrial catalysis. The microenvironment of the enzyme in which the target molecule sits—the binding pocket—contains surfaces with delicately tailored surface properties: water-repelling (hydrophobic) where it will marry up to a region of the target molecule with this same property; water-attractive (hydrophilic) where water molecules might actually be required within the binding pocket to facilitate the reaction; and an external coating to endear the protein to its own environment, whether that be a cell membrane or the watery cytoplasm of the cell. In effect, the enzyme acts as a lock into which only the right key—the enzyme's target molecule—will fit. This lock-and-key principle of enzyme action was first proposed in 1894 by the German biochemist Emil Fischer.

Another major class of globular proteins is the *immunoglobulins*, also known as *antibodies*, which are the warriors of the body's defense system. These mole-cules are able to recognize invaders such as viruses or bacterial products, and bind to them tightly to provide a tag that marks the foreign material for destruction by the immune system's war machines, the lymphocyte cells. That antibodies have this ability might seem near-miraculous, because often they will be required to tag molecules that the body has never encountered before; but the solution is a simple, if perhaps profligate one. The glands of the immune system simply churn out armies of immunoglobulins of diverse and essentially random composition, billions upon billions of structural variants, in the hope that a few of these will happen to have a shape conducive to binding the invader (the antigen). Among so many varieties, the probability of a good match is high, and once successful binding is established, the immune system latches onto that formula for subse-quent refinement and mass production of an antibody. The binding pockets of antibodies have features much like those of enzymes, but in the former case the task is somewhat simpler—merely to hold on to the target, not to enable its modification.

With all this in mind, it is not unreasonable to regard globular proteins as "sculpted" materials. They display features that we might commonly associate with macroscopic materials: they have surfaces, they have elasticity, they have to be carefully packaged, they are made piece by piece on an assembly line. The remarkable fact is that the instructions for this sculpting are all encoded within the protein's *primary* structure, so that once formed, the protein is already programmed to adopt the required shape. This is evident from the fact that if a globular protein is denatured from its native (active) shape by heating, it can recover that shape (renature) when cooled. This is all the more remarkable given the number of alternative chain-folding pathways with which a protein is generally presented: there is a literally astronomical number of choices, many of which give folded structures that are only slightly less stable than the native form of the protein. Yet the chain (usually) manages to avoid getting stuck in these conformations and instead finds its way back to the most stable form. This is truly a property that materials scientists dream of: a material that is self-programmed to adopt a particular, highly complex structure at the molecular level.

Although our understanding of how proteins fold into their complicated globular shapes is still very incomplete, scientists are developing an increasing ability to modify these structures by design. The aims of such studies are manifold: for pharmacologists, the goal might be to develop new protein-based drugs, whereas more fundamental studies are aimed at making polypeptides that mimic the motifs found in natural proteins, to provide model systems for exploring the relation between structure and function. For materials scientists, meanwhile, the ability to tailor proteins raises the possibility of making new peptide-based materials that have some of the characteristics of natural structural proteins—a topic covered in chapter 9.

There are two general approaches to protein design: one takes natural proteins and introduces modifications along the polypeptide chain, while the other plans and builds purely synthetic polypeptides from scratch, which the designers hope will mimic some of the features of natural proteins—this is called de *novo* design. In both cases, the likely structural outcome of a particular amino acid sequence is usually predicted in advance by performing computer-modeling studies. Synthesis of a de *novo*-designed polypeptide can be carried out in several ways, the most common of which are solid-phase and microbial syntheses. In the former, the peptide chains are attached to a solid support, and the amino acids are added one at a time in a controlled, sequential manner. This is fine for making relatively short polypeptides, but becomes tedious for longer chains. In that case, the microbial approach is more effective: the genetic material of bacteria is modified so that they synthesize the designed peptides when cultured in fermentation vats. This approach is usual if one wants to modify natural proteins; in this case, the genes that encode the protein are themselves modified and inserted into the bacteria's genetic material (see chapter 9).

From the many dexterous examples that have now accumulated of *de novo* design, let me pick just one to illustrate how peptide chemists go about building a custom-made "synthetic protein." William DeGrado and colleagues at Du Pont

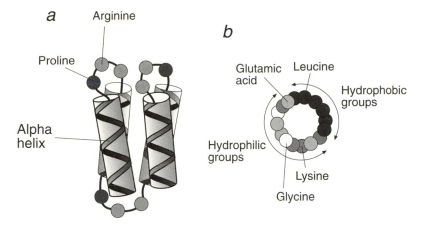

FIGURE 4.5 A peptide chemically synthesized from individual amino acids by William DeGrado and colleagues will spontaneously fold in water into a four-helix bundle, in which four α-helical coils form a cluster (*a*). Each helix has a careful placement of amino acid residues with hydrophilic and hydrophobic side groups, such that all of the former are on one side and all of the latter on the other (*b*, showing a top view down a helix). This encourages the four helices to form a bundle to shield the hydrophobic sides of the coils from the water.

Merck Pharmaceuticals in Wilmington, Delaware, have attempted to design from scratch folded polypeptides with four-helix bundles, in which the polypeptide chain forms four successive α-helices that cluster together, via a weaving back and forth of the chain, into a cylindrical bundle (fig. 4.5*a*). Helical bundles of this sort are common in membrane proteins, which, as discussed later, sit embedded within cell walls. The four-helix bundle in particular is a feature of the membrane protein cytochrome c, which plays a central role in photosynthesis. The formation of α-helical sections of a polypeptide chain is encouraged by the presence of the amino acids lysine and leucine, so DeGrado's group incorporated these into the chains. But they had to figure out how to ensure that these helices clustered together in a bundle. They hypothesized that if they could arrange amino acid side chains along the helices such that one side of the helix became hydrophobic and one side hydrophilic, the helices would come together in water in such a way that the hydrophobic sides lay inside the cluster, protected from water, while the hydrophilic groups were exposed on the outside. They linked the coiled regions together via short chains that could bend back on themselves in a loop (fig. 4.5*a*).

Using solid-state methods, DeGrado's group were able to construct individual polypeptide helices. But to create the full polypeptide, with four helices linked by three loop units, it was easier to make a synthetic gene that encoded the protein's sequence and insert it into the genetic material of the bacterium *Escherichia coli*. They found that the synthetic polypeptide thus produced did indeed adopt the four-helix bundle motif in solution. The team's ultimate goal is to incorporate metal ions into this structure in the hope of attaining catalytic activity like that shown by cytochrome c.

CHAINS OF INFORMATION

The evident importance of proteins in the body's chemistry led researchers to conclude in the early part of the twentieth century that the genetic material—the stuff that transfers hereditary characteristics from one generation to the next— must surely be protein based. That characteristics are inherited from parents has been evident even to the dullest eye—and particularly to livestock breeders and cuckolded husbands—for centuries. Inheritance of characteristics underpinned Darwin's theory of evolution in the 1870s. But only at the beginning of this century was much headway made in unraveling the molecular mechanism of inheritance, based largely on the work of Gregor Mendel, an Austrian monk who proposed that ill-defined "particulate factors" were passed from parents to their offspring. These would later be called *genes*; but it was not until the late 1940s that good evidence was amassed that nucleic acids, not proteins, are the molecular carrier of the genes. The modern understanding of genetics can be traced directly to the solution of the molecular structure of DNA in 1953 by Francis Crick and James Watson, who identified its famous double-helix motif from the crystallographic work of Maurice Wilkins and Rosalind Franklin.

If there existed such a thing as a polymer chemist with no knowledge of biochemistry, he would probably conclude on conducting an analysis of DNA that it was a random copolymer, which is to say that it consists of several (to be precise, four) different kinds of monomer units joined together in a chain apparently at random. These four units are called nucleotides, and each contains three constituent parts: a phosphate group, a sugar (deoxyribose), and a base. The nucleotides differ only in the nature of the base, which may be adenine, thymine, cytosine, or guanine. The phosphate and sugar groups form the backbone of the strand, and the bases hang off the sugars like appendages (fig. 4.6). Two such strands twist around each other in a right-handed double helix.

But our chemist's analysis of both strands of the double helix would indicate that there is more organization to the sequence of nucleotides than can be explained by a process of random assembly. The strands are held together by hydrogen bonds between bases, and while in principle each base could form hydrogen bonds with itself or with any other of the bases, in practice adenine binds only to thymine along the helix, and cytosine only to guanine. Thus, the two strands have precisely complementary structures—wherever an adenine appears on one strand, a thymine appears on the other, and so on.

The sequence of adenine–thymine and cytosine–guanine base pairs is a code. What it encodes is the information required to build a protein. Proteins may be the infinitely adaptable workhorses of the cell, but one thing they cannot do is build themselves. The DNA that an infant inherits from its parents is the most valuable bequest it will ever receive: an instruction kit for making proteins. Since the physical features of organisms are determined by their genetic makeup, and as an organism acquires DNA from both parents, it inherits a mixture of characteristics from both.

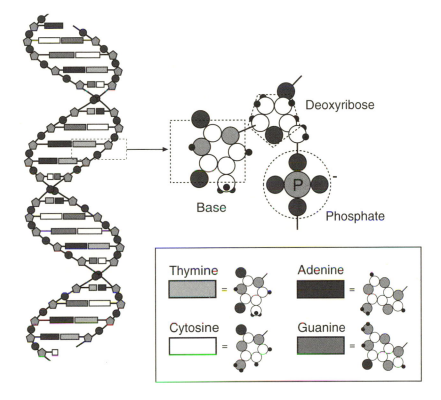

FIGURE 4.6 Deoxyribonucleic acid (DNA) is the basis of life. Aside from a few viruses, the genetic information of all living organisms is encoded in the molecular structure of this molecule. DNA is a linear polymer, in which four different nucleotide bases (thymine, adenine, cytosine, and guanine) are attached to a backbone of phosphate and deoxyribose molecular groups. The order in which the bases appear delineates a code, containing the information needed to put together the protein molecules that do most of the body's biochemical tasks. Two distinct strands of this linear polymer twist around each other in a double helix, held together by hydrogen bonds between the bases that stick out from the backbone. Each base pairs up with a specific partner in this helical staircase: adenine pairs with thymine, and guanine to cytosine.

Reading the Plan

The primary interest of DNA here is not as a material itself but as the blueprint for protein-based materials. Broadly speaking, each protein is encoded by a separate gene on the DNA molecule. Each amino acid in a protein is represented in the corresponding gene by a group of three nucleotides; as there are sixty-four ways to combine four nucleotides in groups of three and only twenty amino acids in natural proteins, this genetic code has some redundancy (some amino acids are encoded in more than one kind of nucleotide triplet). The translation process is mediated by a second kind of nucleic acid, RNA, in which the sugar is ribose instead of deoxyribose and the base thymine is replaced by a similar base called uracil. To translate a gene, enzymes first make a transcript of just that portion of the genetic material in the form of an RNA molecule, which is assembled on the strand of DNA representing that gene. The DNA, which has to be unzipped from its double-helical form for this process, acts as a template on which an RNA version is made by complementary base pairing between the nucleotide bases of the DNA and those of the RNA (fig. 4.7). This is called transcription of the gene.

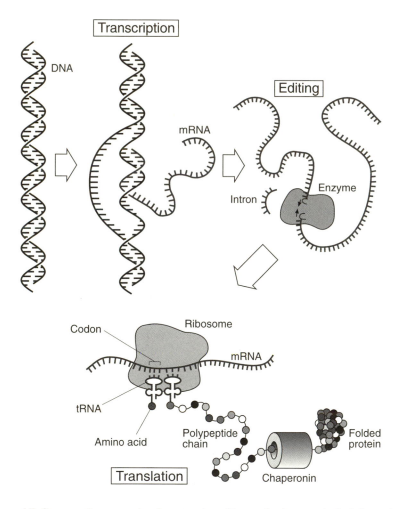

FIGURE 4.7 Genes on the DNA molecule—stretches of base pairs that encode the information to make a single protein—are translated into the proteins themselves in a sequence of steps that involves an intermediary molecule called ribonucleic acid (RNA). First, the DNA double helix is unzipped by enzymes, and an RNA copy of the gene is put together on the template of one strand, guided by complementary pairing between the DNA and RNA bases and orchestrated by enzymes. This is called transcription. In eukaryotic cells the RNA molecule must then be edited—small regions are snipped out and the ends spliced back together. The edited version is called messen-

This RNA molecule, which is called messenger RNA (mRNA), then becomes detached from the gene template and, in eukaryotic cells, travels from the nucleus, where the DNA resides, to a region called the endoplasmic reticulum, where it provides a template for protein synthesis. In eukaryotes the messenger RNA must be edited before it is ready to act as a protein blueprint: in general the coding regions of genes are interspersed with regions that represent genetic gibberish. In the RNA, the coding regions are called exons and the "nonsense" regions introns, and a team of enzymes carefully snips out the introns and splices the exons back together. Why such an apparently wasteful mechanism has been retained in the face of evolutionary pressures is unclear, but there is some suggestion that introns in fact do serve a useful evolutionary purpose, for which reason the common practice of calling them "junk DNA" may be a little unfair.

Protein synthesis, or translation, involves two other kinds of RNA molecule, as well as many enzymes. The messenger RNA becomes bound by a complex entity called the ribosome, a collection of proteins and of ribosomal RNA. Small RNA molecules called transfer RNA (tRNA) search the environs for amino acids and bring them to the ribosome for incorporation into the protein that is to be synthesized on the messenger RNA template. There is a separate tRNA molecule for each of the twenty amino acids, and each one has a group of three nucleotides, called a codon, whose bases are complementary to the triplet of bases that encodes the tRNA's amino acid on the mRNA template. So the tRNAs, loaded with their respective amino acids, dock one by one onto the mRNA template, and the ribosome links their amino acid charges together into a growing polypeptide chain. There is evidence that this chain is folded into its native form, with the assistance of proteins called molecular chaperones, directly as it comes off the ribosomal assembly line (fig. 4.7).

Packing Up Genes

The genetic data bank of human DNA makes for a long read—it contains something like three billion base pairs, in a chain that would stretch to a little under a meter if laid out in a single strand. All of this DNA must be packed into every one of our cells—twice, since we have two copies of the complete genome in each body cell! As the cell nucleus is typically no more than a thousandth of a millimeter across, that is a formidable packaging problem. Admittedly, the DNA double helix is only a couple of nanometers thick, but even so it would be no answer to screw the whole affair up into a tight ball, like a bundle of string, since all of

---

ger RNA (mRNA). This acts as a template on which the corresponding protein is put together by an RNA/protein complex called the ribosome. Each triplet of base pairs on the mRNA encodes a single amino acid in the protein chain; these amino acids are brought to the mRNA/ribosome complex by transfer RNA (tRNA) molecules, which dock onto the mRNA template by complementary base pairing. This process of protein synthesis is called translation. As the protein chain comes off the assembly line, it is folded into its active form, generally with the help of a protein assembly called a molecular chaperone, such as chaperonin.

the chain has to be readily accessible for replication and transcription. Biology's answer, which exploits a hierarchy of packing features, makes DNA in the cell another example of a hierarchical material.

First of all, the DNA in human cells (and in those of all other organisms whose cells have a nucleus, called eukaryotes) is not all in one continuous chain. It is chopped into smaller pieces—forty-six in human cells—called *chromosomes*. The chromosomes have an X-like shape, although in some only two of the four arms are clearly visible. Each species of eukaryote has a characteristic number of chromosomes in the cell's nucleus: cats have thirty-eight, toads thirty-six, and turkeys a mighty eighty-two. There are two of each chromosome in these creatures because they inherit one set from the mother (from the egg) and one from the father (from the sperm). The chromosomes of humans can be distinguished according to the genes that they contain—in general, the same genes are found together on a particular chromosome in all people. But before gene analysis was possible, chromosomes could be identified by the effects of staining them with dyes—certain substances produce a pattern of dark bands that are specific to each chromosome. (It is this tendency to become colored by dyes that gave chromosomes their name.)

Chemical analysis of the chromosomes reveals that they contain not only DNA but proteins too, to which the DNA is tightly bound. This complex of DNA and protein is called chromatin. The proteins in chromatin are there to assist in the tight packaging of the DNA. They are divided into two classes, called histones and (prosaically) nonhistones. Four of the five kinds of histone protein associate together into cylindrical disks, and these bind to the DNA double helix around their edges, producing a double coil of DNA around a histone core—this DNA-histone assembly is called a nucleosome (fig. 4.8). In this way, a thread of DNA acquires a set of nucleosomes along its length like beads on a string. This string of beads then coils up in a left-handed helix, called a solenoid or chromatin fiber, which is about 30 nanometers thick. The fifth histone protein assists in the coiling-up of the fiber.

Once packaged in this way, the length of the DNA strand is reduced by a factor of fifty from its initial, uncomplexed state. But to fit it all into a chromosome, the length must be reduced further by a factor of several thousand. We don't yet understand the details of this higher-order packaging process, in which the 30-nanometer-thick fibers are somehow scrunched up into the roughly 700-nanometer-thick legs of a chromosome (fig. 4.8). This process must be phenomenally well orchestrated, however, because it is clearly highly reproducible (the chromosomes in each cell, and even in each member of the same species, look the same) and because it can be reversed relatively easily when access to the genes in the DNA strand is required. From the perspective of a materials scientist, one can regard chromosomes as functional objects that are assembled in a regulated manner by the hierarchical transfer of information to progressively larger scales. This is something that as yet we cannot mimic in synthetic molecular materials, but like so much else in biology, it provides an "existence proof" that shows what might ultimately be possible if we continue to hone our abilities to manipulate matter.

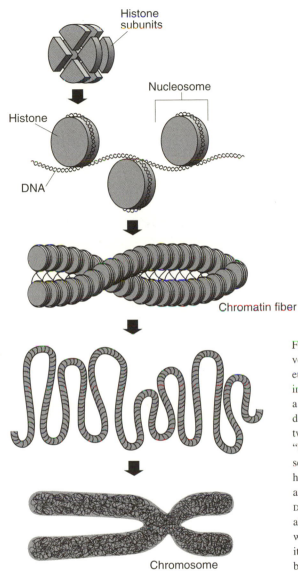

FIGURE 4.8 DNA is packaged very compactly in the cell. In eukaryotic cells it is assembled into chromosomes, which have a hierarchical structure. The double-stranded DNA is wound twice around disklike protein "beads" made from eight separate protein units called histones. The DNA-histone assembly is a nucleosome. The DNA chain, with nucleosomes all along its length, is then wound into a helix which is itself packaged into a compact bundle in the chromosome.

Mass Production

DNA is a substance in great demand. Everyone wants a bit, or preferably a lot. Forensic scientists need it to glean genetic clues about the identity of a criminal, who might have left tiny genetic "fingerprints" behind him in blood or hair follicles. Palaeontologists want to find out more about the constitution of extinct creatures from the sparse and degraded fragments of DNA that can be salvaged from fossils or from amber-encased Cretaceous bloodsucking mosquitoes. (They don't, however, seriously entertain dreams of resurrecting these creatures in *Jurassic Park* style.) Geneticists and drug designers can't get enough DNA: they want to be able to pull out and analyze a particular, rare genetic mutation from a heterogeneous mixture of DNA molecules, or to extract and multiply a particularly effective DNA-based drug from a randomly assembled horde of such molecules.

What is needed in each of these cases is a means to replicate a strand of DNA selectively and accurately. When, in 1983, the American chemist Kary Mullis realized how to do this, it seemed so obvious that he could not at first believe that someone else had not already done it. After all, any cell biologist knew that cells already possess an efficient machinery for replicating DNA. Central to this process is an enzyme called DNA polymerase, which assists in the construction of a complementary strand of DNA on an exposed, "bare" single strand. Mullis and his coworkers at the Cetus Foundation devised a cyclic procedure in which double helices of DNA were first unzipped into single strands, and then complementary strands were built on these with the aid of DNA polymerase enzymes. The new double-stranded molecules were then unzipped for further copying, so that copies of the original double-helical DNA multiplied exponentially. From a single strand, the first cycle makes two; the next cycle makes four; the next eight; and so on. This kind of growth process produces astronomical numbers after a surprisingly small number of cycles.

The only trick that Mullis and colleagues had to build into the procedure was one that got the copying of each single strand under way. DNA polymerase enzymes cannot start from nothing—from a bare single strand. Rather, they need a tiny fragment—a primer—of the new second strand to build on. So the researchers added short primer DNA molecules to the separated (denatured) strands, containing just a few nucleotides. The nucleotide sequences of these primers are complementary to those at one end of the strands that are to be copied, so that the primers bind to these ends ready to be built up by the enzymes over the rest of the strands. This means that one needs to know the nucleotide sequence of the very ends of the initial DNA molecule. But that's all.

Mullis and colleagues developed this idea into the so-called polymerase chain reaction (PCR) (fig. 4.9), which is now used in biotechnology and genetics laboratories worldwide as a means of multiplying minuscule DNA samples into as much as one wants. For this invention, Mullis was awarded the 1993 Nobel Prize for chemistry. To denature the double helices before each cycle of replication, they are simply heated, and are then cooled in the presence of the primer molecules so that these can stick to the ends of the strands. The DNA polymerase enzyme that is used is taken from the bacterium *Thermus aquaticus*, which lives in the hot water around submarine hydrothermal vents; this enzyme is naturally adapted to survive in hot environments, and so is not degraded during the repeated cycles of heating and cooling. PCR can be used to amplify a specific stretch of DNA among many other DNA molecules, because the primer molecules are designed to latch onto the ends of just the required strands (to which their sequences are complementary)—other strands are not copied by the DNA polymerase enzyme, because these others are not primed. PCR now makes DNA a mass-produced polymer.

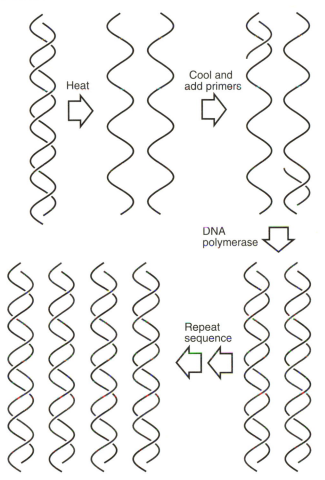

Heat

Cool and
add primers

DNA
polymerase

Repeat
sequence

FIGURE 4.9 DNA molecules can be multiplied by a process known as the polymerase chain reaction. This uses enzymes called DNA polymerases to build up complementary strands from short primer strands attached to single-stranded DNA. By separating the resulting double strands through heat treatment and then adding new primers, the number of strands doubles on each cycle, and copies of the original DNA are multiplied exponentially.

The Invaders

Proteins and nucleic acids make an effective team—the nucleic acids encode the protein structures, and the proteins help to put together the nucleic acids. We can be thankful that living organisms evolved to do more than just repeat this mutual favor again and again—but nothing more is actually *essential*. The simplest organisms that we know contain nothing but protein and nucleic acid: they are viruses, in which a length of DNA or RNA is housed in a protein coat or membrane. Viruses represent the closest thing in nature to nucleic acids as "raw" living material. But they lack the ability to replicate in isolation by metabolizing the substances in their environment; rather, they rely on the protein machinery of cells that they invade to do the job of copying their genetic material. Viruses are able to insert their own DNA into that of the infected cells of the host organism (for RNA viruses, called retroviruses, the RNA is first transcribed "in reverse" to DNA), so that when the host cell replicates its own DNA, the viral DNA is reproduced too. This simplicity of purpose and strategy, as well as their ability to mutate rapidly to evade antiviral treatments, is what makes viruses so terrifying and hard to combat.

The tobacco mosaic virus (TMV), which causes wrinkling of tobacco leaves, is an example of a retrovirus. Under the microscope these viruses appear as rodlike entities; on closer examination, they are seen to be essentially protein tubes with a nucleic-acid core. The elegant structure of TMV holds a peculiar beauty: it is an RNA spiral surrounded by a sheath of protein units that self-assemble in a helical stack (fig. 4.10a). The assembly process can be controlled by acidity (pH). It is driven by the attraction between the RNA core and the protein building blocks, but this can be counterbalanced by electrostatic repulsion between chemical groups

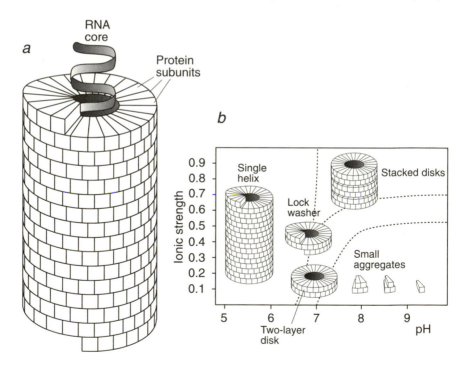

FIGURE 4.10 The tobacco mosaic virus (*a*) is a self-assembling replication machine. It contains a core of coiled RNA surrounded by a coat of protein subunits. The self-assembly of these subunits takes place spontaneously in slightly acidic or near-neutral solution; but at different values of pH, other protein aggregates are formed (*b*. The virus can be regarded as a self-assembling biomaterial.

on the protein units when they lose hydrogen ions to become negatively charged in alkaline solution. So the size of the assembly depends on pH: in highly alkaline solution, the protein units float around on their own, but as the pH is lowered (as the solution becomes more acidic), they start to lose their negative charges and cluster into parts of the helical stack. This is fully formed for pH values slightly to the acidic side of neutral (fig. 4.10b). Materials scientists are now showing a lot of interest in developing materials that assemble themselves spontaneously from elementary building blocks, and they commonly look to viruses for inspiration.

The Boundaries of Life

According to the viewpoint popularized by the biologist Richard Dawkins in his books *The Blind Watchmaker* and *The Selfish Gene*, the fundamental unit of evolution is not the organism, which evolves, diversifies and generally strives for survival in a fight of the fittest, but the gene, whose primary motivation it is to propagate—at the expense of other genes, if need be. This perspective helps us to make sense of a great deal of the vagaries of evolution and the consequences of natural selection; but the idea of a collection of nucleic-acid genes swimming about in some homogeneous soup and competing with one another does not seem greatly to resemble the world we know. And in any case, nucleic acids are scarcely the most stable of molecules and would not fare well if exposed to, say, the harsh environment of the ocean waters. In order to carry out their schemes for world domination, genes need a protective envelope, and that is why we, and bacteria and trees and humpback whales, exist as discrete organisms with our genes packed away in cellular compartments. We may be in some ways at the mercy of our genes, but the packaging—the fabric of the cell itself—is a vital ingredient of the whole.

So while to geneticists life revolves around the duality of gene and protein, to biochemists, physiologists, cell biologists, and immunologists the cell membrane is central to the existence of real organisms. It is a subtle and sophisticated functional material that fulfills many roles. It acts as a barrier to prevent its contents from dispersing and to exclude unwanted foreign materials; but it is also permeable, since the biochemical processes that take place within, such as protein synthesis, require raw materials from without. This permeability is selective, and is able to maintain virtually constant conditions within, despite changes in the surrounding environment. Sometimes this means that the cell wall expels or admits chemical substances in the face of an opposing concentration gradient—for example, admitting a compound to the cell even though the concentration outside the cell is lower than that inside. This is an "uphill" process that requires pumping devices in the cell wall, driven by sources of chemical energy.

The membrane is flexible, but the shapes that it adopts are not wholly arbitrary (sickle-cell anemia, for instance, is a consequence of misshapen red blood cells). In multicellular organisms the walls of neighboring cells stick together in a tissue-specific manner, so that the organism does not merge into a shapeless mass of cells. Certain cells in the immune system are able to recognize foreign invaders and to destroy them, without doing the same to the body's own cells. The visual system contains cells that are sensitive to light (in effect providing miniature optical switches); the cells of the olfactory (smell) system contain switches that respond to specific odor molecules.

All of these functions originate at the cell wall, which has a composite, modular structure in which several "active" components are embedded in a matrix that forms the structural framework. The active components are generally proteins (sometimes modified by the attachment of other chemical groups such as sugars), which perform functions such as transport of chemical species across the membrane, or recognition of other molecules at the cell surface. The matrix of the membrane has various components, but principally it consists of a closely packed array of rodlike molecules called *phospholipids*.

These are members of a class of molecules that chemists call amphiphiles, meaning that they have a liking (*philos*) for two contrasting (*amphi*) environments. In general amphiphilic molecules have a charged (ionic) "head group," which, like inorganic ions, is soluble in water; and a long "tail" composed of chemical groups (typically hydrocarbons, like those in waxy petroleum products) that are insoluble in water. The head group is hydrophilic (water loving), and the tail hydrophobic (water fearing). The head group of a phospholipid contains an ionic phosphate group (PO_4), and the hydrophobic part of the molecule is generally a double tail comprised of two hydrocarbon chains each with a backbone of typically seventeen or so carbon atoms. These two units are linked via a glycerol group, which provides two arms for attachment of the hydrocarbon chains. Such amphiphiles are more accurately called glycerophospholipids. Several other hydrophilic chemical groups may be attached to the phosphate group, such as choline and serine. In the former case, the phospholipid molecule is called phosphatidylcholine (fig. 4.11*a*), for which an older name from biochemistry is lecithin. This is one of the most abundant phospholipids in all cell membranes.

In the cell wall, phospholipids are packed into layers with all the head groups pointing in the same direction. When two such layers are arranged back to back in a *bilayer*, the hydrophobic tails are shielded from the water. The bilayer (fig. 4.11*b*) is the fundamental structural element of all cell membranes.

The membranes of some organisms may contain other kinds of amphiphile, too. Sphingolipids are found in the cell walls of plants and animals, particularly in the cells of the mammalian nervous system. These molecules have no glycerol linking units between the tail and the head; rather, the link is forged by an alcohol group, from which a hydrocarbon tail dangles and to which various hydrophilic groups such as phosphate or sugar molecules can be attached. The cell walls of many more highly evolved animals also contain cholesterol, an amphiphile in which the head group is a humble hydroxyl unit (OH). Cholesterol acts mainly to make a membrane more rigid. In certain situations this rigidity is useful, but when cholesterol accumulates at too great a degree in the cell membranes of arteries, the consequent hardening causes cardiovascular problems.

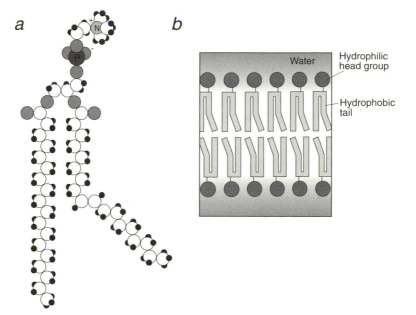

a   *b*

FIGURE 4.11 Cell membranes are composed primarily of phospholipids, of which phosphatidylcholine (*a*) is the most abundant example. Phospholipids have a charged, water-soluble (hydrophilic) head group and a fatty, water-insoluble (hydrophobic) tail. In water, phospholipids
can aggregate into double layers (bilayers) in which the head groups lie on the outside and the
hydrophobic tails are buried within (*b*).

The Cell's Superstructure

In the most primitive single-celled organisms (which is simply to say those that
evolved first, and is not to cast aspersions on their evolutionary success), everything is thrown together in a single membrane sac (fig. 4.12). These organisms,
called prokaryotes, include most bacteria. In more sophisticated forms of life,
however, a more complex cell architecture has evolved. In eukaryotes, which
include some bacteria and everything else (apart from viruses) that lives—plants,
fungi, and all animals—the cell is subdivided into compartments by internal
membranes. The cells of eukaryotes have a central nucleus, a roughly spherical
membrane-bounded compartment within which the genetic material mostly sits.
Replication and transcription of DNA take place in the nucleus, while elsewhere
in the cell other membrane-bounded compartments called organelles, perform a
variety of different functions (fig. 4.12). In one of these, the mitochondrion, most
of the energy-producing metabolic chemistry of aerobic (oxygen-utilizing) cells
takes place. The endoplasmic reticulum is a complex network of membrane folds
and tubes that provides the location for protein synthesis and the transport network for distributing proteins throughout the cell. In the flat membranous bags of

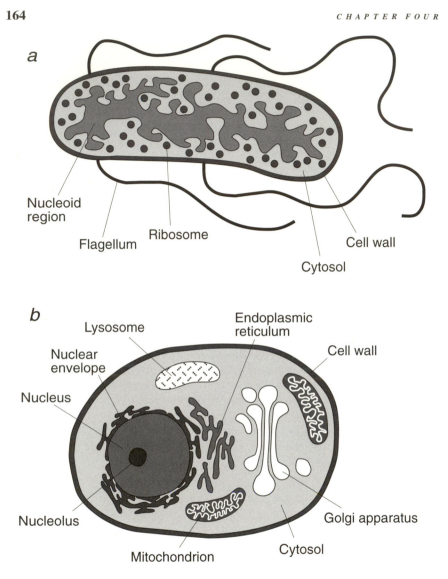

a

Nucleoid region

Flagellum

Ribosome

Cell wall

Cytosol

b

Lysosome

Nuclear envelope

Nucleus

Endoplasmic reticulum

Cell wall

Nucleolus

Mitochondrion

Cytosol

Golgi apparatus

FIGURE 4.12 *a*, A prokaryotic bacterium. The cell interior is not compartmentalized, although the DNA resides in a central nucleoid region. The ribosomes, which are responsible for protein synthesis, float in the cell fluid, called the cytosol. Whiplike appendages called flagella extend from the cell membrane, and act to propel the bacteria through the surrounding medium. *b*, A eukaryotic cell. Membranes divide the cell into several compartments called organelles, each with a specialized function. The nucleus contains most of the cell's DNA, and RNA is synthesized mainly in the dense nucleolus region within the nucleus. Proteins and lipids are synthesized within a network of membranes called the endoplasmic reticulum. The mitochondrion is the location of most of the energy production in aerobic cells. The Golgi apparatus is a collection of flat membranous sacs that play a crucial role in transporting proteins and other molecules around the cell. Lysosomes contain enzymes responsible for digestive processes.

the Golgi complexes, carbohydrates and lipids are stored and attached to proteins, and the modified proteins are dispatched from these compartments to other parts of the cell within vesicles that bud off from the Golgi sacs. In plants, the photosynthetic equipment is housed within an organelle called the chloroplast, the cell's solar-powered energy source. Plant cells also contain compartments called vacuoles, which expand as the plant grows and store vital compounds such as salts, pigments, sugars, and proteins.

We can regard eukaryotic cells as buildings in which the organelles are rooms, the bilayer membranes are walls (sometimes, like cavity walls, doubled up), and the membrane lipids are the bricks. But just as many buildings also have an infrastructure of girders and joists, so the fabric of real eukaryotic cell walls is somewhat more complex than a mere brickwork of lipids. Specifically, their structure is maintained by a web of flexible fibers made of proteins, which together constitute the cytoskeleton. In part, these are support cables that are responsible simply for holding the edifice together and securing the various subcompartments—the organelles—in place. But they also play a more active role in allowing the cell and its components to deform and move, either by contracting or by providing tracks along which components of the cell can be hitched. They are, if you will, "smart" joists that allow the house to change size and shape.

In red blood cells the cytoskeleton is a web of protein strands that crisscrosses the inner surface of the cell wall, attached to the phospholipid membrane through other proteins embedded in the bilayer. The threads of the web are comprised of the protein spectrin, which consists of two polypeptide chains twisted around each other like a rope. The ends of these chains are linked together via actin filaments, in which about twenty actin proteins are joined into short chains, and via ankyrin proteins, which have attachments that are embedded in the bilayer. The actin filaments are not themselves attached to the bilayer but are secured there by other membrane proteins (fig. 4.13). The cytoskeleton of the red blood cell maintains the cell's distinctive shape—that of a disk with a dimple in the center of each face. Defects in the genes that encode the membrane proteins binding the spectrin network give rise to abnormalities in the shape of the red blood cells (spherocytosis, where the cells are more spherical, and elliptocytosis, where they are elliptical), ultimately manifesting itself as anemia.

Actin filaments also constitute a major component of the cytoskeleton of muscle cells, where they play a central role in muscle contraction. Muscle fibers are comprised of long cylindrical cells, which can be up to 4 centimeters long, in which actin strands are lined up along the fiber's axis and interdigitate with thicker filaments of another multistrand coiled protein, myosin (fig. 4.14). During muscle contraction, the myosin filaments pull themselves along the actin filaments with a ratchetlike motion, so that overlap of the interdigitated strands increases.

Clearly, muscle fiber is a highly sophisticated functional material, and our attempts to mimic its contractile behavior, some of which were described in chapter 3, are crude in comparison. In particular, the degree of contraction that muscle

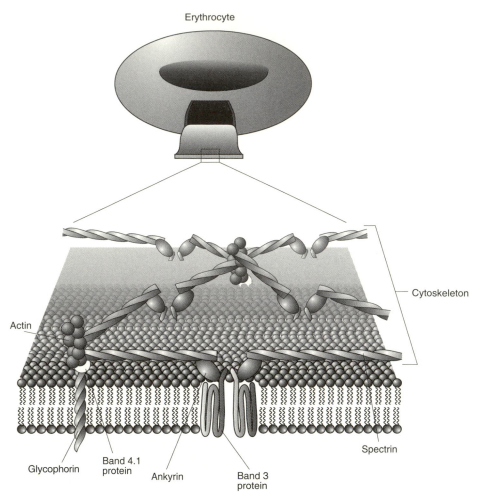

Erythrocyte

Cytoskeleton

Actin

Spectrin

Glycophorin

Band 4.1 protein

Ankyrin

Band 3 protein

FIGURE 4.13 The shapes of cells are controlled by a scaffold of protein filaments called the cytoskeleton. In red blood cells (erythrocytes), the cytoskeleton is a web of helical filaments of the protein spectrin, anchored to the cell wall by other proteins embedded in the lipid membrane.

FIGURE 4.14 (*on opposite page*) Muscle fibers are long, cylindrical-shaped cells packed with filamentary protein structures: thin filaments, made from strands of polymerized actin molecules wound into a double helix and around which in turn a double helix of tropomyosin strands is twisted, and thick filaments composed of α-helices of myosin, coiled around each other in a rod structure. The thin and thick filaments interdigitate, and during muscle contraction the thick filaments are pulled along the thin filaments in a ratchetlike manner (*top*). This motion is effected by the myosin head, which is attached to the myosin α-helices through a flexible hinge unit and which can form a bond to the actin molecules of the thin filament. The motion is driven by chemical energy as a molecule of adenosine triphosphate (ATP) is converted to adenosine

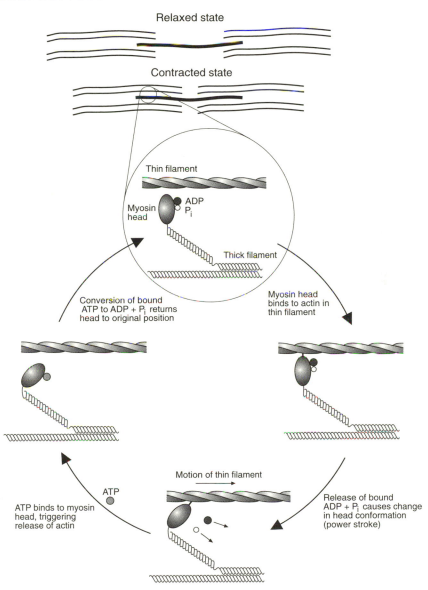

diphosphate (ADP) and a phosphate ion (P_i). Picking up the cycle in the top frame, a myosin head with ADP and P_i attached binds to the actin thin filament. Release of the ADP and P_i then brings about a change in the conformation of the head at the point where it is attached to the myosin α-helix. This conformational change pulls the actin filament to the right. A molecule of ATP then binds to the myosin head, triggering the release of the actin filament from the head. The bound ATP is then converted, by a chemical reaction called hydrolysis, to ADP and P_i; this releases chemical energy and reverses the conformation of the head, returning it to its original position. In this way, the thin filament is gradually pulled to the right along the thick filament.

α-tubulin

β-tubulin

FIGURE 4.15 Microtubules are an important part of the cytoskeleton of many cells. They are cylindrical structures in which many molecules of the protein tubulin are bound together in a helical array. Tubulin itself comes in two varieties—α-tubulin and β-tubulin—which bind together to form a "dimer." Most microtubules are dynamic structures, meaning that tubulin dimers are constantly attaching and detaching from their ends—each addition consumes energy, and is driven by the conversion of guanosine triphosphate (GTP) to guanosine diphosphate (GDP) and phosphate.

is capable of is substantial—and yet the contractile motion does not involve any substantial changes in length of the molecular components, but just the motion of one relative to the other. What is more, like all biological tissues it is self-repairing.

Another important cytoskeletal material is the globular protein tubulin, which can aggregate into hollow cylindrical structures called microtubules (fig. 4.15). This protein is composed of two subunits, α-tubulin and β-tubulin. Microtubules are found in nearly all eukaryotic cells, and are of utmost importance for cell division. When a cell is preparing to divide, it makes a copy of its entire DNA content. This DNA replication takes place in the nucleus and results in a doubling of the chromosome content; rather than two copies of each chromosome, the cell now has four. The cell is then faced with the tricky job of sorting through this collection of chromosomes to ensure that each of the two daughter cells into which it splits gains two of each set.

This sorting process is performed by microtubules. Replication of each chromosome creates a doubled-up version in which the two copies, called chromatids, are linked together near their centers in an X shape. Once this has happened, two clusters of microtubules are assembled outside the nucleus. In each of these clusters, the microtubules radiate outward from a central focus in an anemone-like structure called an aster, ultimately forming a set of strands called the mitotic spindle (fig. 4.16). The X-shaped chromosomes become aligned at the center of this spindle and are split into the two chromatids, which are pulled along the tubules to opposite poles. In this way, a complete copy of the cell's genome is gathered at each pole of the mitotic spindle, and the cell then divides in such a way that one pole remains in each half.

These microtubules are dynamic structures—tubulin units are continually being added and removed. Each addition or dissociation step is an energy-consuming process, in which one molecule of the energy-storage molecule GTP is

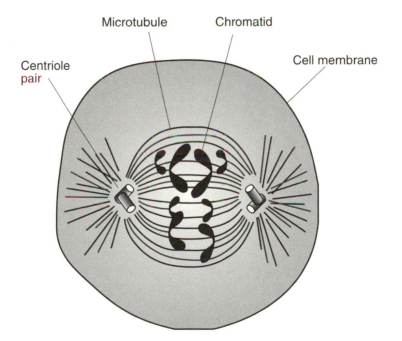

FIGURE 4.16 During cell division (mitosis), microtubules form in two radiating, starlike clusters called asters, centred on two protein structures called a centriole pair. The microtubules grow until those from opposite poles meet and join to form a scaffold called the mitotic spindle. The chromosomes, which have each been replicated within the cell's nucleus before the asters began to form, now become attached at their midpoints to the equator of the spindle. The X-shaped chromosomes then split in half, and the double-lobed halves (called sister chromatids) are drawn along the microtubules toward opposite poles. As they approach the poles, the cell begins to elongate, taking on a peanut shape (this process is not shown here) that eventually divides into two cells, each with a full copy of the chromosomes.

converted to the less energetic GDP. This dynamic assembly and disassembly enables a cell to control the length and the growth direction of its microtubules. Tubulin can also form stable microtubules, whose length remains fixed. Such structures comprise part of the cytoskeleton of eukaryotic cells, where they maintain the cell's shape and act as a kind of transportation grid on which other organelles can be moved around. The cell surfaces of certain single-celled eukaryotes are covered in stable, hairlike microtubule appendages called cilia, which exhibit whiplike motions that propel the cells through a fluid. Cilia are also used by vertebrate animals to transport mucus around the respiratory tract. Bacteria use similar microtubule appendages, called flagella, to propel themselves. These are extraordinary molecular devices: rotary motors embedded in the double membrane of the cell wall, which drive long, helical filaments in circular motions that power the bacteria along like the propellers of ships. Physicists and engineers have now made tiny silicon motors only a few hundredths of a millimeter across, but nothing yet to compare with this.

The Gates of the Cell

The structural components of cell walls thus have a great deal of sophistication. But to survive, cells must interact with and respond to their environment, and they do this primarily through the mediation of the active element of their enveloping fabric—the membrane proteins. These are the crucial channels of communication that allow a multicellular organism to act as a whole rather than as a collection of self-centered individuals. Membrane proteins turn the cell wall into a complex smart material.

Most membrane proteins are so-called integral membrane proteins: deeply embedded in the lipid bilayers, often spanning from one side to the other. They are like icebergs in a lipid sea—the fluidity of the membrane allows the proteins to drift around. This mobility can have important biological roles. Clustering of membrane proteins, for example, can accompany some cell-surface processes such as budding, when a small compartment separates from the cell wall. Other membrane proteins may be bound only weakly to the surface of the cell, via interactions with the lipid head groups or with embedded membrane proteins. These are called peripheral membrane proteins.

Traffic of chemical species across the cell wall is mediated largely by integral, membrane-spanning proteins. Water-soluble species cannot easily pass through the lipid bilayer itself, despite the fluidity of its components, because the interior of the bilayer contains the hydrophobic tails of the lipids, among which water-soluble hydrophilic species are not welcomed. Membrane proteins may provide channels with hydrophilic inner walls, through which ions and other water-soluble molecules can pass. In some cases the membrane protein seems simply to offer a passive hole through which molecules can diffuse from a region of higher to lower concentration; but more often the transport process is tightly controlled. The channel might, for example, admit one kind of species but not others (calcium but not potassium ions, say), or it might permit transport in one direction

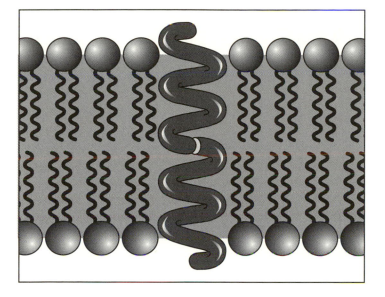

FIGURE 4.17 The membrane protein gramicidin A is an ion channel: it enables metal ions such as sodium and potassium to pass through cell walls. In the cell wall, two linear molecules of gramicidin A coil up end to end to form a helical dimer, which spans the membrane from one side to the other (in fact, the membrane is squeezed together slightly around the edges of the coil, but I have not shown this here). The inside of the helical pore is lined with chemical groups that are conducive to the entry and passage of positively charged ions.

only, regardless of the concentrations on either side. Or its permeability might be switched on and off by other chemical triggers, a phenomenon called gating.

One of the simplest channel-forming membrane proteins is gramicidin A, which allows metal ions such as potassium to pass across the inner membrane of the mitochondrion. It is a peptide that coils up in the mitochondrial membrane; two of these coils stacked end to end span the entire membrane (fig. 4.17). Gramicidin A displays a feature common to such channels: its outer surface bristles with hydrophobic groups, which make it compatible with the lipid tail groups in the interior of the membrane, while its inner surface exposes hydrophilic groups (in this case, carbonyl (CO) groups) which enable the ions to pass. The formation of the channel is a dynamic process—the two halves of the coil come together for just a second or so before dissociating, during which time about ten million ions can pass through the channel. Because gramicidin A will also insert itself into the cell walls of bacteria and flush out their ionic contents, it acts as an antibiotic.

Synthetic Cells

Cell membranes are highly heterogeneous functional materials, containing a host of different biomolecules that carry out a range of tasks. Materials scientists and biochemists have become very interested in building artificial membranes and cells that capture some of these features. Such structures are already proving useful as protective delivery bags for administering drugs to internal organs through the bloodstream; and when joined together, they might be regarded as artificial tissues, with potential biomedical applications. Artificial cell-like structures are also used for chemical-separations technology and for harnessing solar energy, as catalysts and molecular sensors, and even as model systems for studying how life evolved. I will describe some of these applications below and in the ensuing chapters.

For the present time we cannot hope to reproduce all of the functions of real cell walls. In general we are limited to building either "bare" membranes from lipids or other amphiphilic molecules, or membranes with just one kind of active component embedded within it—proteins that selectively bind certain target molecules, perhaps, or ion pumps that create an electrical gradient.

Amphiphilic bilayer membranes are self-assembling materials, the assembly process being driven by the need to bury the hydrophobic tails of their constituent molecules. Under certain conditions, phospholipids can assemble themselves into sacs called vesicles (also known as liposomes) which can be regarded as artificial cells without the functional components (fig. 4.18). Joe Zasadzinski and col-

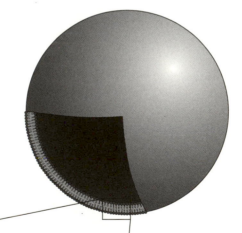

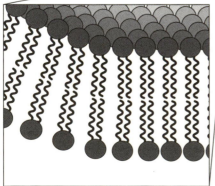

FIGURE 4.18 Phospholipids will form hollow aggregates called vesicles, the walls of which are bilayers.

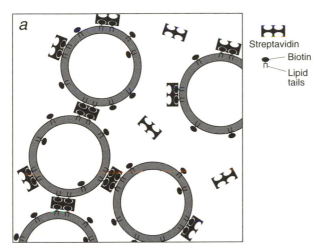

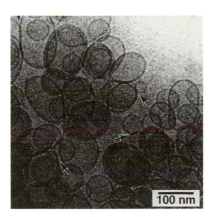

FIGURE 4.19 Vesicles can be bound together by recognition between molecules attached to the head groups of some of the amphiphiles in the walls. Recognition and binding between biotin molecules at the surface of the vesicles and streptavidin proteins in solution (*a*) leads to the formation of tissue-like assemblies of vesicles (*b*). (Photograph courtesy of Joe Zasadzinski, University of California at Santa Barbara; reproduced with permission from *Science* **264**, 1753 [1994].)

leagues at the University of California at Santa Barbara have shown how phospholipid vesicles can be stitched together by lock-and-key interactions between membrane-bound proteins to make artificial tissues. Zasadzinski and colleagues chose to use the protein streptavidin as the lock and the molecule biotin as the key. Biotin is a small organic molecule which acts as a coenzyme—an enzyme's assistant, if you will—in one of the key steps of metabolism, glucose synthesis. Streptavidin binds biotin very tightly. The Santa Barbara team incorporated biotin molecules into vesicles by giving them hydrophobic tails, which buried themselves among the tails of the vesicle phospholipids. They found that when they mixed vesicles containing membrane-bound biotin with streptavidin, the vesicles stuck together like a mass of bubbles (fig. 4.19).

Bilayer vesicles are just one of the diverse kinds of assembly that amphiphilic molecules will form. Other types of aggregate have long been known from experiments on detergents, which are also amphiphilic molecules. Like phospholipids,

FIGURE 4.20 Soaps and detergents contain amphiphilic molecules. Commonly these are fatty acids, which contain a carboxylic acid head group (which can be ionized in water to carboxylate) and a hydrocarbon tail.

the amphiphiles used in soaps and detergents have hydrocarbon tails, but these are generally single-stranded rather than double. The head groups are generally carboxylic acid groups, like those in acetic acid—these can readily lose a hydrogen ion to form negatively charged carboxylate ions in water (fig. 4.20). In detergent science, such amphiphilic molecules are often called surfactants, a condensation of "surface-active agents," which signifies that the molecules do their job at the surface between water and the oily substances they are supposed to remove. Surfactants render globules of grease soluble by burying their hydrophobic tails in the grease and leaving the head groups sticking out at the surface to give the globule a water-soluble coat.

Depending on their concentration in solution, and on other factors such as temperature, amphiphiles—whether they are phospholipids or detergents—will spontaneously assemble themselves into many different structures. The simplest are micelles, globular clusters in which the hydrophobic tails are buried in the inside and the hydrophilic head groups present a water-soluble coat on the outside (fig. 4.21a). In oily solvents, micelles turn inside out, shielding their hydrophilic heads. These aggregates are reverse micelles (fig. 4.21b). At higher concentrations, cylindrical micelles are formed, which look much like micelles in cross section but which are extended, wormlike structures rather than spheres. At still higher concentrations, stacks of bilayer sheets, called lamellar phases, are found. Some of the most intriguing bilayer structures are the so-called cubic phases, in which the sheets are bent into a complex, ordered network of channels. Each of these structures has proved useful in materials synthesis, as we shall see.

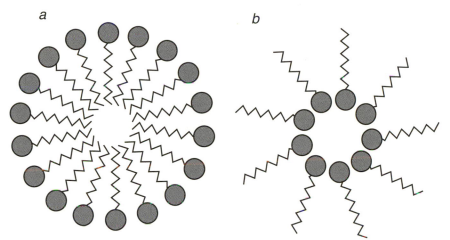

FIGURE 4.21 Amphiphilic molecules will cluster in water to form aggregates called micelles, in which all of the hydrophobic tails are buried within the center of the roughly spherical structure (*a*). In oily solvents, micelles turn inside out to expose the hydrophobic, oil-soluble tails and to bury the hydrophilic head groups. These aggregates are reverse micelles (*b*).

The environment at the molecular scale within artificial lipid membranes is similar to that within a natural cell wall. This means that membrane proteins incorporated into the former may be persuaded that they are still in the latter, so they can function normally. Immobilization of proteins within artificial membranes is of great value to the biotechnological industry, which wishes to exploit the superior properties of proteins in the out-of-context setting of technological devices. Biosensors, for instance, use immobilized proteins as part of a sensing device that detects the presence of the protein's substrate. When the protein binds its substrate molecule, some kind of electrochemical, optical, thermal, or electrical signal is induced that allows its concentration to be monitored. Biosensors are finding wide use in medical diagnostics, particularly for monitoring glucose concentrations in the bloodstream of people with diabetes.

THE THREADS OF THE FABRIC

Cells have a life of their own, as bacteria demonstrate—but multicelled organisms like us need a superstructure, a scaffold that holds us together. Like a suspension bridge, this structure is supported by strong cables, and these are fashioned from *structural* proteins, which have a fibrous nature. Most of the body's strong tissues—skin, muscle, tendon, hair, horn, and claw—are made of fibrous proteins. The range of properties in this list is considerable, but they are derived from only a limited repertoire of fundamental motifs (secondary structures) of the proteins' peptide chains. One such is the α-helix, which is displayed by the protein keratin,

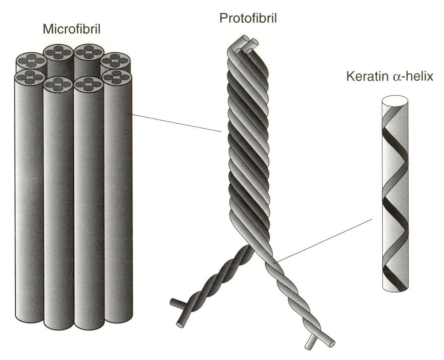

Microfibril

Protofibril

Keratin α-helix

FIGURE 4.22 Keratin forms twisted ropelike fibers called microfibrils from hollow clusters of eight protofibrils with a circular or square cross section. The protofibrils are themselves composed of four keratin α-helices: two pairs of coiled helices wound together in a left-handed supercoil. This hierarchy of coils gives the keratin microfibrils great strength.

the main component of skin, hair and fur, beaks, claws, and fingernails. All of these materials are built up from fibers with a many-tiered, hierarchical structure. Two α-helices of keratin twist around each other in a double helix, and then each of these double helices twists around another to make a four-stranded fiber called a protofibril. While the individual polypeptide helices are right-handed, the double-helical strands wind around each other in a left-handed sense (fig. 4.22). This is not a matter of chance or nature's whimsy; rather, the opposite coiling sense helps to prevent unwinding of the individual strands in the double helices. The same principle of using "coiled coils" that twist in different directions at different structural levels is employed in engineering to strengthen coiled steel cables for suspension bridges.

The protofibrils cluster together in groups of eight, either in a cylindrical or square tubular arrangement, to make a microfibril (fig. 4.22). These microfibrils represent the basic structural element of keratin-based tissues. The toughness of the material is enhanced by cross-linking of the protofibrils and microfibrils via disulfide bonds, which are forged between two cysteine amino acid units in the polypeptide chains. Such linkages are commonly employed in both fibrous and globular proteins to secure certain arrangements of their polypeptide chains. When the number of disulfide bonds between keratin microfibrils is small, the material is soft and flexible, like skin and wool. When the degree of cross-linking is higher, the substance is much harder and tougher, like claw. In all of these materials, the keratin fibers are embedded in a matrix of other proteins. The matrix proteins fall into two groups—one rich in cysteine, the other rich in glycine and tyrosine—and they have a tangled, disordered structure with cross-links binding the chains into a strong net. This fiber-in-matrix structure of keratin makes it a composite material, and it is a structure that materials scientists emulate in many synthetic composites that likewise contain strong fibers dispersed in a disordered matrix.

While on the outside our bodies are mainly keratin, on the inside our structural tissues are largely composed of the protein collagen. This fibrous protein shows even more versatility than keratin. In teeth it forms the fibrous dentine matrix within which a hard inorganic material is crystallized; in tendons (which join muscle to bone) and ligaments (which join bone to bone) it forms a durable and elastic connection; in the cornea of the eye it is transparent; in blood vessels it is elastic; and in gelatin it is a water-absorbent gel. Bone itself is also largely collagen, reinforced with mineral crystals (see p. 198). The differences in properties of each type of collagen are a consequence of differing arrangements of essentially the same polypeptide chains, which are composed largely of the amino acids glycine and proline, along with a significant amount of lysine. Like keratin, collagen is a coiled coil, but in the opposite sense: three *left*-handed helices of the collagen polypeptide chains coil around one another in a *right*-handed triple helix. Three of these triple helices then coil around one another in a right-handed sense to make a collagen microfibril. Several of the proline and lysine units in collagen are modified to contain hydroxy (OH) substituents, making them hydroxyproline and hydroxylysine, respectively. Some of the latter are further modified by enzymes that attach sugar groups, so that collagen is in fact a glycoprotein (that is, a protein that also contains carbohydrate). Other enzymes forge cross-links between neighboring chains via lysine groups; again, the number of these cross-links determine the strength and stiffness of the material. In the cornea of the eye, collagen microfibrils lie together side by side in a highly ordered, virtually crystalline array. This reduces the propensity of the material to scatter light (which occurs primarily from defects, just as it does from cracks in glass), and so the material is transparent. Gelatin is made from collagen by dissolving the triple-helical strands in water and allowing them to unwind by heating the solution: this unwinding is called denaturation. When cooled, the strands then twist back around one another, but different ends of each strand may become caught up in different triple helices, so the result is a tangled web of strands which can hold large quantities of water within its flexible framework: a gel.

While both collagen and keratin are present in skin, its elastic properties, and those of blood vessels, are due primarily to the protein elastin. Like collagen, this is rich in the amino acids proline and glycine. But elastin does not adopt a regular coiled structure, like telephone cable; rather, it forms a looser, flexible coil with

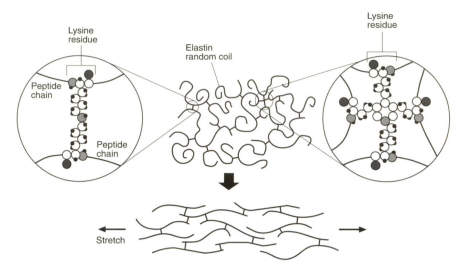

FIGURE 4.23 Elastin is an elastomeric protein that gives tissues such as skin and blood vessels their elasticity. The polypeptide chains adopt a random-coil conformation, in which they remain flexible and loosely coiled. This allows them to behave like springs. The elastin strands are cross-linked into a net via lysine residues, which can join up into either two-way or four-way bridges. When the net is stretched, the random coils unravel; when the tension is released, the chains spring back into the random-coil conformation.

no regular repeat distance, called a random coil. The peptide chain can be un-coiled relatively easily when stretched (fig. 4.23), but the random coil re-forms when the tension is released. Thus the peptide chains act as springs, and the material is elastic (more technically it is elastomeric—see chapter 9). Again it is toughened by cross-linking between chains, either via the same kind of cross-links as are found in collagen or via a cyclic chemical group that joins no fewer than four chains at a single node (fig. 4.23).

Tapping into the Web

There are certain images from nature that resonate so strongly in our experience that they become symbolic or archetypal, cropping up again and again in myth, legend, and metaphor. The tree is surely one of these: a symbol of growth and diversification. And the sea—in its vastness, depth, and fathomless mystery an ideal metaphor for the unconscious mind—is another. So too is the spider's web.

It is not simply that the web is a trap; the spider is not nature's only setter of snares. But there is something about the fact that the web is at once an object of great beauty and symmetry, and a fatal hazard that is all but invisible to its vic-tims, that stirs up both admiration and dread. To many of us the unexpected caress of the silky net is abhorrent even though we could crush the weaver between our fingers; but others will arise at the crack of dawn to capture on film the delicate artistry traced out in glittering droplets of dew (fig. 4.24).

But as well as a natural wonder, the spider's web represents an astonishing feat of materials engineering. How can a thread almost too fine to see with the unaided eye possess such strength—enough to stop a fly-sized projectile traveling at around a meter per second? The mechanical properties of spider silk—another protein-based structural material—seem to become ever more remarkable the more closely one studies them. For fibers of comparable thickness, it takes more energy to break silk than steel or the strongest synthetic polymer fibers such as Kevlar (see chapter 9). Some silks can stretch to up to four times their initial length before breaking. Water may play a crucial role in this resilience. Certain kinds of silk will contract by up to 40 percent when wet: within droplets of water, the silk threads crumple up into a bundle that acts as a kind of windlass which can be drawn out again to absorb the energy of an insect impacting on the web. If the web were simply highly elastic, like a trampoline, there would be a risk of ejecting a captured fly as the threads spring back from the impact. But the fibers are able to dissipate almost three quarters of the energy of the impact as heat, leaving very little to be taken up by the recoil.

Not surprisingly, all of this leaves materials scientists who are concerned with making synthetic fibers looking enviously at spider silk. The inspiration that silk has provided for fiber technology has a long history: Robert Hooke, the father of spring mechanics, suggested in the seventeenth century that the thread of a silkworm might be imitated by drawing out glue into a thread. That

FIGURE 4.24 The spider's web, glistening in the morning dew, has inspired many a poet—and the tremendous toughness and resilience of the threads is now inspiring materials scientists. The water droplets are not merely incidental decoration; they provide the threads with the elasticity needed to absorb the energy of impacting prey without snapping. (Photograph courtesy of Fritz Vollrath, University of Aarhus, Denmark.)

sentiment was followed through three centuries later with the synthesis of nylon, an artificial fiber that can be considered as a kind of synthetic (though inferior) silk.

The protein chains of silk are comprised primarily of the amino acids glycine, alanine, and, to a lesser extent, serine, all of which are compact monomers that leave short side chains dangling from the protein backbone. These three monomers are interspersed, however, with various other amino acids that have bulkier side chains. The precise composition depends on the species of spider, and also on what the silk is being used for. Spiders can produce an astonishing variety of silky materials, each with properties to fit the given task. For web weaving, they make so-called dragline silk; a different kind of thread is used to spin the cocoon that protects the silkworm during its development from a larva to a pupa and thence to an adult. Other varieties include capture silk, glue silk, support fibers for the web (which are different from the main web fibers), attachment silk (which binds the whole assembly to its frame—a branch, say, or the corner of an attic), and wrapping silk to bind the captured prey. Each of these is produced in a different set of specialized glands, and each has a different amino acid composition specified by a different set of genes.

The microstructures formed by these peptide chains in silks are complex and in some respects still imperfectly understood. One thing that is clear, however, is that silk is another example of a natural hierarchical material, with different kinds of structural motif appearing at different size scales. Within all kinds of silk there appear to be small regions, typically 2 to 20 nanometers across, in which the proteins are crystalline, their chains packed together in an orderly, periodic manner. These protein crystallites are embedded in a disorderly tangle of protein chains, like gems in a snake pit. This combination of stiff crystals with a more flexible, softer matrix is a feature of all "crystalline" fibrous polymers, including synthetics like polyethylene and polypropylene. The disordered regions confer flexibility, and in many polymers, both natural and synthetic, the regular sequence of monomer units in the chains is purposely disrupted by the random inclusion of different monomers, to give the final material the desired flexibility. Some silk fibers, such as the dragline silk of golden orb weavers, have further levels of structural organization. They too contain ordered, crystalline regions, about 2 nanometers across (which seem to be composed primarily of protein strands containing just the amino acid alanine), embedded in a disordered matrix of protein strands rich in glycine. Yet also embedded in this matrix are larger "crystals" about 50 nanometers in diameter whose structure is rather perplexing: it contains a fairly orderly arrangement of chains, but there is no exact repetition of any structural unit, as there is in true crystals. Christopher Viney from the University of Washington has proposed that these regions be regarded as "nonperiodic crystals," containing protein chains of several different amino-acid compositions that are nonetheless stacked together in a relatively ordered, "pseudocrystalline" manner (fig. 4.25).

How do the protein strands stack together in the crystalline regions? Each chain adopts a kind of pleated conformation, a zigzagging up-and-down of the amino-

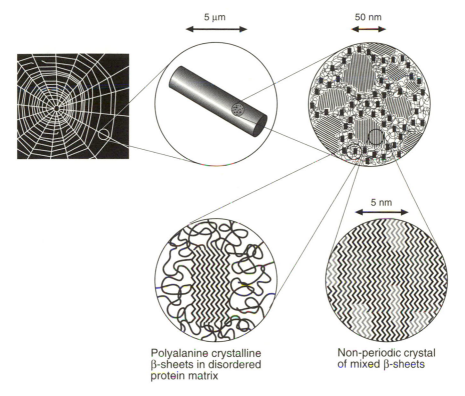

5 μm

50 nm

5 nm

Polyalanine crystalline
β-sheets in disordered
protein matrix

Non-periodic crystal
of mixed β-sheets

FIGURE 4.25 Spider silk is a hierarchical biomaterial, characterized by several different structures on different length scales. Shown here is the structure of dragline silk of the golden orb-weaving spider (*Nephila clavipes*). It contains regions of largely crystalline protein, about 50 nanometers across, embedded in a disordered tangle of protein chains. These "crystalline" regions are not true crystals, as they have no repeating structural element. Rather, the protein chains in these regions are a mixture of polyalanine chains with other kinds of polypeptide—the polyalanine sections are fully ordered into pleated β-sheets (fig. 4.26), but the other polypeptides introduce some disorder. Within the fully disordered matrix, there are much smaller crystallites, about two nanometers across, of stacked polyalanine β-sheets. Not all spider silk is this complex, but all silk threads contain the same general feature of crystalline regions embedded in a disordered matrix.

acid monomers. Neighboring chains are aligned side by side and held together by hydrogen bonds (fig. 4.26). The hydrogen-bonding groups in pleated silk proteins stick out from the backbones first on one side and then on the other, so that each chain can bind to that on either side. This arrangement of polypeptide chains is a common secondary structural motif of many other proteins too, and is called a *pleated β-sheet*. In the silk-fiber crystallites, these sheets are stacked one atop the other in a highly regular array. It is this highly aligned arrangement of polymer chains in the β-sheets that gives silk its high tensile strength. Those parts of the chains containing amino acids with bulky side groups, meanwhile, disrupt this

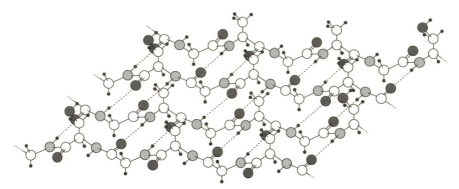

FIGURE 4.26 In the crystalline regions of spider silk, protein strands are arrayed in an orderly, zigzagging arrangement called a pleated β-sheet. Each strand is bound to its neighbors on either side by hydrogen bonds (shown here as dashed lines). In the crystallites, the β-sheets are stacked on top of one another. The β-sheets shown here are those found in the cocoon fibers of the silkworm (*Bombyx mori*), in which the amino acid sequence glycine–alanine–glycine–alanine–glycine–serine repeats along the protein chains.

order and force the chains to degenerate into the disordered tangle of the surrounding matrix.

The trick that the spider must perform, however, is to generate this kind of partly ordered, stiff, water-insoluble structure from an initially disordered solution of soluble protein. This requires a very well orchestrated spinning and drawing process, engineered by some sophisticated spinning machinery. The silk protein is synthesized in a water-soluble form (called silk I) in epithelial cells in the posterior region of the spider's silk gland (fig. 4.27): the amino acid sequence of the protein is specified by a gene, and the chains are put together by an army of enzymes. At this stage the silk solution has a low viscosity, but this increases when the solution is passed from the posterior to a central chamber of the gland, called the lumen, where it is stored for use. The increase in viscosity is an indication that the peptide chains are starting to bind together via hydrogen bonds. From the middle region of the gland the protein is passed down a narrow anterior region which leads to the spinneret, where the threads are excreted. Because of the narrowness of this region, the protein solution becomes highly sheared, and this encourages the polymer strands to line up. Thus, although it is still liquid, the protein chains no longer point in random directions but have a preferred orientation in the direction of shear (that is, along the axis of the anterior tube). The formation of an oriented protein solution at the spider's spinneret anticipates the high degree of alignment in the crystalline regions of the final thread.

At the spinneret, the protein solution loses water rapidly. As it passes through the exit and turns into a silk thread, the protein undergoes an abrupt change to a different state, called silk II. This state is insoluble in water and contains the crystalline regions of stacked β-sheets. The abrupt change is induced by the shear itself as the thread is drawn out, and is an example of what physicists call a phase

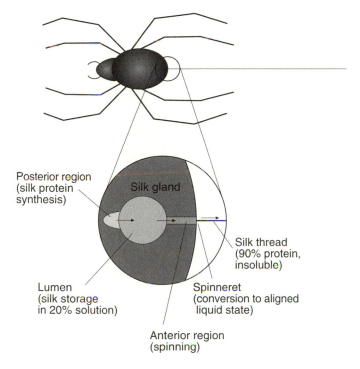

FIGURE 4.27 The silk glands of spiders are polymer-processing units, which generate protein strands and process them into the orderly arrangement found in silk fibers. The proteins are made in epithelial cells in the posterior region, and are stored in a chamber called the lumen. When spinning commences, the silk solution passes into the narrow anterior region, where the protein strands start to become aligned and to stick together via hydrogen bonds. As the silk is excreted from the spinneret, most of the water is shed and the strands line up to form an oriented liquid which is then drawn into partly crystalline, insoluble threads.

transition, of which freezing, melting, and evaporation are familiar examples. So here the spider uses physics to its advantage in finding a way to convert the protein solution in its glands into a thread that will not dissolve when the first drop of dew condenses on it.

PLASTICS FROM PLANTS

The Wood for the Trees

Wood was surely one of the first biopolymeric materials to be pressed into service, being tough and able to support heavy loads yet workable with crude tools. In architectural terms, wood and plant stems are stunningly complex and subtle materials: highly organized, cellular structures that act both as a transport network for fluids and as a strong scaffolding for the plant's solar cells, the green leaves.

A plant's stem carries water and sugar-rich fluids (the products of photosynthesis) between the extremities of root and leaf. These can end up several hundred feet apart in the tallest trees, so tree trunks have to provide a pretty sophisticated plumbing system.

Wood has a hierarchical structure, comprised largely of cellular cavities of different shapes and sizes. Most of the cells in a tree trunk are elongated along its axis. The cells that carry water from the roots upward are called xylem, and these are arranged in concentric layers around the trunk axis. The xylem can be loosely compared with our own arteries, which bear blood from the heart to our extremities; but the tree has no heart to do the pumping. Instead, water is carried up through the xylem by capillary forces, the same forces that draw water into a sponge. This drawing up of water is constantly sustained by its loss at the other end, as the leaves expel it into the air as water vapor—a process known as transpiration. The forces that are generated within the xylem can be enormous; in the tallest trees, the water is "stretched" by tensile forces that would pull it apart into vapor if it were not confined within the xylem.

Tree trunks and plant stems have a second system of cellular channels called the phloem, which carry glucose from the leaves to the rest of the plant. The network formed by the phloem is independent of that formed by the xylem, although the two systems sit side by side. In addition to these vessel-like cells (called tracheids) lying along the trunk's axis, the plant possesses some cells that radiate out horizontally from the axis, conveying fluid outward from the center. These are called ray cells (fig. 4.28). These cell types span a range of sizes and shapes, and together they form a transportation network every bit as complicated as the vents, flues, corridors, and shafts of a skyscraper. Moreover, the hierarchy of structure does not stop at this cellular level; the structure of individual cell walls is a many-layered affair, in which fibrils of the polymer cellulose are woven into matted sheets bound by a matrix of other organic polymers. These sheets are laid on top of one another in the cell walls with their fibrous "grain" facing along different directions, conveying strength and toughness to the wall. Thus, in making plywood we are using (at a larger scale) the structural principles already exploited in the wood itself.

At the molecular level, wood has several components. Fifty percent of the fibrous material consists of strands of cellulose, and a further 30 percent is lignin, a resinous binding matrix. (Notice again the appearance of a fiber-in-matrix composite.) Cellulose is a sugar polymer (a polysaccharide), being made up of chains of glucose molecules in which six carbon atoms and one oxygen are joined into a ring (fig. 4.29). Plants are astonishingly adept polymer chemists, making these rather complex polymers from the simplest of building blocks—carbon dioxide and water. They extract carbon dioxide from the air and fix the carbon atoms, together with hydrogen and oxygen, into the cyclic sugar molecules. This process produces oxygen gas as an unwanted by-product, which the plants expel to our benefit. This is the basis of photosynthesis, an energy-consuming chemical process that is driven by sunlight.

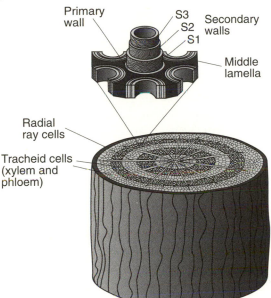

FIGURE 4.28 Wood is a hierarchical material, with structures at many different scales. The stems of woody plants contain concentric arrangements of fluid-bearing cells called tracheids, which convey water to the leaves (in a network called the xylem) and sugar-rich fluids from the leaves (in a network called the phloem). Horizontal, elongated radial cells called ray cells lie at right angles to the tracheids, bearing fluids between the inner and outer rings of the stem. The cells themselves have a multilayered wall structure, containing several fibrous layers with criss-crossing fiber orientations. The primary and secondary walls are made up of cellulose fibers in a matrix of mostly lignin; the material between cells, called the middle lamella, is composed mainly of pectin, hemicellulose, and lignin.

FIGURE 4.29 Cellulose, the main constituent of plant fibers, is a polymer made up of many glucose molecules linked together.

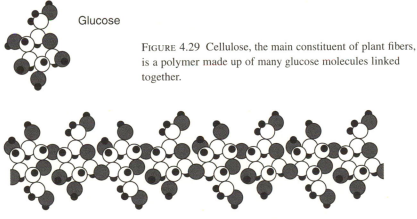

Natural Plastics

With such an abundance of this natural polymer around us, it is perhaps not surprising that cellulose provided the raw material for the dawn of the plastic age. In 1832, Christian Schonbein found that cellulose fibers from cotton can be dissolved in nitric acid to form a material that can be molded and then hardened; this was the compound cellulose nitrate. But it was not until thirty years later that anyone was able to turn this material into a commercial product, in part because it becomes a workable material only when made more pliable. The answer was to add plasticizers such as castor oil or camphor, and the latter was used in the cellulose nitrate material marketed by John and Isaiah Hyatt in the late 1860s under the name celluloid. The possible uses of this tough, hard plastic soon seemed unlimited: it was a good substitute for expensive ivory, it was strong enough to be used for false teeth, and in extruded, fibrous form it could be used in electric-light filaments. The use of celluloid for photographic film was such a hit that the name of the material soon became synonymous with the movie industry itself. It is, however, highly inflammable, as several movie houses found to their cost. In fact, a form of cellulose nitrate containing three nitrate groups on each glucose ring proved to be explosive, earning it the name of gun cotton.

In the 1880s, Count Louis Hilaire de Chardonnet discovered a way to convert cellulose nitrate back to a new form of cellulose that could be extruded into fibers stronger than those of the cotton from which it came. This product, named rayon, established the basis of the synthetic textile industry. A later variant, called viscose rayon, offered superior fiber properties and could be formed into thin plastic sheets, which were marketed as cellophane. Another cellulose-based material, called cellulose acetate, is widely used in the form of fibers, films, and molded shapes, and dissolved in chloroform it provides the "dope" in vehicle paints.

Thus the early years of the plastics industry were dominated by products based on natural cellulose. Today their use is waning, as many new synthetic polymers with better properties have appeared on the scene. But cellulose still has the advantages that it is a natural and renewable product, available at low cost. Today the Sony electronics company uses cellulose produced by bacteria as the diaphragms in hi-fi loudspeakers. Malcolm Brown of the University of Texas at Austin has shown that cellulose fabrics can be obtained from extraordinarily low-technology processes. The bacterium *Acetobacter xylinum* synthesizes fibrous strands of cellulose when fed on a diet of sugar in fermentation troughs, and the strands aggregate into a tangled, fibrous sheet at the water surface. This material, which is fully biodegradable, can be regarded as a kind of natural paper (fig. 4.30). It has the flexibility and absorbance of normal paper, but is much tougher: a sheet no thicker than blotting paper is harder to tear than a telephone directory. Brown says that this material is welcomed by artists as a resilient and aesthetically pleasing alternative to canvas, while its tremendous capacity to absorb water (up to one hundred times its own weight) might find it medical applications as a superabsorbent. Researchers at the Courtaulds company in Coventry, England, have developed a process for creating strong cellulose fibers from wood

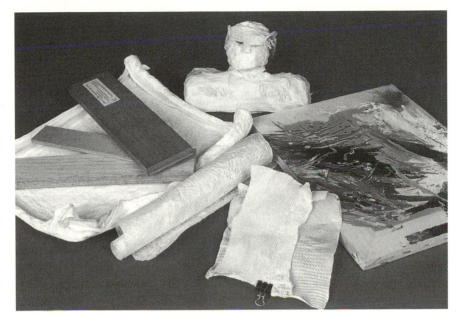

FIGURE 4.30 A tough, fibrous material made of natural cellulose can be produced by growing the cellulose-forming bacteria *Acetobacter xylinum* in a vat of nutrients. The cellulose is deposited at the surface of the water. Its texture is somewhat like rough paper, but it is much stronger and can absorb many times its weight in water. (Photograph courtesy of R. Malcolm Brown, University of Texas at Austin.)

pulp, which is marketed in the form of a textile called Tencel. The fibers are made by dissolving the pulp in an organic solvent and extruding the viscous solution. Tencel is biodegradable and can be fashioned into a variety of fabric textures.

Some bacteria and animals make discriminating use of cellulose-based polymers. The bacteria *Xanthomonas campestris* synthesize a compound consisting of a cellulose backbone with additional saccharides stuck onto every other glucose monomer. This substance, called xanthan gum, is produced commercially by growing the bacteria in huge fermentation vats. The bacteria excrete the polymer, which is soluble in water, so it can be extracted simply by filtering out the microorganisms. Xanthan gum has the property of thixotropy, which means that its viscosity changes when it flows. Shaking the substance turns it from a gel-like consistency to a liquid. This property is useful in food products such as salad dressings, and is also exploited for controlling the viscosity of fluid lubricants used in oil drilling.

Beetles take advantage of the toughness of a cellulose-like material called chitin to fabricate their body armor. This nitrogen-containing polysaccharide is dispersed in a protein matrix to yield the lustrous material that provides beetles and other arthropods with armor plating capable of deterring many a predator's jaws. Prawns and shrimps make use of this protective shell too, forcing diners to go to considerable lengths to extract the juicy flesh from the indigestible mantle. The protein matrix of chitin is commonly cross-linked, and the chitin polymer chains are chemically attached to this matrix too. Some arthropods, such as crabs, reinforce this tough organic composite further with mineral crystals, generally calcium carbonate. The availability of large quantities of chitin literally in the bins of seafood restaurants (or more importantly, in developing countries where crustacea are a staple diet) has stimulated a search for applications; one such may be the use of chitin-derived polymer gels for slow release of drugs, a topic discussed in the next chapter.

The New Face of Plastics Plants

Closely related to cellulose is starch (fig. 4.31), another polysaccharide. But whereas cellulose is resistant to chemical degradation, starch is more easily broken down into its sugar monomers. This was first discovered in 1811 by Constantine Kirchoff, who noted that acids will degrade starch to sugar syrup. Plants use starch as an energy store rather than as a structural material; and indeed we are familiar with the idea that both this compound and its monomeric sugar units can be a rich source of energy for us too.

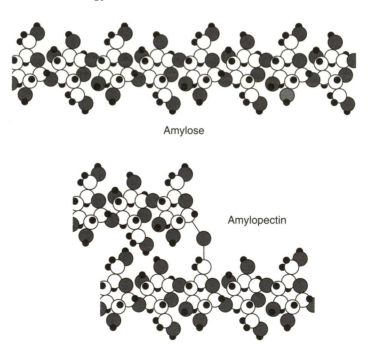

Amylose

Amylopectin

FIGURE 4.31 Starch is made up, like cellulose, of linked glucose molecules; the difference lies in the way that the rings are joined together. There are two main polysaccharide components of starch: amylose, a (mainly) linear chain of glucose units, and amylopectin, which has a complex structure in which chains are cross-linked. Starch acts as an energy reserve for plants, being stored as tiny granules in the roots, seeds, and stems of grains and in potatoes.

Many bacteria store energy in the form of simple polymers called polyhydroxyalkanoates (PHAS), members of a more general class known as polyesters (fig. 4.32). Interest in PHAS has mushroomed in recent years, because these polymers can be extracted from bacteria and processed to form plastic materials with properties similar to those of common synthetic polymers like polypropylene and Terylene. But unlike these synthetics, PHAS are biodegradable materials, being broken down in the natural environment by other bacteria. This realization has blossomed into a vision of a plastics industry in which the materials are made not in chemical processing plants but in vats of bacteria. The possibility of transferring the bacterial PHA-making genes into plants makes for an even more attractive prospect in which plastics are literally grown in green fields.

FIGURE 4.32 Some bacteria store polymers called polyhydroxyalkanoates (PHAS) as an energy reserve, in the same way that our bodies store fat. These polymers, belonging to the class called polyesters, are stored in the form of granules within the bacterial membrane. Shown here is poly(3-hydroxybutyrate) (PHB), formed by certain bacteria from sources of carbon, hydrogen, and oxygen such as sugars or alcohols.

Bacterial PHAS are synthesized in the form of small granules, which the bacteria store up inside their cell membranes for use as an energy reserve if food supplies dwindle. These granules can comprise as much as 80 percent of the dry weight of the bacteria when the bacteria are deprived of an essential nutrient such as nitrogen, oxygen, phosphorus, or magnesium. The PHAS can be harvested by breaking the cells apart with enzymes and extracting the granules in a suitable solvent. PHAS are now produced commercially by feeding bacteria in fermentation vats on relatively cheap carbon sources such as ethanol or milk fat. The precise composition of the PHA produced depends on the composition of the feedstock, making it possible to vary the nature and properties of the plastic product simply by giving the bacteria a different diet.

One of the most commonly produced member of the PHA class is called poly(D-3-hydroxybutyrate), or PHB, which several species of bacteria make naturally from glucose—although in fermentation vats they will be happy to feed off other compounds. PHB, whose structure is depicted in figure 4.32, is a relatively stiff, brittle polymer which melts at around 180 degrees Celsius. A lower melting point would make the plastic more flexible, which is desirable for some applications, and would also make processing easier, since at the relatively high temperature at which PHB melts, it also degrades rapidly. Such a material can be produced simply by providing PHB producers with propionic acid in the feedstock. This compound is transformed by the bacteria into hydroxyvalerate (HV) monomers, which alternate at random with the hydroxybutyrate monomers along the

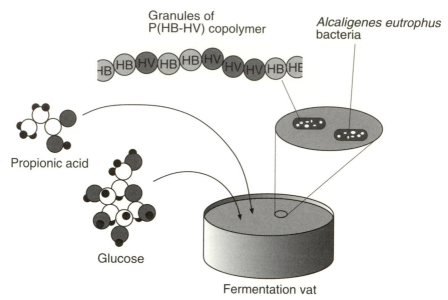

Granules of
P(HB-HV) copolymer

Alcaligenes eutrophus
bacteria

Propionic acid

Glucose

Fermentation vat

FIGURE 4.33 A copolymer containing a mixture of hydroxybutyrate (HB) and hydroxyvalerate (HV) units is marketed by Zeneca Bio Products in the U.K. as a biodegradable, flexible plastic. It is made by feeding bacteria on a diet of glucose and valeric or propionic acid in fermentation vats, and extracting the copolymer from the bacterial cells.

chain (fig. 4.33). Polymers that contain a random alternation of monomer units are called random copolymers. The ICI chemicals company in the United Kingdom markets this P(HB-HV) copolymer commercially as a biodegradable plastic called Biopol, grown by feeding vats of the bacteria *Alcaligenes eutrophus* with glucose and propionic acid.

The genes responsible for PHA synthesis in bacteria can be stitched into the genetic material of another species, turning that organism into a PHA factory too. For example, the PHB-synthesizing genes from *Alcaligenes eutrophus* have been successfully transferred into *Escherichia coli*, the bacteria found in the human gut. This is useful because *E. coli* can be fermented at higher cell densities, leading to improved PHB production. More impressive still is the transfer of the *A. eutrophus* genes into oil-seed rape plants. These plants produce large quantities of vegetable oil; the introduction of PHB genes allows them to divert some of the feedstock into PHB synthesis instead. The British company Zeneca, an offshoot of ICI, has begun to develop PHB plastics from genetically modified rape plants.

All of these plastics are broken down by bacteria in the environment into carbon dioxide and water. A wide variety of bacteria produce enzymes that degrade PHAS, and the plastics will even be degraded by bacteria in the sea. But because bacterial fermentation is not a cheap process for large-scale production, PHAS remain relatively costly in relation to comparable polymers derived from petroleum: they are typically fifty times more expensive than polypropylene, for instance. All the same, ICI believes that the cost will fall dramatically—to perhaps a fifth of the current price—in future years, in which case the special properties of PHAs may allow them to corner at least part of the plastics market.

Sticky Stuff

Two more biopolymers deserve special mention for their continuing value to the materials industry, both of which have been put to the service of humankind for centuries. While we in the Eurocentric Northern Hemisphere have a habit of crediting the Duke of Wellington with the popularization of the rubber boot, the first recorded use of rubber for waterproofing footwear comes from early explorers in Central America, who observed that the indigenous people had already found this use for the water-repellent substance. They extracted it in the form of a gum from tropical trees of the species *Hevea brasiliensis*.

Rubber is an *elastomer*, a material that can be stretched to several times its original length without permanent deformation: it will snap back to its original shape when released. This is distinct from an *elastic* material, which is capable of regaining its initial shape only after a rather small deformation. Confusingly, in common usage "elastic" (as in elastic bands) in fact usually means elastomeric.

Elastomers are polymeric materials whose properties stem from the structure of the polymer chains. As in the case of the elastomeric protein elastin described earlier, the chains adopt a random coiled conformation which can be extended under stress or tension but will coil up again when the tension is released. In other words, the polymer chains are miniature springs. In rubber, the chains are composed of just carbon and hydrogen, but the crucial difference between these and the hydrocarbon chains in plastics like polyethylene and polypropylene is that the rubber chains contain double bonds between some of the carbon atoms: they are unsaturated hydrocarbons. The double bonds create kinks in the chains, preventing them from packing together efficiently and therefore making it hard for crystalline regions to form. For this reason rubber is less stiff than, say, polypropylene—indeed, at room temperature the raw rubber gum remains a viscous liquid. The disorderly chains form tangles of random coils.

Commercial elastomeric rubber products stemmed from Charles Goodyear's discovery in 1839 of a means to stiffen the gum. Goodyear's process is called vulcanization, and it involves cross-linking the polymer chains with sulfur atoms, either individually or in the form of short sulfur chains. Like so many useful discoveries, it was an accident: Goodyear was using sulfur simply as a drying agent, which he hoped would remove the liquid component of the rubber gum. Vulcanized rubber still enjoys wide use, but has the drawback that it cannot be extruded or spun into elastomeric fibers—instead, it is molded into sheets, which are then sliced into strips. Today several synthetic elastomers do better. The most successful of these, called spandex, was first introduced in 1956 and has a greater strength than rubber, as well as being processible into fibers. There are many variants of spandex fibers on the market, all of which are composed primarily of a polymer called polyurethane, whose monomer units are related to the organic molecule urea. Another class of synthetic elastomers, called silicones, are unusual in being silicon-based rather than carbon-based. These materials can be used in molded rubbery products but are too weak to form useful fibers.

NATURE'S HARD STUFF

All of the biomaterials that I have described so far are "organic," in the sense that their molecular components are carbon-based (and usually polymeric) molecules. Structural materials made from these compounds can be extremely tough and strong in tension, but with a few exceptions (such as chitin and wood) they are pretty hopeless under compression—squeeze them and they collapse. In other words, they are soft materials. If you are a seabound organism like a jellyfish, squishiness may be no disadvantage: because of buoyancy, you are not battling against gravity, and in the liquid environment of the sea your slow-motion life does not expose you to sudden, sharp impacts. But big, clumsy landlubbers like ourselves need a robust framework on which to hang our organs and tissues, one that will not sag or buckle under the pull of gravity. We need not only the compliance of organic matter, but the compressive strength and hardness of rock. Which is precisely why we are not just organic, but partly mineral too. Our bones and teeth contain crystals of the phosphate-based mineral called apatite. Mollusks, meanwhile, protect themselves from predators with shells of calcium carbonate, in the form of the minerals aragonite and calcite; and birds use the same materials to protect the developing embryos of their offspring within eggshells. Some marine organisms build themselves elaborate cages of silica, while limpets fashion their teeth from iron oxides.

In a literal sense, "inorganic" can be a misnomer when applied to such hard biominerals, for these crystals are very much the product of living organisms. Whereas in geology minerals are generally deposited by precipitation from solution or solidification from melts, inorganic materials in nature are generally grown within an organic matrix that exerts a strong influence on the shape of the crystals. In bone (where the matrix is primarily the protein collagen), the crystals of apatite are flat plates; in tooth (where the matrix consists largely of the protein enamelin) they are long needles. In shell, two different forms of calcium carbonate (calcite and aragonite) grow within a protein matrix, which determines both the carbonate phase and the crystal shape. The degree of control exerted on crystal growth during biomineral formation (the process generally called *biomineralization*) extends over a wide range of length scales, from the atomic (the organic matrix can determine the way in which ions are packed together in the crystal) to the macroscopic (organics are responsible for the overall patterning of many skeletal and shell shapes).

A Home by the Sea

Perhaps the most striking feature of sea shells is their variety of pattern and form, from the scalloped bowls of bivalved mollusks to the spiral minarets of snails and

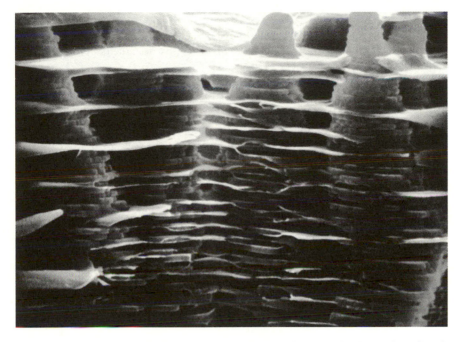

FIGURE 4.34 Mother-of-pearl (nacre) is a composite material containing inorganic (mineral) and organic components. Slabs of the calcium carbonate mineral aragonite are laid down on sheets of proteins and other biomolecules, forming pyramidal stacks of disks. (Photograph courtesy of Mehmet Sarikaya, University of Washington, Seattle.)

whelks (plate 3). These elaborate dwelling places are composed of aragonite crystals—calcium carbonate, generally with a small amount of magnesium—arrayed within a matrix of proteins and polysaccharides. In nacre (mother-of-pearl) the aragonite takes the form of flat slabs stacked on top of each other, with thin sheets of organic material separating successive slabs. The organic sheets are themselves rather complex composites, containing successive layers of chitin, silklike proteins, and acidic polymer molecules that control the way that the aragonite crystals grow. The sheets are laid down first, and the inorganic material is then gradually deposited as expanding circular slabs between the organic layers. These slabs are nucleated one atop the other, so that they take on the appearance of a pyramidal stack of disks (fig. 4.34). This whole assembly is covered with an external layer of calcite.

The aragonite platelets not only have all the same disk-like shape but also have their crystal axes oriented in the same direction. That is to say, within each platelet the ions are stacked together in the same orientation as in all the others, just as the layers of apples in different crates are aligned when the crates are stacked in a warehouse. This correlation between the crystal axes of two crystals that begin growing at different locations shows that the organic matrix is governing not just the shape but also the orientation of the crystals. During shell growth, interactions between the acidic molecules at the surface of the organic sheets and the ions that attach themselves to the growing crystal nucleus somehow guide these ions into a particular orientation, at the same time suppressing the growth along the crystal axis perpendicular to the layers and/or enhancing growth in the parallel directions so as to produce disk-shaped crystals.

How does the organic material exercise this influence? It seems likely that it acts as a kind of template, imposing certain constraints on the way that the inorganic ions stick to its surface. There are several ways in which these constraints can be applied. One is through complementarity of charge: a positive (calcium) ion will prefer to stick to a negatively charged region of the matrix, such as an acidic site containing a carboxylate group. Another is the disposition of these binding sites—the distance between them, and their geometric arrangement. The spacing between ions in a crystal generally differs depending on the direction in which you look at the crystal structure (fig. 4.35a)—that is, it varies for different crystal faces. So if ion-binding sites within the organic matrix are disposed so as to secure ions in an arrangement corresponding to that along a particular crystal face, the preferential growth of that crystal face can be encouraged. The crystal then grows with its axes in a particular orientation (fig. 4.35a). You might recognize this as a process akin to epitaxial growth (chapter 1): the substrate (the organic matrix) acts as a template for epitaxial growth of the crystal.

Control of crystal nucleation and growth in this way has been closely studied in the formation of the shells of the mollusk *Nautilus repertus*. The aragonite is deposited on a pleated β-sheet of a silklike protein, the repeat unit of which almost matches in size the repeat distances between calcium ions in the aragonite crystal lattice (fig. 4.35b). It seems likely that the calcium ions become bound to acidic aspartate units in the protein, which bear a negative charge: two such groups placed with one amino acid between them in the protein chain would lie just far enough apart to bind a calcium ion in a pincerlike grip. In this way, the amino acid sequence of the protein chain can determine where, within the organic matrix, nucleation of aragonite occurs. So ultimately the biomineralization process is *genetically* controlled.

The resulting mineral–organic composite is a mosaic of crystal plates (fig. 4.36, *top*). This structure conveys high toughness against stresses applied perpendicular to the layers (that is, perpendicular to the flat face of the shell). Paradoxically, the origin of this toughness is the relative weakness of the interface between the mineral plates and the organic matrix, which can be pulled apart much more easily than one can disrupt the crystals themselves. These weak interfaces prevent a crack from propagating through the material, as it would through a monolithic slab of the pure, brittle mineral, because the energy of a crack-inducing impact can be rapidly absorbed by the rupture of bonds between the mineral and the matrix. In effect, a crack gets deflected sideways, along the weak interfaces, as it passes through the layers, so that it loses its energy before it penetrates too deeply. The result is a few fractured layers rather than a catastrophic flaw that runs through the whole shell. So while the composite structure makes the shell less hard than a pure crystal, it allows the shell to withstand more battering without simply shattering. This means of toughening materials by incorporating weak

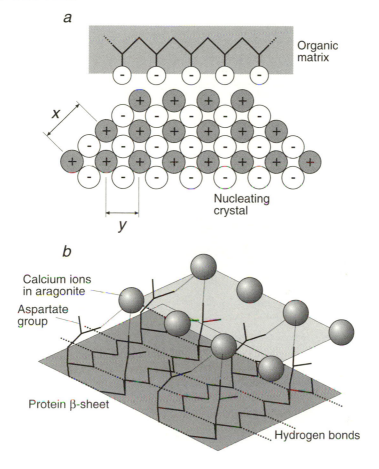

FIGURE 4.35 Most biomineral crystals are deposited on an organic matrix that controls the orientation and structure of the crystals. It is thought that chemical groups at the surface of the matrix may act as a template for the nucleation and growth of the mineral. For example, negatively charged side groups along a protein β-sheet might be arranged the right distance apart to mimic the negative ions on one face of the crystal but not on another face (*a*). Here the negative side groups are spaced a distance y apart, which matches the lattice spacing on the top face of the crystal but not the lattice spacing (x) on the side faces—so the crystal is deposited with the top face facing the matrix. Something of this sort may take place during the formation of the shell of the mollusc *Nautilus repertus*, where aspartate residues (containing negatively charged carboxylate groups) in the protein β-sheet are arranged in more or less the right positions to bind the calcium ions within the aragonite crystal lattice (*b*).

interfaces is frequently exploited in synthetic composites, such as fiber-reinforced ceramics; here microscopic fibers are embedded in a strong ceramic matrix, and the energy of a fracture is expended in pulling the fibers out of the matrix as the crack propagates. Kevin Kendall and colleagues at the ICI laboratories in Runcorn, U.K., described in 1989 a kind of synthetic shell in which slabs of the strong

FIGURE 4.36 The toughness of nacre arises, paradoxically, from the relative weakness of the interface between the inorganic plates and the intervening organic sheets (*top*). Cracks running through the composite material perpendicular to the plates are deflected at right angles along these weak interfaces. Although this causes delamination of the plates, it absorbs the energy of the crack and prevents it from propagating right through the material, as it would through a monolithic crystal. This microstructure can be mimicked to make tough synthetic composite materials in which slabs of the strong ceramic silicon carbide are bound together at weak interfaces by graphite (*bottom*). (Photographs courtesy of Kevin Kendall, Keele University, and Bill Clegg, University of Cambridge.)

ceramic silicon carbide were bonded together by graphite sheets interposed between the slabs, creating a tough, fracture-resistant composite (fig. 4.36, *bottom*).

The shells of birds' eggs are similar in composition to mollusk shells, consisting of calcium carbonate (this time mainly in the crystal form of calcite) laced

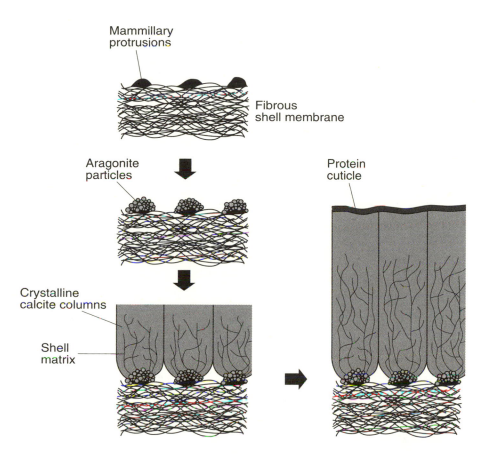

FIGURE 4.37 The formation of eggshell involves the nucleation of inorganic crystals on a membrane of fibrillar proteins around the egg. The nucleation points are defined by protein nodules called mammillary protrusions, and the mineral is first deposited as particles of aragonite with random orientation of the crystal planes. On top of these aragonite piles, columns of oriented crystalline calcite begin to grow upward, interwoven with strands of a protein matrix. Finally, the crystalline columns are overlaid with an organic layer (a cuticle) of glycoproteins, proteins with sugar groups attached.

with an organic matrix. But the microstructure is rather different, reflecting the fact that eggshells must be produced much more quickly than sea shells. The crystals in eggshell are stacked in columns perpendicular to the egg's surface, rather than as slabs parallel to it. This makes the shell less strong and less shatterproof (and after all, the chick has to break out of it eventually), but quicker to grow. The crystals are nucleated around nodules of protein on the surface of a fibrous organic membrane, and the calcite grows in columns up from the nucleation sites (fig. 4.37).

The Meat Rack

Bone is a superb example of a natural hierarchical material and demonstrates how several levels of structure at different size scales can allow a material to fulfill many requirements at once. Bone must be strong but not brittle. It must be rigid and able to bear loads, but must also have some flexibility. It must be lightweight. And it must be able to support a blood supply to the organic tissues within. Rather than list the various levels of structural hierarchy that allow these needs to be satisfied, I'll simply refer you to figure 4.38. The mineral component of bone is basically hydroxyapatite, a crystal comprised of calcium, hydroxyl, and phosphate ions in the ratio $Ca_{10}(OH)_2(PO_4)_6$; but in bone this mineral contains some carbonate ions in place of some of the phosphates. Bone-forming cells called osteoblasts secrete small platelets of this material within a matrix of collagen. The body continually reabsorbs the hydroxyapatite crystals (this is the task of so-called osteoclast cells), and redeposits new material. In this way the thickness and microstructure of the bone can be continually adjusted to accommodate changes in the body's distribution of weight (see chapter 5).

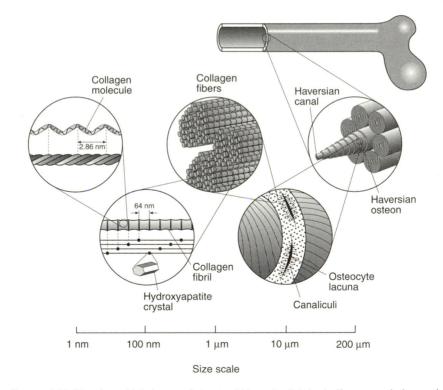

FIGURE 4.38 Bone has a high degree of structural hierarchy. It is basically an organic-inorganic composite material in which crystals of hydroxyapatite are dispersed in a matrix of collagen. The porous structure allows access for blood vessels, and conveys flexibility and low density while maintaining strength.

FIGURE 4.39 Rat tooth enamel contains interwoven fibers of the mineral phase (mainly hydroxyapatite), interspersed with remnants of the organic matrix responsible for directing mineralization in this complex structure. (Photograph courtesy of Paul Calvert, University of Arizona.)

Anyone who has watched wild cats grind up the bones of their prey will appreciate how tough the fabric of teeth must be. Tooth enamel is almost pure mineral—95 percent by volume, about twice the mineral content of bone. Its great strength comes from a structure that is the envy of ceramics manufacturers everywhere. The mineral component, again essentially hydroxyapatite, comes in the form of needlelike fibers that are interwoven like the threads of a carpet (fig. 4.39). This interweaving means that the structure remains strong even if many fibers are fractured. The crisscrossed mineral trellis contains small amounts of protein between the rods, the remnants of the organic matrix in which the crystals were nucleated.

THE ART OF IMITATION

I hope it is clear by now that nature's fabrics are very advanced materials indeed. This is why materials scientists are trying to learn from nature to develop new synthetic materials with sophisticated properties. In the course of this chapter I have already presented some examples of attempts to utilize the constructional principles of natural materials, or to adapt these materials themselves for technological uses—for example, by reconstituting biopolymers into synthetic products like cellophane and rayon, or by hijacking enzymes or whole cells for materials synthesis. I shall discuss other attempts to mimic and adapt nature's materials in later chapters, particularly chapter 9. These efforts are diverse in both their aims and their methods, and they have generally evolved from different directions under the guidance of researchers who would see little in common with each other's work. Yet increasingly in the past decade, materials scientists are appreciating that such studies are collectively defining a new field of endeavor, one that shares something more in the nature of a common philosophy than similar objectives or skills. This field has acquired the name of *biomimetics*: the art of mimicking biology. As scientists come to understand more about the way that natural systems put themselves together, some dominant themes have emerged that serve as a guide to much of biomimetic materials research; here I'd like to suggest a few of them.

Composites. It is rare to find a structural or functional component of an organism that consists of just one material. Cell walls are tremendously elegant composites of many kinds of molecule, including phospholipids, membrane proteins and glycoproteins, and protein-based cytoskeletal frameworks. Bone is an intimate mixture of organic and inorganic materials, primarily apatite in a protein matrix. The soft tissues of the body contain a mélange of structural proteins, each with different mechanical properties. Biomimetic materials tend to be composites too.

Structure and organization on many scales. The paradigmatic materials of the nonliving world—crystalline minerals—tend to have only a single level of organization: the arrangement of atoms in the crystal lattice. The bulk material is made up by simply repeating this structural motif countless times. But in natural materials one tends to find distinct, well-defined structures over many scales of magnification. We saw how this is true for tissue fibers, for instance: the protein molecules have a specific kind of (generally helical) organization, but the structures formed from these helical molecules are themselves organized into strands, bundles, and bundles of bundles. This hierarchical organization conveys many benefits to a material, such as a high strength-to-density ratio or the ability to serve several functions simultaneously. For scientists attempting biomimetic materials synthesis, the challenge here is how to achieve these successive levels of structural organization. The trick to that lies partly with a common element of biomaterials processing, called . . .

Self-assembly. Traditionally, we fashion a material's microstructure the hard way: by putting it there by hand. Thus we saw in chapter 1 how the exquisite microstructures of semiconductor electronics are carved out by electron or ion beams, a method that is now routine and capable of extraordinary feats but which is nevertheless labor-intensive and time-consuming. Nature does not have electron or ion beams at her disposal, but more importantly still, she does not (as far as we can tell) have a guiding intelligence to define the delicate microstructures of natural organisms. Yet she produces such microarchitectural masterpieces as the Oriental terraces of mother-of-pearl, or the soft compartmentalized complexity of the cell and the chloroplast. All of this must be carried out using the simple rules of chemistry, which follow from a handful of fundamental interactions among molecules. We have seen how hydrophobic effects guide cell-like vesicles to assemble spontaneously in solution, and how chemistry alone can construct the entire blueprint for an organism (not to mention putting it together). Self-assembly—the spontaneous formation of ordered and/or complex molecular aggregates—is now becoming a dominant theme of research in modern chemistry, and this in turn is feeding into materials synthesis. The theme will recur particularly strongly in chapter 7.

Control of nucleation and crystal growth. We have seen that, while crystals grow both in the nonliving (geological) and the living (biological) worlds, the two are not the same. In geology, as a rule the crystal phase that forms is the most stable one under a given set of circumstances (such as temperature, pressure, and so forth). The crystalline materials that are formed in biology, meanwhile, have to serve a purpose, usually as a protective, strong material. They have to conform to the shape of the organism, and so usually cannot display the abrupt, geometrical facets of minerals. They have to be sympathetic to the way that the organism grows as a whole, which commonly means that their growth must have a preferred location and direction (mollusk shells grow from their rims, for instance). So crystal growth in biology cannot be left to the whims of inorganic crystal chemistry; it must be controlled and guided, a task that is left to organic molecules and soft tissues.

The use of templates. One of the most common ways in which organisms exercise architectural control over the formation of their structural and functional components is by the use of templates to guide the growth process. Genes themselves are simply templates on which proteins can be grown (via the intermediation of messenger RNA), while organic tissues template the formation of bone and shell. Sometimes the latter involves templating at the molecular level: epitaxy between the molecular structure of an organic membrane and an inorganic crystal deposited at its surface.

The ways in which these and other guiding principles are being exploited in biomimetic materials synthesis are many and, to my mind, dazzling. I shall not attempt to survey these efforts in any comprehensive manner, but I would like to exemplify them here by focusing on one particularly strong theme in recent biomimetic research: the use of cooperative interactions between organic and inorganic materials to make new kinds of patterned composites. This field is literally exploding with potential, and the latest research in this area is demonstrating that, at the convergence of materials science, chemistry, biology, crystal growth and pattern formation, we have the opportunity to make new connections between scientific disciplines that are to the advantage of them all.

Organic Control

The central theme of biomineralization is the growth of a mineral phase within or on top of an organic matrix that controls the nucleation, growth, and eventual size and shape of the crystals. Researchers have attempted to mimic this principle using a variety of synthetic organic matrices. One of the roles of the matrix is to provide ion-binding sites that are more or less in registry with the positions of the ions in the crystal structure of the mineral. Several researchers have been attracted by the potential of surfactant films to provide this kind of crystal template. Surfactant molecules disport themselves on the surface of water with their water-loving heads in the liquid and their insoluble tails protruding from the surface. By varying the density of surfactants in this interfacial film (something that can be achieved simply by squeezing it with a movable barrier), one can vary the packing arrangement and the distance between the head groups on the lower (water-immersed) face of the film. The controlled growth of crystals below surfactant monolayers has been pioneered by Meir Lahav's group at the Weizmann Institute of Science in Rehovot, Israel. Lahav and colleagues showed in 1985 that crystals of the amino acid glycine and of sodium chloride grown from supersaturated solutions at the surface of a surfactant monolayer have a well-defined orientation, with their crystal planes lying parallel to the organic layer (fig. 4.40). Moreover, Lahav and colleagues found that by changing the surfactant they could alter the orientation of the crystals. For example, sodium chloride crystals grew with different crystal faces facing the monolayer when the surfactants were alkyl amines (which have uncharged head groups) or stearic acid (where the carboxylate head groups are ionic). And when the researchers expanded the distance between the head groups in the latter case by giving the surfactants bulkier tails, they could suppress crystal nucleation entirely, presumably because there was then no longer a near-epitaxial relationship between the spacing of the ion-binding sites (the carboxylate groups) and the spacing of the ions in the sodium chloride lattice.

Stephen Mann and colleagues at the University of Bath in England have shown that they can control the orientation and crystal structure of calcium carbonate crystals grown at the surface of surfactant films. For example, at the surface of a carboxylic acid film the calcium carbonate crystallized as calcite with the same crystal planes of all the separate crystallites parallel to the surface of the film, apparently because for this crystal orientation the carboxylate groups of the surfactants mimic the carbonate ions of the crystal both in their spacing and their orientation. But when they grew the crystals using lower concentrations of calcium carbonate in solution, or using amine surfactants instead of carboxylic acids, the researchers obtained a different kind of crystal structure, corresponding to the carbonate mineral vaterite. They reasoned that this change was due to the different constraints on ion/head-group positional matching at the crystal–organic interface when there were fewer calcium ions around or when the head groups were different. These studies show that an organic membrane just one molecule thick can exert a profound effect on the crystallization of inorganic materials.

As well as determining the crystal structure and orientation of the mineral phase, the organic matrices in which biominerals grow influence the final shape

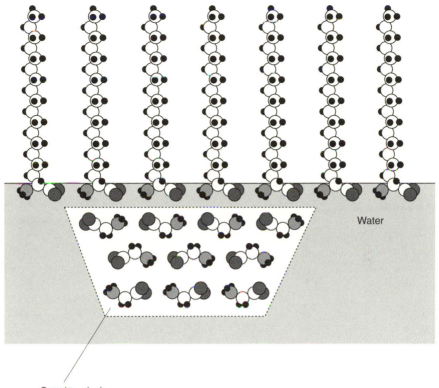

Water

Growing glycine
crystal

FIGURE 4.40 The orientation of crystals grown from solution below a monolayer of amphiphilic molecules can be controlled by the relationship between the atoms or molecules in the crystal and the head groups of the amphiphiles. Meir Lahav and colleagues have grown crystals of the amino acid glycine beneath a monolayer of amino acids with long hydrocarbon chains attached; the latter molecules can form an ordered, close-packed layer at the interface between the solution and air. The spacing between the amino acid head groups of the amphiphiles in the close-packed monolayer was very similar to the spacing between glycine molecules in a specific plane of the glycine crystal (denoted the (010) plane). So the head groups acted as a template to induce crystal growth with the 010 plane oriented parallel to the water surface.

of the mineral materials—crystallization is commonly restricted to well-defined compartments or to specific cells. This is a trick that proves very useful for making inorganic crystals of a predetermined size. There is great interest in making small inorganic particles of a single size and shape, because particles of this sort made from semiconducting materials (such as cadmium selenide) might have useful light-absorbing properties. We saw in chapter 1 how silicon becomes a good light emitter when fashioned into tiny wires a few nanometers across. For this and other semiconductors, the color of the light absorbed and emitted depends on the size of the particles, and so one can tune the light-emitting properties simply by varying the size. Tiny semiconductor particles of this sort might therefore be useful for making optoelectronic and optical display devices, and indeed prototypes of light-emitting diodes containing tiny cadmium selenide particles were reported in 1993.

Researchers are starting to look for inspiration from biomineralization to find ways of making these nanoscale particles. One approach, introduced during the 1980s, is to grow them within tiny "beakers" made by the self-assembly of surfactants into reverse micelles (see fig. 4.21*b*). Stephen Mann's group has used a similar philosophy to make uniformly sized iron oxide particles, using as the mold a natural protein called ferritin, which stores iron ions in bone marrow for eventual incorporation into red blood cells. Ferritin is quite simply an iron repository—the protein is made up of twenty-four molecular subunits, which come together in an aggregate that has a cavity about 12 nanometers across at its center, big enough to hold up to 4,300 iron ions. By supersaturating this encapsulated iron solution, Mann and colleagues were able to trigger the formation of iron oxide particles within the ferritin cavities, generating a fine suspension of protein-coated iron oxide spheres each about 4 nanometers in diameter. These tiny particles are biocompatible and magnetic, and a suspension of them acts as a kind of liquid magnet.

As well as controlling the growth process by depositing the mineral on the *inside* of an organic mold, one can lay it down on the *outside*, so that the organic structure acts as a template. Stephen Mann and Douglas Archibald used this idea in 1993 to make tiny tubes of iron oxide, a few micrometres long and about 100 nanometers in diameter, by using self-assembling lipid tubes as the template. These tubes, comprised of bilayer sheets of lipids curved into cylinders, can form spontaneously in solution. Mann and Archibald deposited crystallites of iron oxide on their hydrophilic surfaces to make miniature composite pipes.

Grand Designs

This templating approach, which exploits the range of architectures that assemble spontaneously from surfactant molecules in solution, is now leading to one of the most dramatic new directions in materials processing: the creation of inorganic materials imprinted with intricate patterns. Not only might such materials find valuable uses in their own right—in biomedicine, chemical engineering, and optoelectronics, for instance—but they may also provide clues as to how nature fabricates her own elaborate biomineral edifices, such as the external skeletons (exoskeletons) fashioned by some of the microscopic denizens of the sea— radiolarians, coccolithophores, and diatoms. These are almost absurdly ornate works of architecture, and both why and how the organisms lavish such attention on them remains largely a mystery. But the synthetic materials made in 1995 by Geoffrey Ozin and colleagues at the University of Toronto in Canada (fig. 4.41*a*–*c*) bear a striking resemblance to these natural biominerals (fig. 4.41*d*–*f*).

Ozin's "artificial radiolarians" are made from an aluminophosphate mineral— not an inorganic material that is found in biominerals—but it seems likely that the patterning mechanism will prove to be quite general. To make their patterned materials, the Toronto group mixed phosphoric acid and aluminum hydroxide

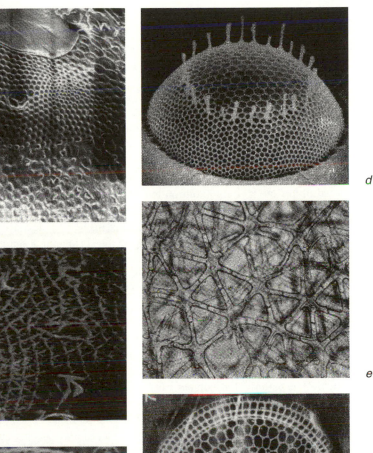

FIGURE 4.41 By using organic materials as a template, Geoffrey Ozin and colleagues have produced synthetic materials with microscopic patterns (*a–c*) that closely resemble those seen on the exoskeletons of marine organisms called radiolarians and diatoms (*d–f*). Ozin believes that the patterns are the imprints of vesicles formed from surfactant molecules and polymer chains present in the reaction mixture. Spherical vesicles might pack together like bubbles in an ordered array on the surface of a growing mineral crystal, and the deposition of inorganic ions in the spaces between the vesicles then creates an imprint of the pattern in the crystal surface. The honeycomb of pits in (*d*) is particularly well mimicked.

together in solution along with a dose of an alkylamine surfactant. A crucial aspect of the synthesis is that the solvent was not water but the organic compound tetraethylene glycol (TEG). Ozin believes that the key to the patterning process lies with the fact that this solvent encourages the surfactants to organize themselves into hollow vesicles.

The aluminophosphate crystallizes in the form of spheres about a millimeter across. Scanning over the surface of these spheres with an electron microscope reveals a landscape of craters like the Moon's surface. The researchers have suggested that the bowl-shaped craters are the imprints of vesicles formed from a mixture of surfactants, negatively charged inorganic ions, and TEG solvent molecules. As these vesicles become packed together, like soap bubbles, they imprint a kind of hexagonal bubble raft on the surface of the growing aluminophosphate crystals. Finer-scale patterns might result from the separation of the surfactants and the TEG molecules into distinct domains in the vesicle walls—the aluminophosphate then precipitates preferentially in the surfactant domains. Ozin's group has seen a wealth of other shapes scattered across the surface of these materials, including tiny hollow eggshell-like features that resemble microscopic barnacles, which are perhaps the mineralized casts of entire vesicles. The idea that vesicles are involved in the formation of these patterns echoes the thoughts of the Scottish zoologist D'Arcy Wentworth Thompson, who speculated in 1917 that radiolarian skeletons are the result of crystallization within a froth of packed vesicles.

Under the electron microscope, one can see that the aluminophosphate is deposited in crystalline layers about three nanometers thick—it is a lamellar material. So the patterned structures can be compared to a papier-mâché model, in which the features are fashioned from parallel layers of fabric glued together. We can identify a rich hierarchical structure here: the millimeter-scale spherical shape of the particles, the micrometer-scale disks and bowls on the surface, the submicrometer-scale pores and striations, the nanometer-scale lamellae, and finally the atomic-scale arrangement of ions.

Stephen Mann at Bath has made patterned materials of calcium and magnesium carbonate (aragonite) that consist of convoluted ("reticulated") networks of mineral branches reminiscent of a thorny hedgerow (fig. 4.42a). Mann and his colleague Dominic Walsh made these structures also using a surfactant-based reaction medium to effect the patterning, although their procedure is rather more complex than Ozin's. The solvent is a mixture of oil (tetradecane) and water, along with a surfactant. The water is supersaturated in calcium bicarbonate and spiked with magnesium ions.

Mann and Walsh spread a few drops of this mixture on the surface of copper or brass and washed the film in a hot organic solvent. This washing step causes the oil/water mixture to separate into microscopic droplets of oil, stabilized by the surfactant and surrounded by the supersaturated aqueous solution. This mixture is called a microemulsion. Dissolved carbon dioxide in the solution then gradually diffuses out, causing the calcium carbonate to precipitate around the oil droplets. Washing removes the oil and surfactant, and the remaining mineral phase preserves a cast of the labyrinthine channels between the oil droplets.

Mann and Walsh have created hollow shells from their reticulated aragonite films (fig. 4.42*b*) by casting them on the surface of microscopic, gold-coated polystyrene beads, which they subsequently dissolved away using acetone and ethanol. Placed side by side with a coccolithophore shell (fig. 4.42*c*), Mann's hollow spheres are a convincing imitation of nature.

These aragonite networks have a structure not dissimilar to bone and might therefore serve as bone substitutes. As I show in the next chapter, there is a great interest in materials that will provide temporary replacements for bone, acting as scaffolds that are gradually superseded by the new growth of real bone. And materials with microscopic, uniform pores, like those seen in Ozin's patterned

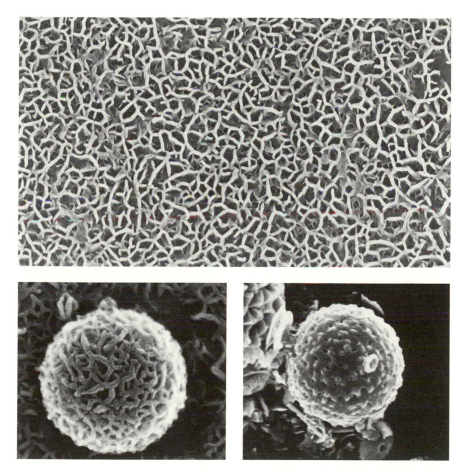

FIGURE 4.42 Chemistry imitates nature. Crystallization of the carbonate mineral aragonite within an emulsion of oil and water can generate a porous, reticulated solid (*top*). The crystal provides a cast of the structures present in the reaction medium. By casting this porous film on the surface of microscopic beads, a hollow porous sphere can be synthesized (*bottom left*) which resembles the shell of a coccolithophore (*bottom right*). (Photographs courtesy of Stephen Mann, University of Bath.)

materials, are in much demand for filtration and separation technologies. But it is hard to predict at this stage what applications these patterned structures might find, since we are still only just beginning to explore the range of patterns that can be made. Their most immediate value is in showing just how far we can go in controlling the structure of materials over a range of length scales by the marriage of the organic and inorganic worlds. As the need for such control becomes ever more urgent in the development of new advanced materials, we can hope that clever chemistry might replace complicated, expensive, and labor-intensive technologies. And nature is surely showing us the way.

Spare Parts

BIOMEDICAL MATERIALS

> Now, for most folks one pair of legs lasts a lifetime, and that must
> be because they use them mercifully, as a tender-hearted old lady
> uses her roly-poly old coach horses. But Ahab; oh he's a hard
> driver. Look, driven one leg to death, and spavined the other
> for life, and now wears out bone legs by the cord.
>
> —Herman Melville, *Moby Dick*

Most of the human body can now be replaced with artificial parts. Whereas once prosthetic replacements were engineered to be inert, now there is an increasing trend toward bioactive or bioadaptive materials that cooperate with living tissues rather than ignoring them. The ultimate goal is the growth of new organs, which will come from a marriage of molecular biology and materials science.

REPLACING parts of the human body with artificial parts is an ancient practice: the earliest known artificial leg, made of wood and metal, dates from 300 B.C. The crudest prosthetic limbs, like Captain Ahab's, do not require particularly sophisticated materials, since their main function is merely a mechanical one—they do not carry out a vital, life-supporting role, and they do not come into direct contact with the internal workings of the body. But replacing vital organs such as the liver, kidneys, or heart is another matter. All of these organs come into intimate contact with the blood stream, and so are subject to the scrutiny of the immune system. This is why replacing organs is not a simple matter of stitching a healthy organ from a donor in place of the faulty or damaged organ: if the immune system does not recognize the cells of a tissue as native to the body, it initiates an immune response that causes inflammation and ultimately destruction of the foreign tissue. These problems of "immune rejection" dogged early attempts at organ transplants, but advances in understanding its mechanism in the 1940s and 1950s led to a recognition of the need to match tissue type as closely as possible between donor and recipient. Drugs were also developed that helped to suppress the immune response. Today, transplants of kidneys, livers, hearts, lungs, and pancreases can be guaranteed a very high success rate; something like 20,000 heart transplants and hundreds of thousands of kidney transplants have now been performed.

But the big problem is supply. In the United States, deaths through liver failure outnumber the number of liver donors by about ten to one, and while around 100,000 people each year need an organ transplant of some kind, only a quarter of these will ever receive one. Several attempts have been made to transplant organs from monkeys into humans, but all have been unsuccessful, and the very idea raises formidable ethical problems. So too does the proposal to use organs from more distantly related animals such as pigs, first "humanizing" them by giving the animals human genes that produce a protective protein coat on the animal's cells to minimize immune rejection.

When transplants are not possible, the failure of some organs is dealt with by reproducing their function with devices outside the body—kidney dialysis machines that remove toxins from the blood, for example, or "artificial lungs" that aerate the blood—or by drugs, such as insulin administered to replace that which a faulty pancreas no longer produces. But the ideal would be to equip the patient with an artificial device that can be inserted into the body surgically, where it does its job continually without the need for constant trips to the hospital. We can now contemplate the prospect of such synthetic organs—made either from natural tissues, synthetic materials, or a mixture of the two—which will carry out the biological function of the original while deceiving the body into believing that they *are* the original. These artificial organs will revolutionize medical care.

Quite aside from the issue of finding nontoxic materials that have the right physical properties to reproduce the function of an organ, the problems with any artificial device that comes into direct contact with the bloodstream are much the same as those encountered in organ transplants. Even if the foreign material is an inert, sterile, synthetic substance, its presence does not go unnoticed. The body responds to it as if it were healing a wound, for example by depositing new tissue on the surface. Excessive deposition of scar tissue on an implanted device can have untoward consequences such as blockage of blood vessels. And if an artificial device is implanted that penetrates the skin, it is essential that a permanent seal be established at the point of penetration; otherwise, infection will inevitably develop.

Organ replacement is probably the most dramatic prospect that the new biomedical materials offer; but there is a much greater need per capita for biocompatible materials for the treatment of wounds and of damaged hard and soft tissues. Materials for assisting bone healing or for replacing damaged or degraded bones (such as the highly porous and brittle bones that result from osteoporosis) are in tremendous demand. Replacement of the ball-and-socket hip joint in elderly people is particularly common, and has traditionally involved the use of strong inorganic materials such as stainless steel or titanium and its alloys. But these materials have a lifetime of just ten to fifteen years before they fail. If the body can still generate healthy, strong bone, the ideal material for treating bone damage is one that can be slowly replaced by real bone. Artificial skin, meanwhile, is needed for grafts applied to burn victims. Over the past few decades, materials like this have become increasingly available.

In the past, researchers developing biomedical materials for these applications took the view that the ideal was a highly inert material, which interacts as little as possible with the body and so minimizes the chances of interfering with the growth and function of the cells with which the material comes into contact. But nothing foreign in the body can truly pass unnoticed, and so there is now a growing tendency toward making materials that are *designed* to interact with the body—it is sometimes better to exploit this inevitable interaction than to pretend that it is not there. Thus, for example, temporary bone substitutes are designed to promote bone growth and to be biodegradable in the body, so that the material acts as a scaffold for the re-formation of healthy bone, and vanishes as natural bone develops.

These new biomedical materials can replace most of the purely structural components of the body: bones, skin, ligaments, arteries. They can provide biodegradable supports for tissue regrowth and repair; they can be used to encapsulate foreign cells or drugs, allowing them to be delivered to specific organs without being destroyed in the body; they can perform specialized physical functions, such as artificial lenses for eyes; and they can even be assembled into synthetic organs such as artificial hearts.

Together, these developments make it possible for us to begin to regard the human body as a mechanism in which faulty parts can be replaced on demand from an off-the-shelf selection of parts. Put this way, the possibility may sound cold and even unsettling, raising specters of Frankenstein-style creations. But that, in the age of genetic engineering, is something that our societies must face up to and discuss. As it becomes possible to do more and more to aid ailing human bodies, we must ask ourselves increasingly what is desirable or acceptable and what is not. Is it right that artificial skin should be developed, at considerable cost, for cosmetic surgery? What is the role of advanced biomedical materials in developing countries that do not have the resources even to treat diseases or injuries that in richer nations would be trivial or eradicated? When options are limited by cost rather than by the question of what is medically possible, how are the choices made? These are not necessarily questions for those working on biomedical materials, but they are questions that will arise from that work and which the medical and the broader communities must address urgently. They are a reminder of the ever present truth that scientific advancement does not take place in a vacuum.

Many of the studies described in this chapter, like much of biomedical research of any sort, involve experiments on animals at some stage in their development. That poses an ethical dilemma both for the researchers and for society more generally. By discussing this work, I do not intend to offer an endorsement of animal experiments; the issue is much debated elsewhere, and needs to be. I would suggest, however, that the question is not merely that of whether we should trade the lives or the suffering of humans against those of animals. We must also look carefully at the nature and origin of the conditions that we are aiming to treat. Much research in the United States is conducted on heart disease, and indeed it is one of the biggest causes of death in the country. But how often is it a disease exacerbated by the choices we make, by the kinds of lives we choose to lead? Should we aim to justify animal experiments related to heart disease in the same way as those related to, say, hereditary conditions that cause pain, suffering, and death to children, or those that arise from the social conditions in which millions find themselves in the developing world? To thread our way through the complex ethical questions that these considerations throw out, I believe that we need to know what is currently possible in medical science. My hope is that this chapter can make some small contribution to informing that debate.

PASSAGES OF THE HEART

Heart disease is the scourge of the affluent countries—a reminder of what may lie in store when other illnesses can be more readily suppressed, and of the drawbacks of the lifestyles and diets of the Western world. Many heart conditions are caused by the strain imposed on the heart by obstructions to the flow of blood, particularly in the main passages of the circulatory system. One of the most common is atherosclerosis, a condition in which deposits of coagulated blood material build up on the inner surfaces of arteries and blood vessels, leading to blockages. Atherosclerosis in arteries more distant from the heart, such as those in the legs, can cause gangrene due to poor blood supply to extremities such as the feet. Often, the only effective treatment for the condition is to replace the affected arteries entirely.

In some cases, artificial arteries can be constructed surgically using tissue grafts from the patient—this ensures that there will be no immune or toxic response to the implants. But it is not always possible to extract from the patient graft tissue of sufficient size and with the right properties for making blood vessels. So there is a great deal of demand for materials that will provide artificial arteries.

This is partly a problem of finding a material with the right mechanical properties: artificial blood vessels must be tough but have a flexibility that enables them to adapt themselves to the shape of the surrounding tissues and organs, yet without buckling or kinking (as this would block the blood flow, like the kinking of a garden hose). In the nineteenth century the Frenchman Alexis Carrel was prepared to sacrifice flexibility in order to make artificial blood-vessel substitutes from tubes of metal and glass. Clearly these will do not more than the bare minimum of providing a channel for containing blood flow, and no one would relish the prospect of wandering around with glass tubes inside them. But today's synthetic arteries are made from polymers, primarily the polyester called Dacron (also used in clothing fabrics) and the nonstick material polytetrafluoroethylene (PTFE). Dacron arteries are fashioned from fibers of the polymer knitted together into tubes with porous walls. This knitted texture provides the right mechanical flexibility, but it also means that the vessels have small holes through which blood can leak. To seal the gaps, the woven fibers are first treated with blood that is clotted by heating, or with the protein albumin—the main component of egg white—which is also transformed by heat into a coagulated mass. When the vessels are inserted into the body, both of these sealant materials act as biodegradable matrices into which cells migrate and deposit the structural protein collagen, producing a natural/synthetic hybrid structure. Eventually, the artificial vessels acquire a smooth, "natural" lining of collagen, called a pseudointima.

Although Dacron fibers can be easily woven into the textured vessel walls, they have the slight drawback that they induce inflammation when first implanted. PTFE is harder to fashion into a suitable porous texture, but as an expanded porous foam (the same form, in fact, as is used for the "breathable" waterproof material Goretex) it can be made into tubes which have the advantage that their fluorinated, nonstick surface deters cells from adhering and growing over the vessels as scar tissue. Polyurethanes have also been investigated as potential materials for artificial arteries, because they are highly flexible, durable, and (for reasons not yet understood) very compatible with blood. There are some worries, however, that as polyurethanes are gradually broken down in the body, toxic degradation products are generated. Some attempts are being made to combine the advantageous properties of several of these materials, for example by depositing thin layers of PTFE on the surfaces of polyurethane or polyester vessels.

Aside from the need to get the strength and mechanical properties right, the main challenge to making artificial blood vessels is to reduce the formation of clots, or *thrombi*, on their surfaces. Thrombosis, the deposition of coagulated blood material, begins to take place on artificial blood vessels within minutes of their contact with blood, and threatens to block them very rapidly unless some preventative action is taken. The formation of a thrombus involves two related and interdependent processes: coagulation of some of the proteins in blood, and deposition of cells called platelets, which stick to each other. As coagulation proceeds, it enhances platelet deposition, and the two processes thus reinforce each other to develop a clot. The problems associated with thrombosis mean that at present artificial arteries can be made with diameters no less than 6 millimeters; tubes narrower than this become clogged too quickly. Even for large vessels, thrombosis must be avoided by administering the anticoagulant protein heparin into the patient's bloodstream.

But what engineers of artificial vessels would really like to be able to do is to avoid thrombosis in the same way that natural vessels do. Rather than simply having a surface that passively discourages the deposition of clot material, blood vessels resist thrombosis by actively combatting it. They are lined with so-called endothelial cells, which have a varied armory for conducting this battle. The cells produce and release heparin (to prevent coagulation) and prostacyclin, a protein that prevents platelet deposition; and they capture from the bloodstream an enzyme called thrombin, which promotes coagulation, taking it literally out of circulation. To develop a material that can do all of these things is currently too much to ask; but by mimicking one or some of them, researchers hope to achieve

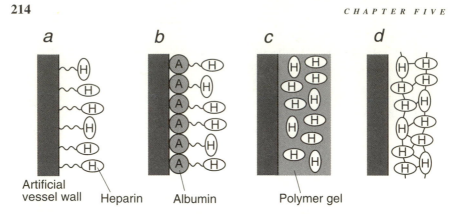

Artificial vessel wall   Heparin   Albumin   Polymer gel

FIGURE 5.1 The anticlotting agent heparin can be immobilized at the surface of artificial blood vessels to suppress the formation of clots (thrombi) that block the vessels. Heparin molecules may be attached in several different ways, including tethering via polymer strands (*a*), tethering via albumin protein molecules at the vessel wall (*b*), dispersal in a layer of polymer gel (*c*), and cross-linking to form a polymerized heparin film (*d*).

some measure of resistance to thrombosis. One favored route is to attach heparin molecules to the surface of the artificial vessels, and a variety of approaches to this have been tried. Heparin molecules might be attached directly to the polymer wall by chemical bonds, or they might be tethered there via strands of other polymers. They might be entrapped within a porous polymer coating on the inner surface of the vessels, or they might themselves be cross-linked into a polymer film that lines the walls (fig. 5.1). There have been similar attempts to immobilize prostacyclin on the polymer walls.

But perhaps the best solution would be to encourage endothelial cells themselves to grow on the artificial surfaces, to give them a natural clot-resistant lining. Endothelial cells will be formed if the body can be fooled into thinking that the artificial vessel is a real one, by giving it a lining that mimics the natural layer underlying endothelial cells in real vessels. This natural layer is called the basement membrane, and it overlies the main fabric of the vessel. Some success has been achieved in making artificial membranes lined with a collagen gel doped with proteins that encourage endothelial cell growth and adhesion, but thicker and more uniform layers of cells are needed before these artificial vessels begin to resemble their natural counterparts closely.

While artificial blood vessels may help to alleviate the stress placed on the heart, a weak heart is still a weak heart, and liable to failure. Ultimately the only solution to a heart condition may be a transplant or, one might hope, replacement with an artificial substitute. Given the physiological and symbolic significance traditionally accorded to the heart, it comes as something of a surprise to find that it is actually one of the easiest organs to consider replacing with an artificial device. In fact, prototype artificial hearts are already in existence, and replacement of component parts of the heart—the valves controlling blood flow in and out—has been carried out for decades. Indeed, because the blood flows through

PLATE 1 Blue laser light will offer greater data storage densities in optical recording media (such as compact optical and magneto-optic disks) than are available from the current generation of infrared lasers based on gallium arsenide and related II–V alloys. The shorter wavelength allows these lasers to read and write smaller bit sizes. Lasers that emit at blue wavelengths are now becoming available, thanks to wide-band-gap II–VI alloys such as zinc selenide (used in the device shown here) and new III–V alloys such as gallium nitride. (Photograph courtesy of Arto Nurmikko, Brown University, Providence, Rhode Island.)

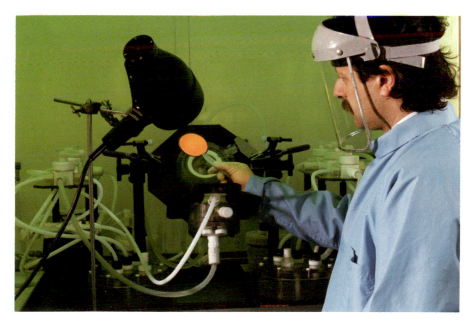

PLATE 2 Silicon will emit visible light when etched electrochemically into a mesh of "quantum wires" just a few nanometres thick. In the silicon disk shown here, about 80 percent of the material has been removed from a surface layer by etching, and the porous material emits orange light (photoluminescence) when held under an ultrviolet lamp. (Photograph courtesy of Leigh Canham, Defence Research Agency, Malvern, U.K.)

PLATE 3 Mollusk shells not only are sophisticated composite materials but can be ornately decorated with colored pigments produced by cells in the protein outer mantle. These patterns are frozen "chemical waves" generated by complex, nonlinear interactions between cells during shell growth, which modulate the production of pigment. (Photograph courtesy of Hans Meinhardt, Max Planck Institute for Developmental Biology, Tübingen, Germany.)

PLATE 4 Nickel-metal hydride batteries are a strong contender in the rechargeable battery market. The Ovonic nickel-metal hydride cells shown here are produced by the Ovonic Battery Company, a subsidiary of Energy Conversion Devices Inc. in Troy, Michigan. (Photograph courtesy of Stanford Ovshinsky, Energy Conversion Devices.)

PLATE 5 The emission-free electric bus is on its way. Ballard Power Systems Inc. in Vancouver, Canada, has created a prototype bus powered by a hydrogen fuel cell, which converts the energy of hydrogen combustion directly to electricity. The only product of the reaction is water. This commercial prototype holds sixty passengers and its on-board hydrogen tanks give it a range of 250 miles. Demonstration fleets of the Ballard bus are now being tested by commercial transit operators. (Photograph courtesy of Ballard Power Systems Inc.)

PLATE 6 Solar panels based on amorphous silicon, marketed by the Sanyo Corporation in Japan, are fashioned into traditional Japanese roofing tiles, providing environmentally sensitive domestic power generation while blending with the surroundings. (Photograph courtesy of Sanyo Corporation, Osaka, Japan.)

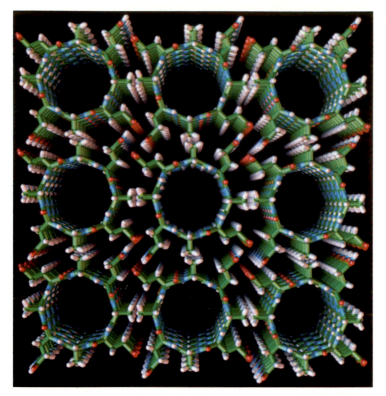

PLATE 7 Down the tubes? These tubular structures are made from cyclic peptide molecules, which self-assemble into stacks that pack together in a regular, crystalline array. The channels are 7 to 8 angstroms across. In this computer representation of the crystal structure, green represents carbon, blue nitrogen, red oxygen, and white hydrogen. (Image courtesy of M. Reza Ghadiri, Scripps Research Institute, La Jolla, California.)

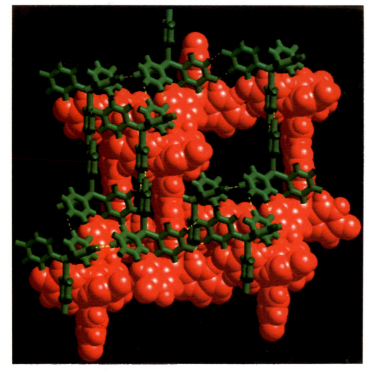

PLATE 8 Tetrahedral molecules with "sticky ends" that form hydrogen bonds will self-assemble into an open-framework structure with the same connectivity as the diamond crystal lattice. To fill the large voids in the crystal structure, a second identical network interpenetrates the first. Here one network is shown as red balls, which give an indication of the size of the constituent molecules, while the other is represented as green sticks, which shows more clearly the geometry of the molecules. Even with interpenetration, the crystal retains large, roughly square channels. (Image courtesy of Jim Wuest, University of Montreal, Canada.)

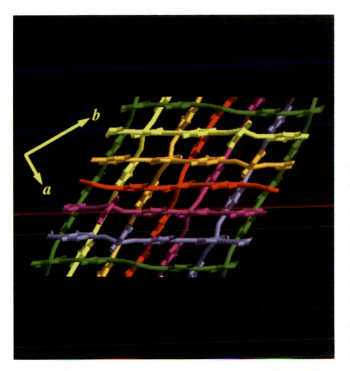

PLATE 9 In the hinged open-framework crystal synthesized by Jeffrey Moore and coworkers from silver ions and organic molecular building blocks, no fewer than eight distinct networks interpenetrate—here they are shown in different colors. (Image courtesy of Jeffrey Moore, University of Illinois.)

PLATE 10 An aerogel can be up to 99 percent empty space; it is a kind of "frozen smoke," with a highly porous structure in which tiny particles less than a nanometer across aggregate into a tenuous web. The materials are translucent and have potentially useful properties such as slow sound-transmission speeds and low thermal conductivity. Here scientists from Lawrence Livermore National Laboratory in California are holding a cylinder of an ultra-low-density aerogel made from silica. (Photograph courtesy of Steve Wampler, Lawrence Livermore National Laboratory.)

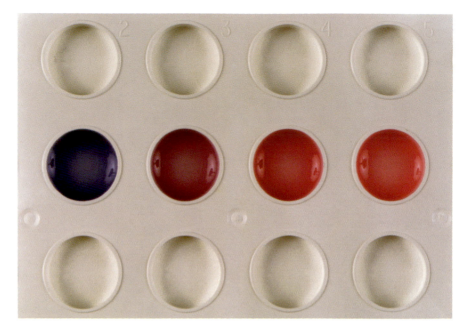

PLATE 11 Films of cross-linked amphiphilic molecules with diacetylene groups in their tails are richly colored, because the polydiacetylene backbone absorbs visible light strongly. The films shown here, synthesized by researchers at the Lawrence Berkeley Laboratory in California, contain sialic acid groups bound to their surface, to which the influenza A virus will bind. In their pristine state the films are blue (*left*), but when the viruses become attached to the surface, changes in the conformation of the cross-linked amphiphiles bring about a color change to red. Increasing numbers of viruses have been added from left to right. These films provide a kind of litmus test for sensing the virus. (Photograph courtesy of Deborah Charych, Lawrence Berkeley Laboratory.)

PLATE 12 Light-emitting diodes can be made entirely from plastic. The light-emitting material in the device shown here is a polymer based on poly(*p*-phenylene vinylene), which emits yellow-green light when electrons and holes are injected into it from electrical contacts on either side. The top electrode is made from the conducting polymer polyaniline; the back

PLATE 13 The surface of copper at the atomic scale. Here steps and ledges correspond to the edges of layers of atoms, while single peaks are lone atoms poking above a flat atomic plane. The vertical scale is exaggerated to make these features clear. The ripples that spread out from these surface irregularities are standing waves of electron density: because the surface electrons have a wavelike nature, they are scattered from the surface features like ripples in a pond. The scanning tunneling microscope is able to reveal these electron waves, since the current that flows through the microscope's needlelike tip depends on the local electron density below the tip. (Picture courtesy of Don Eigler, IBM Almaden, San Jose, California.)

electrode is a thin film of calcium metal, protected from oxidation by a transparent polymer layer. This flexible, plastic LED can be bent repeatedly, and even folded back on itself in a 180-degree bend, without detriment to the light-emitting behavior. (Photograph courtesy of Alan Heeger, UNIAX Corporation, Santa Barbara, California.)

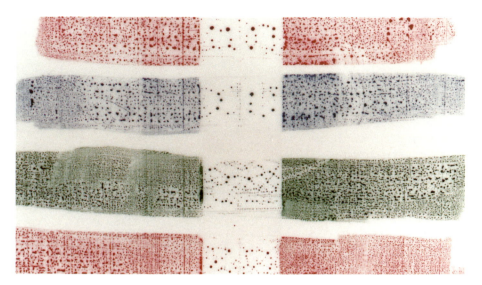

PLATE 14 A nonstick surface coating developed by chemists at the Dow Chemicals Company rejects all attempts to mark it. Paint and marker pens (as shown here) will leave droplets that can be simply wiped off the surface. To either side of the strip of (transparent) nonstick coating in these images is Teflon, which in comparison is more clearly marked by the ink. (Photograph courtesy of Donald Schmidt, Dow Chemicals Company, Michigan.)

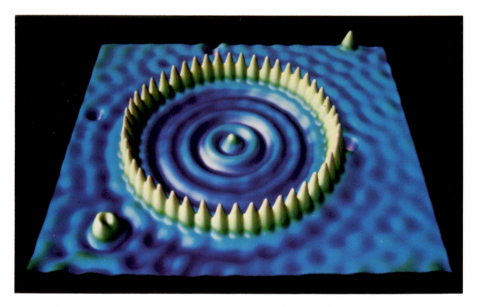

PLATE 15 Don Eigler and colleagues have used the scanning tunneling microscope to drag forty-eight iron atoms into a circular corral on the surface of copper. Surface electrons in the underlying copper are confined within the stockade, since they bounce back from the irregularities of the walls. This sets up standing waves in the corral, evident as concentric ripples in electron density. (Picture courtesy of Don Eigler, IBM Almaden, San Jose, California.)

passages of much greater cross-sectional area in the heart than in blood vessels, in some ways there are fewer problems associated with thrombosis in making artificial hearts than in making synthetic replacements for the vascular system.

The heart is basically a simple pump. It is accessed by four valves that open in response to blood flow in just one direction—two admit blood to the heart, and two let it out again. These valves can fail to open or become structurally damaged by certain diseases. Addressing heart-valve failure by remedial surgery on the damaged component is very difficult and often impossible, so the only way of treating such problems is usually to replace the valve altogether. Replacement with valves taken from other hearts—both from those of dead humans and from pig hearts—is sometimes effective, but the transplants do not tend to have a very long lifetime before they tear or are degraded. The earliest artificial heart valves, developed in the 1950s and first used in the early 1960s, consisted of stainless steel hinged disks within channels of silicone rubber. It was necessary to administer anticoagulant drugs to prevent these valves from becoming stuck or blocked by clots; this treatment has the drawback that it can also suppress the beneficial effects of blood clotting, leading to a risk of hemorrhages. Most advanced valves today have their sealing disks made from a special kind of carbon called *pyrolytic carbon*, a kind of disordered form of graphite. This material is sometimes deposited as a coating onto disks of pure graphite or of metal. Pyrolytic carbon is strong and resistant to mechanical wear, and it also has the crucial property of being resistant to thrombus formation, the most common problem encountered in valve replacement. The disks are attached by a hinged arm (often of a metal such as titanium and perhaps coated with pyrolytic graphite) to a fabric ring made of a polymer such as Dacron or PTFE, which is sewn to the tissue of the heart-valve opening (fig. 5.2).

Besides replacing faulty valves with artificial ones made from purely synthetic materials, it is also possible to construct artificial valves from biological tissue, much as surgeons can perform reconstructive surgery on damaged limbs or on

FIGURE 5.2 Artificial heart valves are commonly made from disks (occluders) coated with pyrolytic carbon, which is resistant to the development of blood clots on its surface; the metal hinges are commonly made of titanium. The white outer ring is sewn to the valve opening of the heart. (Photograph courtesy of F. J. Schoen, Harvard Medical School.)

wounds. Tissue from cow hearts has been used to make these "bioprosthetic" components, as well as tissue from the patients themselves (connective tissue in the thigh can be used, for example). This approach has the advantage that the material used is naturally biocompatible, so that thrombosis is less of a problem; but instead, complications can arise from calcification, whereby deposits of hard, bonelike minerals form on the surface of the replacement valves and cause them to stick and tear. This process can be combatted by administering drugs that inhibit calcification, or by incorporating such drugs into the bioprosthetic device so that they are slowly released. But more needs to be known about the factors that allow crystals to form on the tissue surfaces before this problem can be ameliorated entirely.

A complete artificial heart, driven by pneumatic pumping, was developed by researchers at the University of Utah in the 1970s, who licensed their technology for worldwide marketing to the Symbion company. In 1981 Symbion acquired approval from the U.S. Food and Drug Administration for implantation of their device, called the Jarvik-7, into humans. The first transplant was made in 1984, and the patient survived for 619 days. But other transplants were less successful, with survival in one case extending for only 10 days. The Jarvik-7 and other artificial hearts are now generally used just as temporary cardiac replacements while the patient awaits a transplant of a natural organ. A total of 272 patients have received a Jarvik-7 transplant up to 1995, while a handful of others have been given temporary support from other kinds of artificial hearts.

The considerations involved in choosing the fabric for the heart cavity are rather similar to those that crop up in making artificial arteries—in particular, there is a need for flexibility and durability, as well as thrombus resistance. The Jarvik-7 heart is made primarily of a polyurethane called Biomer, reinforced with a mesh of Dacron (fig. 5.3). There is some advantage in giving the blood-contacting surface of artificial hearts a pseudointima lining of collagen, the growth of which is encouraged by giving the material a rough, textured surface that actually promotes the sticking of platelets and blood proteins—the very thing that needs to be avoided in more constricted spaces such as arteries and valves. The textured surface also traps fibroblasts (cells that produce collagen), and so the fabric gradually acquires a lining of this protein. The fact that the lining thickens the walls does not matter because the heart cavity is large.

Jarvik-7 is pumped by air, which inflates a polymer diaphragm within the Biomer housing. The diaphragm is also made of Biomer, but it must be highly resistant to wear in the face of repeated inflation and deflation. To confer such resistance, thin layers of graphite are inserted between successive layers of Biomer: the graphite acts as a lubricant, reducing wear and tear on the polymer.

As we all well know, a real heart responds to the body as well as passively pumping blood. When adrenalin is injected from the adrenal gland into the bloodstream, for example, the blood vessels dilate and the heart beats faster, increasing the supply of oxygen for metabolism and preparing the body for exertion. An artificial heart that can reproduce such behavior would involve much more sophisticated biomedical technology than is currently at our disposal.

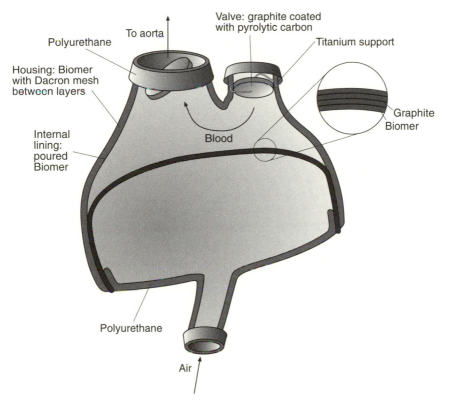

FIGURE 5.3 The Jarvik-7 heart, manufactured by the Symbion company, is a complete, air-pumped heart replacement device. The walls of the main body are made of a polyurethane called Biomer, developed by the Ethicon company in New Jersey; this material shows good blood compatibility, in that blood clots do not readily form at its surface. A Dacron mesh is sandwiched between Biomer layers to convey strength and rigidity. The air-pumped diaphragm is also made of Biomer, this time interleaved with graphite for lubrication. The valves are of the tilting-disk design, made from pyrolytic carbon deposited on graphite.

Blood Supply

"Please give blood" implores the billboard poster, and I consider it. Being of a squeamish nature when it comes to such things, I cannot regard the prospect with great enthusiasm, yet I know how important adequate blood supplies are to the medical profession: during major operations a patient can lose up to a quarter of the total circulating volume of blood, which must be replaced by transfusion. And concerns about the possibility of transmitting viral diseases such as hepatitis and AIDS through blood transfusions, exacerbated by tragic examples of such contamination in France, India, and Japan, have added to the pressures faced by blood banks. Will blood ever be available in limitless quantities as a synthetic factory product?

We can think of blood as a functional, colloidal biomaterial (indeed, it might help my squeamishness if I were more able to do so). It is functional in the sense that it is the medium that distributes oxygen to the body tissues, where it is used in metabolism. And it is colloidal in the sense that it is a suspension of small particles in water—most notably the erythrocytes or red blood cells that harbor the protein hemoglobin, which binds oxygen when it diffuses through the lung walls and gives it up again where it is needed. Erythrocytes are disklike cells about a tenth of a micrometer across which are filled with hemoglobin. Where the concentration of oxygen in the surrounding fluid is high (as in the lungs), hemoglobin has a high propensity to bind oxygen, but where it is low (as in body tissues) this binding tendency diminishes and the hemoglobin will give up its oxygen. The oxygen released is then swept up by myoglobin, a protein related to hemoglobin which resides in the tissues and acts as their oxygen repository.

Making artificial blood is much more than a matter of finding a fluid that will bind oxygen in a similar manner to hemoglobin. As it would be mixed with real blood, an artificial substitute must have the same flow properties. It must pass through the blood-filtering organs with equal facility, and must not be toxic. It must be stable not only in the body but also on the shelf, in storage. It must not inhibit coagulation, nor induce clotting. In short, it must be able to fool the body (which is far from easily fooled) that it is the real thing.

These are demanding criteria, but several avenues of research into artificial blood are now beginning to approach a viable product. Safe and effective blood substitutes for long-term use remain some way off, however, and it is not yet clear which, if any, of the substitutes currently being explored will fit the bill.

There are currently three leading candidates for an artificial blood substitute, all of which incorporate oxygen-binding compounds carried as a colloidal suspension in water. In two of these, the oxygen molecules are carried by hemoglobin molecules like those in erythrocytes, either modified in some way or encapsulated within artificial cell-like sacs. Hemoglobins themselves can be isolated from sources other than human blood, so this is not simply (as it might at first appear) a matter of making poor blood from rich. But although hemoglobin is soluble, we cannot get a blood substitute merely by dispersing free hemoglobin molecules in water, because real blood contains enzyme proteins that break it down. "Naked" hemoglobin is also toxic to the kidneys; and in addition, a hemoglobin solution lacks the crucial flow properties of blood, which are characteristic of a colloidal suspension rather than a simple solution.

One way to make colloidal, hemoglobin-containing suspensions is to cross-link the hemoglobin molecules into larger polymeric particles. This confers some resistance against enzymatic biodegradation. The most promising product from this line of attack seems to be one in which groups of four inter-linked hemoglobin molecules are themselves bound together by a cross-linking agent called pyridoxal phosphate, yielding colloidal suspensions that have similar oxygen-binding characteristics to real blood. Other polymers, such as nylon and polyethylene glycol, have also been used as cross-linking agents. This approach was initially thought to be too prone to the problem of toxicity, and although some of these hurdles have now been overcome, it seems that cross-linked poly-hemoglobins are still prone to complications—including constriction of blood vessels and depression of blood pressure, the triggering of immune responses, and sometimes an unwillingness of the hemoglobin complexes to give up their oxygen, which leads to their being oxidized themselves and to the consequent release of toxic free radicals.

In 1994 Kenneth Suslick and his student Mike Wong from the University of Illinois at Urbana-Champaign hit on a promising new approach to making artificial blood from cross-linked hemoglobin. They found that by passing ultrasonic waves through a hemoglobin solution, they could create hollow microspheres of hemoglobin in which around one million of the molecules were welded together by chemical bonds. Ultrasound induces cavitation—the formation of tiny air bubbles in the solution which then implode, releasing intense heat. Suslick and Wong found that chemical reactions stimulated by this heat appeared to "spot-weld" hemoglobin molecules arrayed around the spherical walls of bubbles, creating roughly spherical polyhemoglobin spheres about a micrometer across (fig. 5.4). The cross-linked hemoglobin retains its ability to take up oxygen, and the microspheres can also hold dissolved oxygen inside. All told, a suspension of the microspheres is able to carry more oxygen than natural blood, and Suslick believes that the hemoglobin capsules will be resistant to enzymatic attack. These attributes, and the fact that the microspheres can be stored far longer than natural blood (which has a shelf life of about 35 days), bode well for their use as a blood substitute. But the true potential will become clear only after intensive clinical trials, which are now underway.

Another way of protecting hemoglobin from degradation, while also suppressing its toxic effects, is to put the molecules within artificial cell-like membranes such as liposomes (page 172). Thomas Chang of McGill University in Montreal first explored this approach in 1964 by encapsulating hemoglobin inside vesicles of nylon and other polymers; today liposomes made from mixtures of phospholipids and cholesterol are being explored as synthetic erythrocytes. This approach seems so far to provide the most problem-free blood substitute, no doubt because it comes the closest to mimicking real red blood cells. All the same, complications can arise from the tendency of liposomes to induce clotting.

The third candidate for artificial blood does not employ hemoglobin at all, but instead uses a purely artificial class of compounds for oxygen transport. These are perfluorocarbons, polymers containing fluorine and carbon atoms, which are capable of dissolving large amounts of oxygen. Their oxygen-transporting capacity thus derives from a purely physical effect—solubilization—rather than from chemical binding, as in hemoglobin. The perfluorocarbons are insoluble in water, but they can be suspended as a colloidal emulsion, rather like milk, in the presence of surfactants. To prepare these suspensions, the perfluorocarbons are mixed with other organic compounds and dispersed as droplets of about the same size as erythrocytes by adding surfactants such as glycerol or phospholipids and subjecting the mixture to high-frequency sound waves. The most common perfluorocarbon used is perfluorinated decalin, which, in a mixture with the compound

0.25 µm

FIGURE 5.4 Hemoglobin molecules can be cross-linked into hollow microspheres by exposing a solution to ultrasound. The cross-linked hemoglobins not only retain their ability to bind oxygen, as the free molecules do in blood, but this ability is actually enhanced by some kind of cooperative process. So far, these hemoglobin microspheres look to be a promising candidate for artificial blood—they do not appear to have significant toxic side-effects, and a suspension of them can carry more oxygen than real blood. (Photograph courtesy of Ken Suslick, University of Illinois at Urbana-Champaign.)

perfluorotripropylamine, represents the basis of the first commercially available blood substitute of this sort, called Fluosol.

The main problem with perfluorinated blood substitutes is that they are retained in filtering organs such as the liver and the spleen, resulting in the swelling of these organs over long periods. It is possible that the extreme stability of the perfluorinated compounds, which is an advantage insofar as it prevents the molecules from being broken down in the bloodstream, hinders the biodegradation that would otherwise prevent their accumulation in cells. Another complication is that some perfluorocarbons can become attached to erythrocytes, making the blood cells inflexible and impairing their flow through capillaries; the end result of this is rather similar to the effects of sickle-cell anemia, diminishing oxygen delivery to the tissues.

Notwithstanding these problems, perfluorocarbon blood substitutes have already shown some clinical success. They have been used, for example, in transfusions for patients whose religious beliefs prevent them from being able to accept donated blood. And the very high oxygen uptake might make them especially useful in situations where there is an urgent need to enhance oxygen delivery to the tissues—in the treatment of anemias, for example, or of oxygen starvation of the brain following a stroke.

STRUCTURAL REPAIRS

The main function of bone is to hold us up—it is a scaffolding on which the soft stuff is draped. So one's first instinct in searching for materials for bone replacement might be to look for strong, rigid, and wear-resistant materials. And indeed, most implants for bone replacement today are metals—stainless steel, titanium, and titanium and cobalt alloys. These materials are resistant to fracture (metals are strong without being brittle) and to corrosion. Strong ceramic materials such as aluminum and zirconium oxides (alumina and zirconia) are also used when the bone replacements are subjected only to compressive (squeezing, not stretching or bending) stresses—these materials are more brittle and, while strong under compression, are liable to crack under tension. Sapphire, a very hard gemstone, is an impure form of alumina which has been used for tooth implants. As a hard-tissue substitute, alumina is usually used in the form of a dense, polycrystalline material, which combines high strength with good wear resistance and biocompatibility. The good resistance to frictional wear makes alumina especially well suited for replacement hip joints, in which an alumina ball rotates in a plastic socket (fig. 5.5). These units have typical lifetimes of up to thirty years.

But strength is not the only criterion for a good bone replacement; in fact, too much strength can be a drawback. Bone, like most biological materials, is responsive to its environment—when a bone is stressed, the bone-producing cells called osteoblasts are stimulated into generating more bone. In this way the skeleton is able to fine-tune itself to suit the loads that it must bear. If a disability or a peculiarity of one's occupation places more than average stresses on specific bones, then they will be able to grow to accommodate that. So if a bone is replaced by a metallic counterpart that is stronger than the original bone, the replacement will tend to bear a greater proportion of the load, shielding the surrounding skeleton from its normal stress levels.

The mineral phosphate material of natural bone is slowly but constantly dissolved (resorbed) by the body, and in the normal course of events it is replenished by fresh bone synthesized by the osteoblasts—this gradual recycling ensures that our bones do not gradually deteriorate with age, at least while the osteoblasts

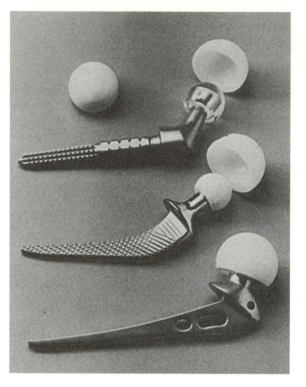

FIGURE 5.5 Its inertness and good resistance to mechanical wear make aluminum oxide (alumina) an ideal material for the ball component of hip-joint replacements. In the examples shown here, an alumina ball is attached to a metal stem (which is inserted into the bone), and rotates in a cup of polyethylene. (Photograph courtesy of J. B. Park, University of Iowa.)

remain healthy and active. But shielding of stresses by strong metal implants discourages this process of replenishment, because the sheltered bones feel no "need" to keep themselves strong. Thus, bone replacements cemented to neighboring bone may become loose over time, as the surrounding bone is resorbed. For this reason, the lifetime of metal bone replacements is not more than twenty years—likely to be enough if the replacement is given to an elderly person, but not a young one.

Considerations of this sort have made it increasingly clear that the best bone implants are those that approximate the properties—not only the strength but also the density, flexibility and tissue-contacting characteristics—of real bone. Naturally, the most desirable way of all to repair damaged bone is to regrow natural, undamaged bone in its place. For joint replacement, this is not yet feasible; if large volumes of bone are removed, the body is simply not able to grow an entirely new piece of bone—a new hip joint, for example—in its place. And in any event, bone replacements are commonly needed for the very reason that the process of bone growth has become unreliable—in elderly people, for example, it slows to a standstill and the old bones become porous and brittle. But when all that is needed is for new bone to seal over a fracture, it becomes possible to use temporary bone sutures to hold the fragments in place while the crack heals. A decade or so ago, such sutures would be made from metal, and were anything but temporary: bolted into place over the fracture, they would stay there even when the break was healed, unless removed by surgery. Motorcycle racing drivers would often end up walking around with a skeleton loaded with metal plates. But bone sutures can now be made from materials that, while strong enough to support the break while it heals, are biodegradable over time and are eventually dissolved away. Degradable supports made from polymers of lactic and glycolic acid are now in widespread use for bones that do not carry major loads, and are broken down harmlessly into carbon dioxide and water (fig. 5.6). Resorbable bone replacements have also been developed from inorganic materials that are very similar to natural bone, so that the bone-dissolving cells (osteoclasts) can break them down and slowly replace them with real bone. Such materials include calcium phosphate salts (similar in composition to the apatite in real bone) and calcium sulfate (plaster of Paris). These materials are not, however, very strong, so their application tends to be limited to bone that does not carry much stress, such as repairs to jaw or skull fractures.

For long-term replacement of load-bearing bones, inert materials with mechanical properties (such as strength and flexibility) that are more similar to those of bone have been developed from polymers reinforced with carbon fibers. But it has become increasingly clear that the ideal bone substitute is not a material that interacts as little as possible with the surrounding tissues, but one that will form a secure bond with the tissues by allowing, and even encouraging, new cells to

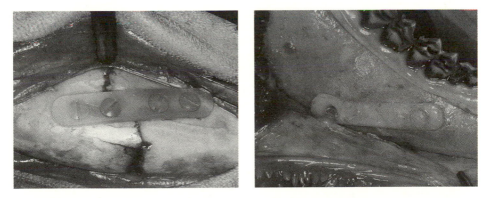

FIGURE 5.6 The healing of fractured bones can be assisted by biodegradable sutures—plates that hold the two parts of the bone together while new bone is produced at the fracture. The copolymer of lactic and glycolic acids has been used in this capacity, as it is slowly broken down to carbon dioxide and water within the body. The image on the *left* shows the freshly implanted polymer suture across a jaw-bone fracture. That on the *right* shows the fracture 11 weeks later: the bone has healed, and the suture has been largely degraded. (Photograph courtesy of A. J. Bennings, University of Groningen.)

grow. One way of achieving this is to use a material that is inert but porous, so that new tissue, and ultimately new bone, can grow into the pores and help to prevent loosening and movement of the implant. Healthy tissue can grow only in pores that are at least around a tenth of a millimeter wide; otherwise, the new tissue cannot develop an effective blood supply. A preponderance of pores this big can make the implant material rather weak, however. Moreover, even tiny movements of the implant (which are hard to avoid) can cause complications by cutting off the blood supply to the tissue in the pores, leading to inflammation. And if porous metals are used, the large surface area created by the pores makes the metal rather more reactive than a monolithic slab, and so corrosion sets in more readily.

Some of these difficulties can be avoided by using porous ceramic materials instead of metals, or by giving porous metals a protective ceramic coating. Mineralized natural coral has been used as a master from which molds are made for casting bone replacements—the highly interconnected, uniform pores of corals are ideal for this purpose. Porous bone replacements have been made in this way from hard ceramics such as alumina and titania, as well as from polymers such as polyurethane, silicone rubber, and poly(methyl methacrylate). And as coral skeletons are themselves made from a reasonably biocompatible material (calcium carbonate), fragments of coral itself have been used in bone surgery on damaged vertebrae. But the problem of weakness remains for these porous materials, and it becomes ever greater as the material ages and small cracks develop. In the long term, porous implants may be effective only if made from a ceramic such as calcium phosphate which will be gradually resorbed and replaced with real bone.

Get Active

The most promising approach to ensuring a strong, long-lasting adhesive interface between a synthetic bone replacement and the surrounding tissue involves so-called bioactive materials, which aim to mimic natural bone to the extent that tissues will bond to their surface. This adhesion can ultimately lead to the laying down of new bone at the interface between the implant and natural bone, such that the two are held together very firmly by chemical bonds. In effect, bioactive substances aim to persuade the body that the implant is not really a foreign substance at all but is real bone.

The first bioactive bone-replacement materials, developed by Larry Hench and colleagues at the University of Florida at Gainesville, were based on normal (silica) glass. Hench found that by mixing into the silica glass certain amounts of sodium, calcium, and phosphorus oxides, he could produce a material (christened Bioglass) to which natural bone would form chemical bonds in the presence of bone-forming tissue. Whether or not bonding occurs is highly sensitive to the composition of the glass: if there is less than five times as much calcium as phosphorus, or if there are very small amounts of alumina present, bonding does not take place. The composition also determines whether or not the glass can form a stable, cohesive interface with soft tissues—some Bioglasses can bond only to

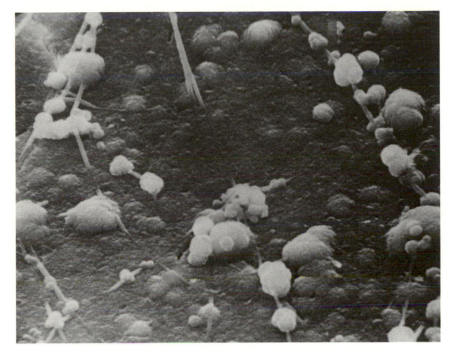

FIGURE 5.7 Bioglass is a ceramic bone-replacement material to which natural tissues and bone will bind. The formation of a secure interface between the prosthetic material and natural tissue is a characteristic of bioactive materials. The binding of soft tissue to Bioglass is initiated by the formation of collagen fibrils around clusters of the biomineral hydroxylcarbonate apatite at the surface of the Bioglass implant. (Photograph courtesy of Larry Hench, Imperial College, London.)

bone and not to soft tissues such as the collagen-rich ligament tissue that connects muscles to bone. In all cases, the bonding seems to be initiated by the formation, on the Bioglass surface, of a layer of the mineral hydroxylcarbonate apatite (HCA), a carbonate-containing calcium phosphate formed at natural bone–tissue interfaces. At a bone–Bioglass interface, the HCA mediates the joint like a layer of glue; at a Bioglass–soft tissue interface, fibrils of collagen become interwoven with clusters of HCA crystals (fig. 5.7).

Bioglasses can be made stronger by growing small crystals of ceramic materials within them. These glass-ceramic composites have all of the bond-forming properties of the bioactive glasses themselves. Bioglasses and glass-ceramics have now been used to replace bones in the ear and the backbone, and in dental surgery on the jaw.

Calcium phosphate ceramics can also be made bioactive—that is, capable of forming stable bonds with bone. I mentioned earlier that calcium phosphates can be used for bone-replacing implants that are eventually resorbed by the body. Such materials are not themselves bioactive, since the interface that they form with other tissues is not stable in the long term; rather, the materials are essentially biodegradable. But calcium phosphates can adopt a wide range of different crystal structures, depending on factors such as the temperature and acidity of the solution from which they form and the amount of water incorporated into the crystal lattice. Under typical physiological conditions (and if the acidity is not too great), the most stable crystal phase is hydroxyapatite (HA), the form of calcium phosphate most commonly found in natural bone. Ceramic implants of HA are bioactive, capable of bonding chemically to bone. Such implants can be formed by compressing a powder of HA into the desired shape and then heating the compaction to weld the grains together (a process called sintering). Bioactive HA implants are widely used for ear and dental bone replacement. Osteoblasts deposit crystallites of natural HA at the bone–implant interface whose crystal planes are aligned with those of the implant material, creating a smooth and effective joint.

These glass and ceramic materials are strong but rather brittle, and so in the long term they are susceptible to fatigue and failure by cracking. So composites of bioactive ceramics and metals or other ductile materials have been developed with improved strength and lifetime: the main load-bearing material is metal, but this is given a coating of the ceramic (often hydroxyapatite) to allow it to bind with surrounding bone. Several criteria must be met simultaneously by these composites: as well as being able to form a good interface with bone or soft tissue, they must be strong and long-lasting but not so strong or unyielding as to cause degradation of surrounding bone by the kind of stress shielding mentioned earlier. Bioglass/metal composites have been made that incorporate stainless steel or titanium fibers into the glass matrix; unlike monolithic metal components, the fibers convey enough flexibility to give the material mechanical properties similar to natural bone. Conversely, bone-replacement composites have been made from an inert, biocompatible and flexible matrix such as polyethylene stiffened with grains of a bioactive ceramic such as hydroxyapatite. This composite has much of the toughness of the pure polymer, but bends and flexes like real bone, to which it can bond by virtue of the exposed hydroxyapatite at the surface. It seems likely that the ideal bone replacement will not be any single material but will emerge from composites of this sort, which can blend the advantages of several different materials. That should come as no surprise, since as we saw in the previous chapter, nature embodies the same philosophy in materials such as shell, horn and indeed bone itself.

TISSUE ENGINEERING

It is all too easy to discover how delicate is the barrier between our internal machinery and the world outside, and all too often we might regret not having the tough hide of a buffalo or the hard carapace of a beetle. Skin was never meant, however, to confer protection against the slings and arrows of outrageous fortune—for that we must rely on our wits—but rather to shield us from the more insidious advances of infecting agents such as bacteria. Thus, one of the major concerns when the skin is damaged is not that we will leak but that the wound will become infected. The second main role of skin is to prevent us from drying out: our machinery needs to stay well lubricated. Yet of course skin does allow outward moisture transport, and for this reason it is highly porous. Such transport provides an important cooling function, since the evaporation of sweat carries heat away with it.

Skin is a highly complex and "active" material, containing in its lower layer (the dermis) a dense network of nerve and blood vessels. Reproducing a material of this complexity has so far eluded biomedical engineers. Early attempts to develop purely synthetic polymers as skin replacements were unsuccessful in preventing infection or avoiding immune rejection. The only truly effective answer, it seems, is to grow real skin—and this is, of course, something that the body does quite readily to heal wounds. But when large areas of skin are damaged, for example by severe burns, the danger of infection while the wound heals is substantial. Several medical companies now market temporary skin replacement materials for situations of this kind, made from polymers that not only afford some protection against infection but also provide a scaffold on which the new skin can grow. In effect, the scaffolding stands in for the tissue itself while it is growing. The scaffolds will ideally be biodegradable at such a rate that they are removed neither significantly faster nor more slowly than the new tissue regrows.

This approach is part of a more general enterprise in biomedical materials research, loosely termed *tissue engineering*. The idea behind this emerging new field is that, rather than making materials that mimic those found in nature, one finds materials that assist and encourage nature to make her own tissues. The advantages are obvious: new tissues grown from healthy, fully functioning cells do not suffer from toxicity or incompatibility problems such as thrombosis. When the tissues are grown from the patient's own cells, there are no problems of immune rejection either; indeed, the process is then essentially that of natural healing, but with assistance or in the presence of temporary support structures.

One of the first biodegradable scaffolds for skin regeneration was developed during the 1980s by Ioannis Yannas at the Massachusetts Institute of Technology (MIT). Yannas made a porous polymer film from collagen fibers from cows, combined with a polysaccharide polymer from shark cartilage. Both of these natural materials are attacked and broken down when grafted to the skin of humans, but the polysaccharide slows down the rate of degradation sufficiently to allow the porous film to support new skin growth before it disappears.

Yannas and colleagues coated the porous films with silicone rubber to make them impermeable, so that when the skin substitute was sutured in place over a wound, bacteria could not get at the wound from outside and fluids would not be lost from inside. Tissue-forming cells called fibroblasts migrate from the patient's connective tissues into the porous material and start to generate a new dermis. Left to its own devices, however, this regeneration of skin would not be complete. After a few weeks, it was necessary to transplant small pieces of epidermis from elsewhere on the patient's body to the surface of the developing dermis (this

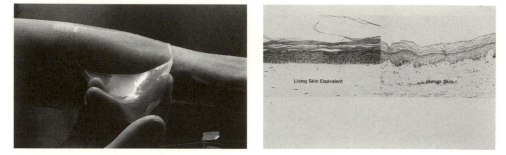

FIGURE 5.8 Artificial skin can be made by growing skin (epidermal) cells within a biodegradable polymer mesh in a culture medium. Graftskin, a synthetic skin produced by Organogenesis, is grown in this way on a collagen support (*a*). It has all of the layers of different cell types as real human skin (*b*: Graftskin is on the left, human skin on the right.) (Photographs courtesy of Organogenesis, Canton, Massachusetts.)

required that the protective silicone layer first be removed). The epidermis transplants create a complete new epidermis within a few days.

An alternative to the supported growth of new skin on the patient's body is to grow it in a culture vessel, in which a scaffold is "seeded" with epidermal cells from the recipient and these cells are allowed to multiply and colonize the scaffold by being supplied with the nutrients they need. The preformed skin can then be grafted onto the patient's body. This approach has been adopted by scientists at the Harvard Medical School, who used small skin samples to grow sheets of synthetic layered epidermis in culture in the presence of substances that accelerated the rate at which the epidermal cells grew and multiplied. The U.S. company Biosurface (now Genzyme Tissue Repair) commercialized this process. Another U.S. company, Advanced Tissue Sciences, began in 1989 to develop a synthetic skin called Dermagraft, grown in culture on a meshlike scaffolding of a lactic acid–glycolic acid copolymer which is nontoxic and biodegradable. This product is intended for treatment of severe burns and skin ulcers, and is now in advanced clinical trials. Organogenesis, a Massachusetts-based company, markets a similar product called Graftskin (fig. 5.8), grown on a collagen gel matrix. Because both of these products are essentially natural skin, they can be expected to show much the same kind of response to toxic or inflammatory substances as would normal human skin. For that reason, the synthetic skin can be used to test cosmetic and health products without danger to humans or suffering to animals. The Procter and Gamble company now uses synthetic skin to test some of its products.

Researchers at MIT have explored several different ways of making tissue scaffoldings from copolymers that incorporate lactic acid. For example, Robert Langer has collaborated with chemists to make supports of a lysine–lactic acid copolymer with chemical groups attached to the lysine molecules that allow cells to stick to them. The sticky groups consist of a short chain of three amino acids—arginine, glycine, and aspartic acid. Cells stick to one another by forming bridges with this same three-component peptide chain: it dangles from membrane proteins embedded in cell walls, and is bound by an adhesive protein called fibronectin. These adhesive scaffolds are another example of a bioactive material.

In addition to guiding skin growth, polymer scaffolds have been used for growing other kinds of tissue, such as the strong tendon, ligament, and cartilage tissues that connect bones and muscles in joints. As we saw in the previous chapter, these materials are made largely from the elastic filaments of fibrous proteins. Early attempts to develop synthetic substitutes were directed at finding artificial polymers with comparable mechanical strength and resilience, and among the materials investigated were Kevlar (chapter 9), Dacron polyesters, PTFE fibers, and pyrolyzed carbon fibers. These largely inert materials have varying degrees of biocompatibility, but none of them provides a perfect substitute for the natural material. A semibiodegradable scaffolding for connective tissue has been made by researchers at the University of Medicine and Dentistry of New Jersey, who coated pyrolyzed carbon fibers with polylactic acid. Once the fibers are implanted, the body begins to break down the lactic acid polymer and replace it with cells that produce collagen, the principal component of natural tendons and ligaments. After almost a year, the carbon fibers are fully surrounded by new connective tissue (fig. 5.9). But the carbon fibers themselves are not biodegradable, so

FIGURE 5.9 Scaffolds for regrowth of tendons have to be strong enough to support tension as the tissue develops. It is hard to find a fully biodegradable material that is strong enough, although Harold Alexander, J. Russell Parsons, and coworkers of the University of Medicine and Dentistry of New Jersey at Newark have made a partly biodegradable composite from carbon fibers (which are not degraded) surrounded by polylactic acid (which are). After around nine months, the polylactic acid is fully broken down and the remaining carbon fibers are surrounded by natural tissue. (Photograph courtesy of J. Russell Parsons, University of Medicine and Dentistry of New Jersey.)

these fibers stay in place and run the risk of triggering an inflammatory response in the surrounding tissue. The ideal would be to make a scaffolding that is fully degradable as the new tissue is deposited within it. The difficulty here, relative to the scaffolds used for growing skin, is that the framework must be strong enough to do the job of the tendon while the new tissue grows—this is why the carbon fibers are needed, because polylactic acid alone is not generally strong enough (although attempts are being made to develop a stronger version). A way around this problem would be to grow the tissue outside the body in culture vessels, so that the scaffolding does not have to support any load. A weaker but fully bio-degradable scaffolding could then be used, and the tissue could subsequently be implanted surgically. Robert Langer and the MIT group have shown that their scaffolds can be used for growing cartilage tissue, which connects load-bearing joints, in culture vessels.

The Organ Growers

Replacing the damaged structural fabric of the body—skin, bone, arteries—is one thing, but replacing its functional devices is quite another. As we saw earlier, the heart is perhaps unusual among bodily organs in that its function is, in the crudest sense, simply mechanical, and so it is realistic to think of making a mechanical device that will reproduce this function. But the kidneys, the liver, the pancreas, the brain, the hormone-producing glands all have very sophisticated functions that involve a degree of chemical processing. We can copy the function of the kidneys to some extent by using dialysis membranes that filter out toxins from the blood, but this is rather crude mimicry which is not without its complications. Dialysis machines are now portable and usable in the home, but they are not "artificial kidneys" in the sense of being devices that can be implanted in the body.

The fact is that nothing works better for blood purification than kidney cells, nor better for insulin synthesis than pancreatic cells, and so forth. So while it may one day prove possible to make purely synthetic and biocompatible devices that carry out the same function as these natural ones, many researchers are presently taking the line that a more practical route to artificial organs is to reproduce their function with healthy natural cells acquired either from the patient or from elsewhere.

Bringing "foreign" cells into contact with the blood of the recipient is really a form of transplantation, except that only a small amount of tissue, not an entire organ, is transplanted. so this carries all the attendant problems of immune rejection, and the foreign cells will be broken down unless protected. One solution is to house the cells in a device that provides the nutrients required to sustain the cells and gives access to the circulating blood plasma of the patient but which filters out the (large) molecular components of the immune system, such as anti-bodies. This strategy has been used by researchers at the University of Massachusetts Medical School to treat diabetes using pancreas cells from a nondiabetic animal. One type of diabetes (called insulin-dependent diabetes) is the result of the inability of the pancreas to synthesize insulin, a protein hormone that controls glucose transport. Insulin is produced by conglomerates of cells within the pancreas with the baroque name of islets of Langerhans. The researchers created a kind of artificial pancreas in which blood circulated through a tubular membrane containing islets of Langerhans. The membrane contained very small pores, which made it permeable to blood plasma, nutrients, and to insulin but not to antibodies. So the islets could inject insulin into the bloodstream while remaining shielded from an immune attack. When implanted in diabetic dogs, this device provided them with the insulin needed to stabilize their blood glucose at normal levels. But problems with blood clotting on the membrane meant that the lifetime of the device was limited to far less than a year.

An analogous artificial liver has been developed by researchers at the University of Minnesota. They trap pig liver cells within hollow, tubular fiber membranes and circulate the blood of liver-failure patients across the fibers. The liver cells, called hepatocytes, are immobilized within a collagen gel, so that a nutrient-rich medium can be passed down the fibers to sustain the cells without washing them away. An array of many fibers is encased in a cylindrical shell, and the blood is circulated through this cell and is detoxified by the hepatocytes (fig. 5.10). Similar devices have been used to keep liver patients alive while awaiting transplants.

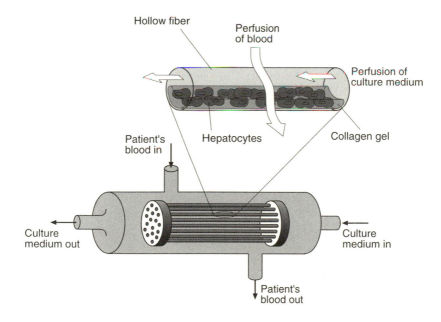

FIGURE 5.10 Patients with liver failure can be sustained by an artificial liver developed by researchers at the University of Minnesota and the Regenerex company in Minneapolis. The device contains pig liver cells (hepatocytes), which remove toxic substances from the blood. The cells are immobilized in a collagen gel housed inside hollow, permeable fibers—the entire assembly contains many thousands of these fibers. The patient's blood is perfused around the fibers inside a cylindrical container, allowing the hepatocytes to detoxify it. The cells themselves are sustained by a nutrient-rich culture medium fed through the fibers.

A drawback of multicomponent devices of this sort is that it is not at all easy to implant them safely and reliably into the body; they may function perfectly well as external machines, to which a patient's bloodstream may be connected, but they are likely to be too big, too bulky, or too thrombus-inducing to allow them to act as genuine artificial organs. A more promising approach to the long-term use of foreign healthy cells is to transplant them inside a soft, protective coat that shields them from the immune system of the recipient. This idea was first demonstrated in 1933, when an Italian scientist named Vincenzo Bisceglie showed that tumor cells from a mouse could be encapsulated in a membrane of nitrocellulose and inserted into a guinea pig. Inside the polymer membrane the foreign cells lived for longer than usual, showing that the membrane was affording some protection against an immune response. In the late 1970s Anthony Sun, then at the Connaught Research Institute in Ontario, developed this approach for the transplantation of islets of Langerhans to treat diabetes. Sun and colleagues suspended individual islets in droplets of an alginate gel, a polysaccharide produced by seaweed. They gave each gel droplet a tough but porous polymeric coating by transferring it into a solution of the polypeptide polylysine—this reacts with the gel to form a thin membrane over the droplet's surface. The researchers then liquefied the gel by extracting the calcium ions it contained (these could pass to and fro through the porous membrane), thereby obtaining individual islets suspended in a liquid inside a spherical membrane.

Sun injected these insulin-generating microcapsules into diabetic mice, and found that they reduced the mice's blood glucose to normal levels for several months without being attacked by the immune system. But they sometimes caused inflammation of the host tissue, followed by the formation of scar tissue around the implant. In effect, the body treats the implant as a kind of wound. The formation of scar tissue can be suppressed, however, by covering the surface of the microcapsules with polymer "hairs" that push away other proteins and cells, preventing them from binding to the surface.

It is not easy to control the size of these microspheres, and they are typically substantially larger than the islets they enclose—while the latter might be 50 to 200 micrometers across, the microspheres have diameters of around 300 to 500 micrometers (fig. 5.11). This means that much of the microsphere's internal space is wasted. If the islets could be shrink wrapped, their performance would be much better, because it would then be easier for oxygen and nutrients from the surrounding blood to reach the islet cells once they have passed through the membrane. With this in mind, Jeffrey Hubbell of the California Institute of Technology and colleagues have synthesized membranes directly on the surfaces of the islets. They suspended islets in a solution of polyethylene glycol with light-sensitive cross-linking groups attached to the polymer chains. Cross-linking was induced on the islet surfaces by shining green light through the solution. The cross-linking was restricted to the islet surfaces by first staining them with a dye that absorbed the light. In this way, each islet acquired a gel-like permeable polymer coat about 10 micrometers thick.

With such modifications, these encapsulated islets look highly promising as a long-term alternative to insulin injection for diabetics. These and related encapsu-

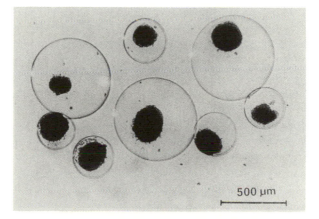

FIGURE 5.11 Insulin-producing cells from the pancreas called islets of Langerhans can be transplanted for the treatment of diabetes by enclosing small aggregates of the cells within thin, spherical polymer membranes, which confer protection against the body's immune system. Shown here are islets encapsulated within membranes of alginate (a polysaccharide) and polylysine, which is permeable to nutrients and insulin but not to the large protein enzymes that would break down unprotected foreign cells. These transplants are able to reduce blood glucose in diabetic animals. (Photograph kindly supplied by Robert Langer, MIT).

lation techniques are also being explored for transplanting other kinds of cell. Sun (now at the University of Toronto) has experimented with microsphere-entrapped liver cells; and Patrick Aebisher at the Vaudois University Hospital Center in Lausanne, Switzerland, has shown that transplanted brain cells, placed inside hollow fibers or small capsules of an alginate/polylysine polymer similar to that first used by Sun, show promise for treating neurological diseases such as Parkinson's disease, which is caused by the failure of cells in the brain to produce dopamine. This molecule allows neurons to communicate with one another (such compounds are known as neurotransmitters). There is some evidence that the Parkinson-like symptoms seen in rats treated with neurotoxic (neuron-destroying) agents are alleviated when the rats receive membrane-encapsulated brain cells that produce dopamine.

There can also be some virtue in suppressing neurotransmission—for example, when the signal transmitted is largely that of pain suffered by patients with terminal cancer. Cells from cows that release the neurosuppressant molecule catecholamine can alleviate pain when injected, in protective membranes, into the spinal fluid of such patients.

Some researchers are trying to grow entire replacement organs from transplanted cells cultured in a polymer scaffold that supports new tissue growth. If this becomes possible, a few healthy cells taken from an otherwise defective organ could be used as the seeds for a new, immunocompatible organ. Robert Langer and colleagues have used their biodegradable lactic acid–glycolic acid copolymer scaffolds to grow cohesive assemblies of liver cells, which might ultimately serve as new organs that can be implanted into the bodies of the hosts from which the seed cells were taken. "Eventually," says Langer, "whole organs such as kidneys and livers will be designed, fabricated, and transferred to patients."

Such "regrown" organs need to acquire a blood supply to provide oxygen and nutrients. This will develop automatically if the cell scaffolding is implanted in the body during the early stages of its colonization by cells—blood vessels grow from the surrounding tissue, just as they do into new tissue that grows during wound healing. But the survival of the new cells depends critically on the efficient and rapid development of this supply system, particularly for tissue types (such as the liver) that require a substantial supply of blood. To encourage the growth of a vessel network, Linda Cima at MIT has used as a scaffolding made from a porous polymer doped with a protein that stimulates this process, called angiogenic growth factor. As the polymer scaffold is broken down, the protein is released and the growth of the blood supply is promoted. An alternative approach is to let the blood supply grow *before* new cells colonize the scaffold. The scaffold is implanted while empty of cells, and new blood vessels creep over it like ivy. Once this supply is in place, the scaffold can then be seeded with the cells that will multiply into new tissue. Eyal Ron, working at the Genetics Institute in Massachusetts, has used this principle to encourage bone growth, by doping a degradable polymer scaffolding with a growth factor called bone morphogenic protein, which stimulates bone-forming cells to move into the scaffolding and deposit new bone.

Living Factories

Besides growing new tissues and organs, researchers can culture cells to produce all manner of useful biomolecular products. Devices such as the one described earlier in which liver cells were cultured for blood detoxification are called *bioreactors*: vessels in which cells are kept alive outside their host organisms by a supply of nutrients. We encountered other examples of bioreactors in the previous chapter, in which the cells were bacteria (which do not *have* a host organism) and the aim was to let the bacteria produce polyhydroxyalkanoate polymers. Such devices, which use bacteria to synthesize some complex molecule (often a drug of some kind), are the most common manifestation of bioreactors—they are, if you like, biological chemical factories. Researchers are now hoping to use bioreactors to make more complex biological products, in particular blood. This, of course, is the ideal solution to the demand for synthetic blood, since one gets not a substitute that satisfies only some of the requirements but the real thing. The blood type produced in these bioreactors would depend simply on the choice of blood-producing cells.

In the body, blood is made by cells in the bone marrow, so in principle it should be possible to create bioreactors in which cultured bone marrow cells produce blood. Such devices do not yet exist, but significant advances have nevertheless been made in the business of culturing bone marrow cells. This alone promises to be tremendously useful, because these cells are needed by cancer patients undergoing chemotherapy. This treatment destroys bone marrow, and so diminishes the patient's ability to generate blood. The hope is that some bone marrow cells could be extracted from the patient before treatment, multiplied in culture, and then returned to the patient after treatment to replenish the stocks lost by chemotherapy. But to sustain and multiply bone marrow cells in culture media is a tricky task, and only very recently has the technology reached the stage of clinical trials, which are being conducted on a device developed at the University of Michigan in collaboration with the Michigan-based company Aastrom Biosciences.

While the ingenuity of materials scientists has already yielded valuable, purely artificial substitutes for the body's tissues, and while no doubt even better products lie in store, it seems likely that tissue engineering and the growth of cells in culture will ultimately be the way to supply complex biomedical materials and substances. We have become used to the idea that, while trees can grow new branches and toads can grow new legs, our own self-repair mechanisms are much more limited. But it is not impractical to imagine a time when we will be able to multiply just about any of our cells into replacement organs, tissues, and body parts that can then be grafted into place without fear of immune rejection.

And taking this vision a step further, we might imagine combatting a suddenly discovered genetic disease by taking the genetically defective cells, performing genetic surgery (cutting out the faulty genes and stitching in functional ones, a technique called *gene therapy*), and multiplying the "healed" cells to form an organ with fully functional genes. Gene therapy offers the potential to cure genetic diseases at their root, by putting right the genetic defect that causes the disease (I shall say more on this topic shortly). If genetically defective cells can be extracted and treated to correct the genetic defect, it might then be possible to culture these cells until they develop into a fully functional and fully compatible organ or tissue which could then be implanted in the patient. Already there are plans to combat hemophilia in this way by genetic remediation of the relevant cells in bone marrow.

SPECIAL DELIVERY

There is something unnatural about the way that we take medication. I'm not talking about the fact that many drugs are purely synthetic substances designed to artificially stimulate or inhibit a particular biochemical process in the body—that, too, is unnatural, but often, at the present stage of medical research, unavoidable. No, it's the method of administering the drug that I'm thinking of. We might take our pills twice a day, morning and evening, or after or before food. But it would be peculiar if the body conducted all of its processes in this spasmodic, pulsed fashion—a shot of insulin in the morning to see us through to lunch, a quick release of dopamines to allow us an intense but gradually fading spell of thinking, bursts of antibodies every now and then like sporadic bombardments of our body's invaders. In short, there is a notable discrepancy between the way the body maintains its chemical balance naturally and the way that we have traditionally supplemented or adjusted this balance with drugs.

Often this does not matter too much, but sometimes it places a strain on the body to have to cope with a sudden, intense input of a drug which then gradually fades to insignificance, when what is really needed is a moderate but constant supply over a long period of time. Insulin for diabetics is a particularly important example—the diabetic has to estimate how much she might need to balance her sugar intake over the next six hours or so. A gross miscalculation might have serious physiological consequences. But maintaining a steady level of insulin in the bloodstream would suffice to cope with most dietary habits.

Of course, another problem with pill-taking is that it relies on the conscientiousness, not to say simply the memory, of the recipient. If you are anything like me, that is not a very reliable option. It does not matter too much if I forget to take my dose of cough mixture, but if the drug is instead to treat hypertension, or if it is insulin or a birth-control pill, the consequences can be more serious. And of course there is usually a period of at least six hours during which we can't trust to oral administration, because we are asleep.

Another limitation of our traditional drugs is that their methods of administration are so crude. Although a drug may be intended to do its work in just one organ, we might just pop it into a convenient orifice and hope it finds its way there. Sometimes accurate targeting of the right tissues is crucial: in particular, drugs designed to kill cancer cells should not knock out healthy cells at the same time. The amount of a drug that needs to be administered is generally far greater if it is taken in some indiscriminate fashion, such as orally, rather than delivered directly to the relevant part of the body. For example, the amount of the contraceptive steroid progesterone that must be taken orally to provide three days prevention against conception would suffice to guarantee this for over a year if administered directly in the uterus.

In recent years, the delivery of drugs to the body has become a fine art, and these shortcomings are being superseded. Researchers have been able to develop materials that allow for a great deal of control over the way that drugs are released, for example by ensuring that a drug finds its way into the bloodstream at a specific, near-constant rate for a specified period of time. These drug delivery systems promise to make medical treatment a less hazardous and also much less intrusive affair, and indeed to extend the range of ailments that one can treat with drugs.

Internal Mail

A simple model for helping us to think about controlled drug release is a plastic bag with many pinholes. Fill the bag with water, and it will be released gradually through the holes. To alter the rate at which the bag emptied, we can alter the number of holes, or their size. Efforts to make controlled release systems can be regarded as attempts to make miniature bags of this sort, which can be taken orally or intravenously without toxic or other side effects from the bag itself. Often this means that the bag must be biodegradable.

One approach to drug delivery was mentioned briefly in the previous chapter: the use of the cell-like artificial vesicles called liposomes as protective shells. The typical procedure here is that the liposomes are formed from lipid molecules in the presence of the drug, so that they encapsulate a certain amount of drug molecules, and they are then passed into the bloodstream (generally intravenously) where they carry the drugs to the intended targets while protecting them either from being broken down by the body or from exercising their influence in unwanted places. Phospholipid liposomes are not toxic and do not induce strong immune responses, but they do have the drawback that they are not particularly stable—they are degraded relatively easily by the body's defense system, and also do not last long in storage.

Because they attach themselves to cell walls, liposomes are particularly effective for delivering drugs directly to cells rather than into the bloodstream. They have been exploited for conveying anticancer drugs, such as the natural compound doxorubicin, to cancer cells while reducing the toxicity of the drug to healthy cells; and they are attractive candidates for gene therapy—for conveying a fragment of DNA, a "replacement gene"—to cells that lack the gene in their own DNA. One can imagine (although this has not yet been achieved) incorporating into the phospholipid membrane of liposomes membrane proteins or other molecules that will be recognized and bound by receptors on the surfaces of specific target cells, thus allowing highly selective delivery of the liposome's contents. One of the principal virtues of liposomes as drug-delivery agents is therefore their potential for targeted delivery.

What they are less good at is *controlled* release—the steady, slow release of drugs into the body. Here the timescales needed are often just too long— liposomes survive for only a limited time in the body before enzymes break them down, whereas it is often desirable to achieve controlled release over several days or even several weeks or months. This kind of prolonged, constant release can be provided by mechanical devices, pumps that inject a solution of the drug into the body. But that is a cumbersome and expensive answer, even though pumps small enough for implantation are now available. Most research on controlled-release systems now focuses on polymers, which act in a manner more akin to that of the perforated plastic bag.

The holes in these polymer drug-delivery systems, which allow the drug molecules to pass out of them, are generally gaps in the tangled web of chainlike polymer molecules. It is possible to exercise a considerable degree of control over the size of these gaps, and thus over the permeability and release rate of the polymer. In general, the delivery system contains either drug molecules dispersed throughout the polymer network, like peas among spaghetti, or a reservoir of the drug encapsulated in a polymer membrane through which the drug can slowly diffuse. In either case, the system might comprise either a single unit that can be inserted or implanted in the body, or a collection of tiny microspheres, like little granules or bubbles each only a fraction of a millimeter in diameter, which can be injected or ingested (fig. 5.12).

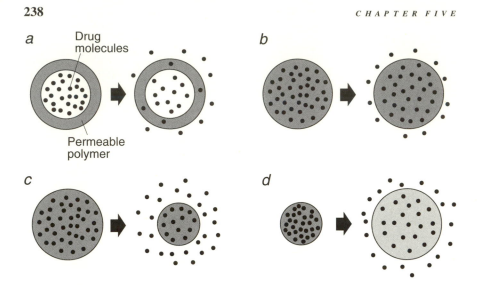

FIGURE 5.12 Polymer microspheres can be used to deliver drugs in a variety of ways. Such microspheres are typically a few micrometers across; loaded with drug molecules, they are injected into the bloodstream. The drug can be encapsulated in a central cavity within hollow microspheres, from where it slowly escapes through the permeable polymer wall (*a*). The drug may be dissolved or dispersed within the polymer itself, escaping either by diffusion (*b*) or by gradual dissolution or biodegradation of the polymer (*c*). A further possibility is to use an environmentally responsive polymer like those discussed in chapter 3; the drug is dispersed in the polymer but can diffuse out when the polymer encounters conditions that cause the network to swell (*d*).

Silicone implants are commonly used as slowly permeable polymer drug-delivery systems. An example is the Norplant system, which is a silicone capsule about the size of a match stick containing a birth-control drug. Six of these devices implanted under the skin of the forearm will deliver a contraceptive drug over a period of five years—the long lifetime is a consequence of the resistance of silicone polymers to degradation.

But sometimes biodegradability is advantageous for controlled release—gradual dissolution of the drug-containing polymer matrix can provide an alternative to diffusion for slow release of the drug. Polymer microspheres that operate by this mechanism can be regarded as miniature, slowly dissolving pills, with the advantage that the chemical properties of the polymer can be fine-tuned so that the dissolution rate can be varied, perhaps from hours to months. It is of course essential to ensure that the breakdown products are not toxic, and most systems of this sort make use of the lactic and glycolic acid copolymers mentioned earlier. A drawback of these materials, however, is that they break down everywhere at once—not just at the surface of the microspheres but inside too, with the result that the degradation process does not resemble the gradual dissolution of a sucked gobstopper but the crumbling of a honeycombed Whopper. This makes it hard to control the rate at which the drug within the polymer matrix is released, and can cause irregular pulses of the drug (called dose dumping), which can be dangerous when the drug is toxic at high dosages. There is consequently much interest in developing polymer microspheres that will dissolve at the surface only. Promising in this regard are copolymers containing so-called polyanhydrides, disks of which have been used with success to administer anticancer drugs called nitrosoureas to patients suffering from brain tumors. Nitrosoureas are toxic in high concentrations, and conventional intravenous treatments risk toxic side effects in other organs. Implanted, nitrosourea-containing polyanhydride disks are able to deliver the drug directly to the brain at a safe, constant rate.

Rather than ensuring a constant release rate, sometimes quite the opposite is desirable in a drug-delivery system. If it is necessary simply to increase or decrease the delivery rate by degrees over the lifetime of the delivery system (for example, during the time it takes for a microsphere to be fully degraded), it may be possible to preprogram that property into the chemical or material makeup of the system. Edith Mathiowitz and colleagues at Brown University have been able to prepare microspheres with uniform, concentric shells of different polymers (fig. 5.13); one can imagine making such onionlike spheres in which each layer

FIGURE 5.13 In these polymer microspheres developed by researchers at Brown University in Rhode Island, the outer coat and inner core are made from different polymers. Multiwalled microspheres fashioned from polymers with different rates of degradation would allow drugs, held within the polymer matrix or in a central cavity, to be delivered to a schedule that is specified by the sphere's composition. (Photograph courtesy of Edith Mathiowitz, Brown University.)

has a different dissolution rate, so that the pattern of release of drugs immobilized in the polymers varies in a predetermined way.

But a more difficult challenge is to get the pattern of drug release to respond to physiological changes within the body—for example, to vary the output of an insulin-delivery system as the sugar level in the bloodstream changes. Robert Langer's group at MIT has attempted to achieve this by making a polymer-based delivery system that is essentially a magnetically operated pump. They made a composite material consisting of an elastic copolymer in which were embedded tiny magnetic beads. They loaded this material with drug molecules within the pores of the polymer network. When the material was exposed to an oscillating magnetic field, the magnetic beads jiggled about, which had the effect of squeezing the pores and forcing out the drug. The rate of release could be increased by a factor of thirty by applying the oscillating field. Langer and colleagues postulated that diabetics could be given implants of the material loaded with insulin, and could carry small magnetic devices, say of the size of a wristwatch, which could trigger enhanced release on demand. This magnet/polymer composite system can be regarded as a smart material of the type discussed in chapter 3. Other stimulus-responsive drug-release systems are being developed that can be activated by ultrasound, light, electricity, acidity, or temperature.

Smarter still, however, would be a system that does not need an external stimulus at all to respond to changes in blood sugar, but which senses and adapts to these changes all by itself. These systems would then not have to rely on the user noticing the need for a greater or lesser supply of the drug and acting accordingly. An ingenious system of this sort has been devised by Joseph Kost, Buddy Ratner, and Tom Horbett at the University of Washington. It consists of a polyamine membrane loaded with insulin and also containing the enzyme glucose oxidase. As the name suggests, this enzyme oxidizes glucose—specifically, it converts it to a compound called gluconic acid. This acidic molecule reacts with the amine groups of the polymer membrane, awarding them with a hydrogen ion that leaves them positively charged. These charged amine groups then repel each other electrically, causing the membrane to swell, which leaves more space between the polymer molecules for the insulin to get out. So the result is that when glucose concentrations in the surrounding fluid increase, the membrane releases more insulin.

Many polymer membranes contain pores through which small drug molecules can diffuse, but which are too small to allow the passage of larger molecules. This posed a challenge when, in the 1970s and 1980s, new drugs began to be developed that are based on large, proteinlike (peptide) molecules. These molecules are broken down rapidly if simply injected or ingested into the body, and so creating delivery systems that would allow them to reach their targets, and to be released at a steady but moderate rate, became an important problem. It was widely believed at first that polymer delivery systems would not be equal to this task—the few polymeric materials that would allow large molecules to diffuse through them, such as polyacrylamide gels, gave too great a rate of discharge and could also damage tissues.

But in 1976, Robert Langer and colleagues found that certain polymers, generally ones that were highly hydrophobic (water-repellent) such as copolymers of ethylene and vinyl acetate, could be mixed with powdered proteins and formed into microspheres that would release the proteins at a steady, slow rate, persisting sometimes for up to one hundred days. There seemed to be no limit to the size of the large molecules that could be released controllably in this way, nor to their nature: proteins, nucleic acids, and polysaccharides (sugar polymers) could all be used. As the, say, polymer/protein composite forms, the two types of compound separate into regions typically a few tenths of a micrometer across, just as the oil and water separate out in salad dressing. When the protein began to diffuse out from the surface of this composite, a tortuous network of pores is left behind and other proteins farther within the material find their way out through this network of very large pores.

Langer and others have since learned how to control the size and shape of the pores in these materials, and thereby to tailor the rate of release to anything between a day and three years. In 1989, a controlled-release system of this sort—microspheres made from a safe biocompatible copolymer of lactic and glycolic acid—was approved by the U.S. Food and Drug Administration for use with a large-molecule peptide drug that combats prostate cancer. This was the first polymeric controlled-release system for peptide-based drugs to find medical approval, and it now provides the most widely used treatment for advanced prostate cancer.

Skin Deep

Most drugs, like most poisons, are effective only when taken internally—the skin keeps them out. This, after all, is what skin is for: resisting infection and contamination by foreign substances. But some drugs can be administered through the skin, an option that—when feasible—is convenient and stress-free (no nasty needles are required), and which can be more effective than oral delivery (since the latter often incurs significant drug degradation in the liver). Delivery of drugs through the skin (so-called transdermal delivery) generally involves the application of an adhesive patch to an appropriate area of the body, which releases the drug from a reservoir in the patch through a membrane that contacts the skin (fig. 5.14). The materials requirements for transdermal patches can be rather extensive: the patches might incorporate, for instance, a tough protective outer coating, a layer to retain moisture in the drug reservoir, a membrane that controls the rate of release from the reservoir, and an adhesive material to attach the patch to the skin. Generally, all of these materials will be polymers with properties tailored to each application.

The first transdermal drug-delivery system released the drug scopolamine, which combats motion sickness. Patches placed on the chest are now widely used to administer nitroglycerin for treating angina, and other systems are available for the treatment of hypertension and postmenopausal conditions. It may one day become possible to administer contraceptives and tumor suppressants this way. As well as requiring the identification of suitable polymeric materials, expanding

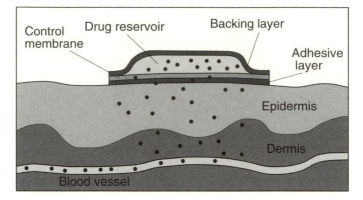

FIGURE 5.14 As many as ten different polymers can be incorporated into a transdermal patch, each serving a different function. The drug molecules diffuse into the skin at a rate determined by a control membrane between the drug reservoir and the skin surface. Ultimately they pass into the blood vessels beneath the epidermis and are carried through the bloodstream.

the range of drugs that can be delivered depends on finding ways to enhance drug transfer through the skin. Mild electrical stimulation and the use of ultrasound show some promise, and the drugs themselves can sometimes be modified chemically to allow better transmission. It is not unreasonable to suppose that the hypodermic needle may one day become redundant, as drug treatment becomes a matter of nothing more than sticking on a Band-Aid.

Controlled release is not only desirable for drugs that are used to combat illnesses. As I suggested above, for example, controlled-release systems may soon be invaluable for gene therapy. Another area in which the potential benefits would be tremendous is vaccination. Some kinds of vaccine cannot be administered in a single shot—two or more are needed at different times to provide the necessary amount of antibodies. Tetanus vaccine is one such, and in the developing countries the need for several injections poses a serious problem, as many people do not come back again after the first. This leads to over half a million deaths each year of babies who have contracted tetanus from their mothers. Polymer delivery systems are being explored that might one day be able to deliver a steady dose of vaccine over several months or even years from a single shot.

THE NEXT PHASE

The human body has a finite capacity for self-repair—we can grow a new fingernail, but not a new leg. Medical science has always worked within these constraints—repairing or promoting wound healing where it is possible, transplanting when supply and current surgical skill permit, and fashioning artificial prosthetic parts when all else fails. But we can now see that biomedical materials today offer the prospect of something else altogether: the fabrication of new parts, grown from the recipient's own cells on a scaffold of materials that work with nature to make the feat possible. A new leg will never grow by itself; but the marriage of cell biology with materials engineering may yet make that astonishing goal feasible. A scaffold of biodegradable or bioactive materials will be seeded with the respective cells, and bit by bit, tissue by tissue, the bones, muscles, arteries, nerves, and skin of a new leg will take shape until a complete limb has been created that can then be surgically transplanted. This, at least, is the dream. It is certainly too early to say whether it will happen—all one can say is that this is the direction in which biomedical materials science is now moving, and that we are beginning to understand what will be required. There will surely be intervening technologies, in which for instance hybrid organs containing synthetic and natural tissues are fabricated and implanted in the body.

Medicine is not just about repair, of course, but about maintaining well-being. Often that requires an awareness of what the body is doing—a warning system for things that might go wrong. Traditionally this may mean little more than checking the pulse and standing on the scales, but in the future it may be possible to produce a full map of the body's composition and state of health. New sensing technologies may be devised that will trace minute amounts of toxins, and alarm systems carried on a wristwatch may provide round-the-clock monitoring of those at risk from diabetes, from heart conditions, from viral infection. And I have not spoken about the decreasing scale of surgery, about micromachines that perform operations through a pinprick in the skin. All of this has a premium in materials.

As these technologies evolve, so, too, will biomedical ethics have to adapt. This is a complex area, one that is becoming increasingly important as medical science finds at its disposal things that would have been unimaginable to doctors and surgeons at the early part of the century. The ethical dilemmas posed by gene therapies have been much trumpeted, but there are surely ethical questions to be asked too about the replacement of body parts, and the prolonged life expectancy that this might bring. This is all uncharted territory—I do not believe that we have ready-made moral road maps to guide us. We are surely learning how to work medical wonders, and at the same time we must urgently learn how to practice them with responsibility and humanity.

Full Power

MATERIALS FOR CLEAN ENERGY

> I was just saying today that if we didn't do something to control
>
> them motorcars, they'll wipe out the whole lot of us.
>
> —Flann O'Brien, *Faustus Kelly*

Energy production is the principal source of human pollution, generating wastes that cause global warming, acid rain, urban smog, and radioactive contamination. But the development of new materials for efficient batteries and solar cells now promises a cleaner way to power the world.

ONE DAY in December of 1991, the citizens of London awoke to find the city enveloped in a dirty brown haze. The grimy sky was not something that you would notice except from a panoramic viewpoint—from the top of a high-rise block, say. But if you were asthmatic, you discovered soon enough that the air that day was scarcely fit for breathing. Pollution levels had soared; at one pollution-monitoring center the readings had gone off the scale. It later transpired that at least 160 people had died from respiratory problems as a result of the poor air on that single day.

This is not an isolated case, of course—it will be a familiar scenario to residents of any large urban development, from Los Angeles to Tokyo. In cities around the world, people are choking to death on air pollution. London itself had seen even worse—the infamous "pea-souper" smogs of Victorian London created a murk that blocked visibility at arm's length, and the city remained at peril from these noxious shrouds until the late 1950s. But whereas these were caused primarily by sulfur dioxide released by coal burning in domestic fires, the smogs of the modern city have a new culprit: the car. The brown color of these hazes is caused by nitrogen oxides, particularly nitrogen dioxide, which are emitted from automobile exhausts. Although these gases are also produced by coal-fired power stations, over half of the amount released into the air over Britain comes from car fumes. Nitrogen dioxide levels there increased by about 35 percent between 1986 and 1991, and just about all of this rise was due to car emissions. When exposed to sunlight, nitrogen dioxide undergoes chemical reactions that produce other pollutants, including ozone and peroxyacetyl nitrate, which cause eye irritation, respiratory problems such as asthma, and crop damage. All things considered, the car has a lot to answer for.

And of course, that is only part of the story. Car fumes are primarily carbon oxides: carbon monoxide, a poisonous gas, and carbon dioxide, the primary agent of potential global warming. Over the past few decades, lead from gasoline has found its way into all the nooks and crannies of the world, as far as the snows of Antarctica. And to anyone who lives in an urban environment, the sheer noise of the motor car—the persistent roar of the streets, the steady growl of distant traffic—is a pollutant in its own right.

But what is to be done? Every few years, one hears promises of salvation on the horizon, in the form of the electric car. This will be nonpolluting and noiseless, almost as environmentally friendly as a bicycle. Yet still we wait. Will the electric car ever become a commercial reality, before we have sent all of the available fuel reserves up into the atmosphere?

Enthusiasts are now proclaiming that the day of the electric car is all but upon us. And the evidence is beginning to look persuasive: General Motors has produced an electric vehicle, externally indistinguishable from those that fill our roads, which can run for well over 200 miles without needing to be recharged and without losing an acceleration of 0 to 60 miles per hour in 8 seconds. They say it will be as affordable as a vehicle powered by a dirty old internal combustion engine, and that it is so smooth and so quiet that, once you've driven it, you'll never want to go back to a gasoline pump again.

That there is a demand for such products is undeniable; in fact, there is a *requirement*. The state of California has decreed that about four thousand electric vehicles must be sold by 1998. By 2003, 10 percent of cars produced in the state by all major companies must be emission-free. The only way to eradicate noxious exhaust emissions is to power the vehicle electrically. And in the face of the vast polluting power of today's vehicles, motor companies are desperate to regain their credibility by developing and promoting electric cars.

The success of these ventures hinges on one crucial question: is it possible to develop a battery that is up to the job? The list of demands is daunting. Not only must the battery be able to store sufficient energy to sustain long drives without stopping for recharging; it must also provide the necessary power to ensure that there is no significant loss in acceleration. Recharging must be quick and easy, and inexpensive too. The battery must have a long life, both in action and on the shelf, and this means that repeated recharging cycles must not degrade its performance. It must be able to withstand the hammering to which most cars are subjected. It must not be full of toxic or otherwise environmentally hazardous substances. It must be competitively priced.

The problem of developing an electric vehicle is therefore largely a problem of developing adequate battery materials. Needless to say, transport is just one area that stands to gain from improved battery design: batteries that are cheaper, lighter, and more powerful are in demand for applications ranging from digital watches to spacecraft power systems. Each of these applications tends to have its own requirements: for some, low weight is important, for others high power is the main concern. And in each case, the identification of suitable materials is central.

Batteries provide just one route to cleaner power sources; another is solar energy. Both involve the conversion of a form of energy into electricity; in the former case, it is electrochemical, and in the latter it is light. Moreover, the two fields are somewhat interdependent: the way in which the availability of solar energy varies with time—it is on-line only during the day—does not generally coincide with the time dependence of the demand for power, so batteries that capture and store solar energy are essential for its potential exploitation. And yet that potential is vast—the Sun provides the Earth with ten thousand times more energy than we need to keep our entire civilization running, and we anticipate that it will do so for billions of years to come. But so far solar energy has not found large-scale application. Yet as fossil-fuel supplies continue to be consumed at an ever faster rate, and the threat of global warming looms as a result, the demand for effective ways to harness sunlight becomes increasingly urgent. With it comes a need for materials that convert light to electricity efficiently. Can new materials save the day?

POWER PACKS

When you think about it, the fact that the icon of the twentieth century—the automobile—runs on the same basic process that Neolithic people used to cook their meat is almost risible. It is an absurdly inefficient process. The energy is provided by nature in a concentrated form, an organic compound (wood for our ancestors; gasoline for our cars) that is stable in air only because the kick needed to convert it to more stable chemicals is marginally too great. By providing that kick, with a spark from a flint or from a spark plug, we set off the combustion process, and the organic material burns away giving out heat in all directions. We then scramble to gather as much of that heat as we can before the fuel is exhausted.

Batteries seem to provide an altogether more civilized alternative. They provide energy on demand, storing it up not in the form of a messy, hazardous compound dug out of the bowels of the Earth but as substances that yearn to pass electrons one to the other. By keeping those chemicals apart, we store up the electrons' energy like storing water at the top of a hydroelectric dam. Providing a pathway "down" which electrons can flow from one compartment to the other is like opening the taps on the dam; but we can close them off again at will, so that "water" remains at the top ready to unleash another burst of power. If the level of the "water" gets too low, well, we can top it up again. And all of this can fit into a package the size of a nickel.

The battery is a device that converts chemical energy to electrical energy. It relies on the differing propensities of chemical species to acquire or donate electrons—something that lies at the heart of all of electrochemistry. And indeed, all of chemistry, since in a sense it is all concerned with the movement and exchange of electrons, whether they be whirling around an atom's nucleus, performing figures of eight in molecular orbitals or trailing through the energy bands of solids. Whenever a chemical bond is formed or broken, electrons change hands. They are chemistry's currency.

The fundamental process underlying the operation of a battery is simply stated: it involves a chemical reaction in which electrons are transferred from one chemical species to another. In a battery this process is carried out in two half-reactions—one that involves the loss of electrons and one that involves their gain. The battery is an electrochemical cell divided into two half-cells, and the reaction proceeds when these are connected together by an electrically conducting pathway. The passage of electrons from one half-cell to the other corresponds to an electric current. Each half-cell contains an electrode in contact with the reacting species: that which passes electrons into the circuit when the battery discharges is called the *anode* and is the negative terminal, while that which receives electrons is called the *cathode*, and is the battery's positive terminal (fig. 6.1). The electrical

FIGURE 6.1 A battery is a kind of electrochemical cell, in which a current (a flow of electrons) is generated by electrochemical reactions at two electrodes. These are called half-cell reactions: one generates electrons, and one consumes them. The process is illustrated here for a cell in which hydrogen gas is converted to hydrogen ions at the negative electrode (the anode, made of platinum) while silver ions are converted to neutral silver atoms (silver metal) at the positive electrode (the cathode, made of silver coated in insoluble silver chloride). At the anode, molecules of hydrogen (H_2) split into atoms (platinum metal facilitates this process) and each of these leaves an electron on the anode, passing into solution as positive hydrogen ions (H^+). At the cathode, silver ions (Ag^+) pick up an electron to form neutral silver atoms. The electrical circuit is completed by the liquid electrolyte in which the electrodes are immersed, in this case hydrochloric acid; the electrolyte contains dissolved ions (H^+ and Cl^-), which can carry a current through the liquid. So long as hydrogen gas is supplied at the anode, this cell will continue to produce a current until all of the silver chloride on the cathode has been converted to silver metal.

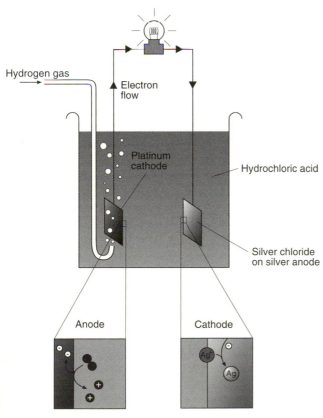

Hydrogen gas

Electron flow

Platinum cathode

Hydrochloric acid

Silver chloride on silver anode

Anode

Cathode

Half-cell reactions

circuit is completed by a so-called electrolyte, an electrically conducting sub-stance placed between the two electrodes which carries a flow of charge between them. In "wet" cells, the electrolyte is a liquid containing dissolved ions, whose motion generates an electrical current; "dry" cells contain an electrolyte that is altogether less sloshy—for example, a solid with mobile ions, a rubbery ion-conducting polymer, or a porous solid saturated with an ionic solution.

Heavy Metal Power

One of the mainstays of the battery industry is the lead–acid battery, so called because it involves an electrochemical reaction between lead metal, lead dioxide, and a sulfuric acid electrolyte. The nature of the electron transactions here are as follows: at the negative electrode, atoms of metallic lead give up two of their electrons each, to become doubly charged lead ions, denoted Pb^{2+}; and at the positive electrode the lead in the dioxide, which is in the form of ions with four positive charges (Pb^{4+}), acquires those two electrons to become Pb^{2+} ions too (fig. 6.2). As this reaction proceeds, both the lead anode and the lead dioxide get converted into lead sulfate ($PbSO_4$). If this reaction were allowed to run to com-pletion, the lead anode would dissolve away completely and the battery would run out of power. But it can be recharged by reversing the process, which involves connecting the battery's negative terminal to the negative terminal of another power source, supplying it with electrons, and similarly allowing the positive terminal of the source to provide an electron sink for the electrons acquired at the battery's positive terminal. Then the Pb^{2+} ions pick up electrons at the battery's negative terminal and get converted back to neutral lead atoms, which are depos-ited on the electrode; and the Pb^{2+} ions at the positive terminal give up electrons to reform Pb^{4+}, regenerating lead dioxide. Thus, the lead–acid battery is an energy store that can be repeatedly filled and emptied; such a device is called a *storage battery* or *secondary battery*.

The electrochemical reaction that takes place in these devices was first reported by the Frenchman Gaston Plante in 1859, long before it was developed into a battery process. There are several different designs of lead–acid batteries now in use. Generally these contain spongy lead anodes immersed in a paste of lead dioxide and an electrolyte of dilute sulfuric acid (fig. 6.2). But because of the presence of lead, the devices are relatively heavy, and the use of both lead (a toxic metal) and sulfuric acid (a corrosive agent) means that disposal must be handled carefully. A further drawback is that because on each charge–recharge cycle the lead electrodes are partly dissolved and then reconstituted, they tend to degrade to an unusable state after only a few cycles. On the other hand, lead–acid batteries are relatively cheap. The batteries that provide the electrical power for sparking, for headlamps and taillights, and for other electrical accessories in today's auto-mobiles are of this type. More advanced forms of lead–acid batteries, which de-velop sufficient power to drive a vehicle, cost considerably more—typically about $10,000 at present. This may seem a lot, but it is cheaper than many of the alternative batteries under development, and several automobile companies

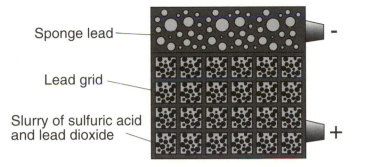

Sponge lead

Lead grid

Slurry of sulfuric acid
and lead dioxide

−

+

Anode   Lead dioxide   Cathode

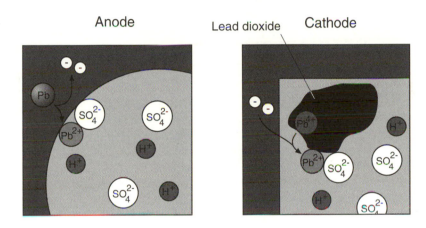

Half-cell reactions

FIGURE 6.2 The lead–acid battery is one of the most common kinds of storage battery. The anode is generally a grid of lead metal pasted with a slurry of lead dioxide and sulfuric acid (H_2SO_4—this is the electrolyte). The cathode is porous, spongy lead metal. At the cathode, lead ions with four positive charges (Pb^{4+}) within the lead dioxide particles pick up two electrons from the electrode and become doubly charged ions (Pb^{2+}). These ions combine with the sulphate ions in the electrolyte to form insoluble lead sulphate ($PbSO_4$). At the anode, neutral lead atoms lose two electrons to form Pb^{2+} ions; these again form $PbSO_4$. Because they contain lead and acid, these batteries are heavy, toxic, and potentially corrosive if they leak.

have developed vehicles powered by cells of this type despite their drawbacks in other respects. General Motors has built a two-seater sports car that can travel for up to 135 kilometers without the need to recharge its lead–acid battery, and which has a top speed of 120 kilometers per hour. GM plans to put these vehicles into commercial production in late 1996. Electrosource, a company in Austin, Texas, has produced a lead–acid battery that can undergo nine hundred charging cycles, and which will power a van for at least 125 kilometers before needing to be recharged. The electrodes are composite structures in which a fiberglass skeleton is coated with lead, reducing the weight by around 40 percent relative to devices of comparable performance. Electrosource believes that its device will be available for less than $3,000, which would make it a strong contender in the marketplace.

In some batteries the discharging process leads to irreversible changes in the nature of the components, so that once discharge is complete the battery cannot be recharged. These are called *primary batteries*, and are strictly "one-shot" devices, thrown away once their power is expended. An example is the mercury cell, which is an efficient device for long-term, slow discharge—an attribute that is valuable for applications such as powering hearing aids. The electrochemical reaction in the mercury cell is very simple: at the anode, which is made of zinc, atoms of zinc give up two electrons to form Zn^{2+} ions, and at the cathode, which is a compressed pellet of mercuric oxide (HgO), mercury ions (Hg^{2+}) take up the two electrons to form mercury metal. The zinc ions pass into an electrolyte of potassium hydroxide absorbed in a porous medium (fig. 6.3)—thus these are dry cells. Mercury cells and other primary batteries are valuable in certain niche applications, but because they are not rechargeable they will never be in the running for powering electric vehicles.

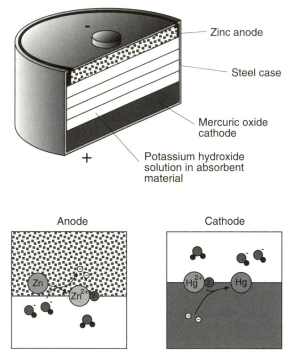

Half-cell reactions

FIGURE 6.3 The mercury battery is an example of a primary battery, which is not rechargeable; once the half-cell reactions have gone to completion, the cell must be discarded. It is a "dry" cell, since the electrolyte (potassium hydroxide) is absorbed in a porous material. The half-cell re-

ADVANCED BATTERIES

Brimstone Power

The search for rivals to lead–acid batteries as power sources for electric vehicles hinges largely on the identification of an electrochemical system that stores a large amount of energy and delivers high power while using lightweight materials and being rechargeable repeatedly without deterioration. The energy storage capacity is measured in terms of a specific energy density—a measure of the total energy output (in watt hours) per kilogram of battery mass. One of the front-runners is the sodium–sulfur battery, which has high efficiency, low density, and a good recharging lifetime. A sodium–sulfur battery developed by Asea Brown Bovery Ltd in Heidelberg, Germany, delivers a power of 100 watts and a specific energy density of 85 to 90 watt hours per kilogram of battery mass (which I shall compare later against an internal combustion engine) and can withstand nearly six hundred recharging cycles. A version produced by Silent Power Ltd in Runcorn, U.K., can undergo eight hundred cycles, albeit at the cost of a slightly lower power-to-weight ratio. The Ford Motor Company has produced a test fleet of fifty-three minivans powered by sodium–sulfur batteries. These vehicles, called Ecostars and based on Ford's European Escort vans, offer a speed of 120 kilometers per hour with a range of 160 kilometers before recharging.

For all this, the battery is a rather peculiar one by electrochemical standards. For a start, both of the electrodes are molten, while the electrolyte is solid. One of these molten electrodes (the anode) is sodium, a metal so reactive that chemists have to treat it with great care even in the solid state; the other (the cathode) is molten sulfur, the brimstone of volcanoes. At the former electrode, sodium atoms from the liquid metal donate an electron into the circuit to form sodium ions; these ions then pass through the solid electrolyte, a ceramic material called beta-alumina, to the positive electrode where sulfur atoms pick up electrons to become sulfide ions. The sodium and sulfide ions combine to make sodium sulfide. Both the high reactivity of sodium and the liquid state of the electrodes work in the battery's favor: the former allows the battery to develop a high voltage and thus a high power output, and the latter means that when the battery is recharged, the electrodes are regenerated in exactly the same state as they started from. In addition, both sodium and sulfur are light elements, so the battery has a high power-to-weight ratio.

But there are disadvantages, the greatest of which is that the device has to be heated to 370 degrees Celsius to keep the electrodes molten. This power con-

---

actions basically involve the conversion of zinc atoms to zinc ions (Zn^{2+}) at the anode (generating two electrons) and of mercury ions (Hg^{2+}) to mercury atoms at the cathode, consuming two electrons. The hydroxide ions of the electrolyte (OH^-, shown here using the same convention for representing atoms as in previous chapters) are involved in these reactions, however: they are transformed into water (H_2O) and oxide ions (O^{2-}) at the anode, while precisely the reverse happens at the cathode.

sumption partially offsets the battery's output. In addition, the battery must be encased in thermally insulating material, negating some of the advantage of the lightweight substances from which it is made. And the battery can withstand only a few heating and cooling cycles, so the high temperature must be maintained at all times. Finally, the cost is considerable: the Ecostar batteries cost $46,000 each. Although sodium–sulfur batteries were once considered very promising for electric-vehicle applications, these drawbacks have proved critical, and almost all efforts to commercialize this technology have now ceased.

So there is ample room for other contenders, and no shortage of those. The nickel-cadmium battery, which uses an electrochemical reaction first proposed in 1899, has been set to work in several prototype electric vehicles, such as Fiat's Panda Elettra (from Italy), Nissan's FEV (Japan), and Renault's Zoom (France). These vehicles claim a competitive top speed (between 110 and 120 kilometers per hour) and range (between 104 and 150 kilometers), and Fiat has planned a production run of five hundred of these vehicles. Nickel–cadmium batteries are already in widespread use as rechargeable cells for low-power applications, such as flashlights. They have electrodes made of hydrated nickel oxide ($NiOOH.H_2O$; the positive electrode) and cadmium (the negative electrode), in contact with an electrolyte of potassium hydroxide solution. During discharge, cadmium metal is converted to cadmium hydroxide at the positive electrode, with loss of electrons to the circuit, and nickel hydroxide is formed at the negative electrode, with a corresponding uptake of electrons. The main drawback of these batteries for large-scale use in vehicles is the presence of cadmium, which is both toxic and expensive. Some researchers believe that in the long run these problems will undermine the bid of nickel–cadmium cells in the electric-vehicle market.

Electric Rockers

Ensuring that the electrodes maintain their integrity over many discharge–recharge cycles is one of the principal problems in the design of secondary batteries: dissolving and then redepositing a solid electrode will not, in general, return it to its original state. For this reason, there is a lot of interest in exploring battery designs in which the electrodes are simply hosts for charge-carrying species that pass back and forth through the electrolyte like electron ferries calling between two ports. One such device, the lithium battery, appears to be one of the most promising candidates for rechargeable batteries in the long term, according to the U.S. Advanced Battery Consortium (USABC), an organization jointly funded by the U.S. government and by power industries.

In the lithium battery the positive electrode is made of a material that will take up lithium ions in the gaps between atoms in the crystal structure, without that structure being significantly perturbed. When the lithium ions are removed, the gaps relax slightly as if deflated. Lithium ions are very small, and also very light (lithium is the lightest metal there is), and so can fit into the spaces between atoms in a variety of materials. The resulting materials are known as intercalation compounds. Lithium ions will form intercalation compounds with many so-called transition-metal chalcogenides, compounds of transition metals and sulfur, selenium or tellurium which have layered structures—here the lithium ions merely slip between the layers. These materials, and in particular molybdenum or titanium disulfide, have been explored as potential electrodes for lithium batteries. But the most promising positive-electrode materials are oxides of manganese, nickel, and cobalt. In these oxides the ionic crystal lattice has a relatively open structure, with spaces or channels between the ions into which the small lithium ions can fit.

In the earliest lithium batteries the negative electrode was lithium metal, which will dissolve into lithium ions as the metal atoms each give up an electron. The lithium ions are then ferried to the positive electrode through an ion-conducting electrolyte of some kind. This electrolyte is generally a lithium salt such as lithium perchlorate ($LiClO_4$), dissolved in an organic solvent such as propylene carbonate or dispersed in a polymer such as polyethylene oxide. The first commercial lithium batteries, which appeared in the 1970s, were primary cells—nonrechargeable devices whose positive terminals were made of materials that can incorporate lithium ions: polytetrafluoroethylene (PTFE), manganese dioxide, or copper sulfide. They are used to power cardiac pacemakers, for which they can provide a suitably slow discharge rate.

Rechargeable lithium batteries are a more recent invention: they were first marketed in the 1980s for use in portable electronic equipment such as laptop computers and portable phones. These too used lithium metal (sometimes in an alloy with aluminum) as the negative electrode, and lithium-ion intercalators as the positive electrode. But rechargeability brings attendant problems: lithium electrodes tend to suffer from a short life cycle, not only because of the usual problems associated with dissolution and regeneration of a metal electrode but because lithium metal is extremely reactive and tends to undergo corrosion reactions at the interface with the electrolyte. These not only degrade the performance of the cell but—most seriously of all—pose a safety hazard. The early attempts at commercializing rechargeable lithium batteries were dogged by fires caused by batteries spontaneously bursting into flame. The Canadian company Moli Energy went bankrupt when lithium-metal batteries marketed by them in the 1980s proved hazardous—a cellular phone powered by one of their batteries caught fire and injured its user. These problems effectively crushed lithium-battery technology for a decade.

Now rechargeable lithium batteries are back in force, because of successes in replacing the problematic lithium metal electrode. The trick has been to find a material that, like the positive electrode, will take up lithium ions in an intercalation compound. One can't simply use the *same* material as that in the positive electrode, because then there would be no driving force for the lithium ions to make the journey from one electrode to the other. Rather, there must be some energetic benefit to the ions moving from the negative to the positive electrode (the process that takes place during battery discharge). When this happens, electrons must also pass from the negative to the positive electrode, to maintain a balance of charge. So the flow of lithium ions through the electrolyte is accompanied by a corresponding flow of electrons through the external circuit—the battery generates a current.

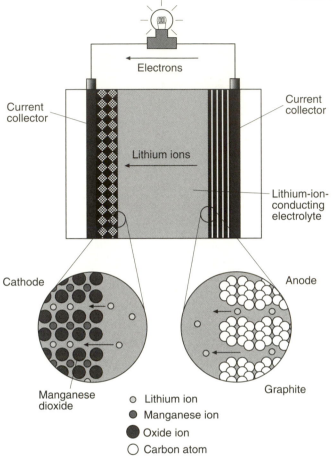

Electrons

Current collector

Current collector

Lithium ions

Lithium-ion-conducting electrolyte

Cathode

Anode

Manganese dioxide

Graphite

○ Lithium ion
● Manganese ion
⬤ Oxide ion
○ Carbon atom

FIGURE 6.4 In the "rocking-chair" rechargeable lithium battery, lithium ions are ferried back and forth between two electrodes that can take up the ions in spaces within their crystal structures. The cathode is made from a transition-metal oxide, such as cobalt or manganese dioxide, and the anode generally consists of graphite, which can incorporate lithium ions in the gaps between the sheets of carbon atoms. The passage of lithium ions from anode to cathode during discharge is accompanied by a corresponding flow of electrons through the external circuit to maintain charge balance. To recharge the batteries, a reverse voltage is applied across the two terminals, drawing the lithium ions back into the graphite anode.

The negative-electrode material most studied is graphite, which consists of sheets of carbon atoms linked into hexagons, like chicken wire (see fig. 8.2 on page 317). The sheets are held together relatively weakly, and lithium ions can slip between them easily to form an intercalation compound. Because graphite binds lithium ions rather less strongly than the transition-metal oxides used for the positive electrodes, it is energetically favorable for these ions to pass from the graphite negative electrode to the positive electrode (fig. 6.4). During recharge,

FIGURE 6.5 The first rechargeable rocking-chair lithium batteries to become commercially available were developed by the Sony Corporation in Japan. They have an operating voltage of 3.6 volts, three times that of rechargeable nickel–cadmium cells. (Photograph courtesy of the Sony Corporation.)

the lithium ions are drawn back into the graphite. Thus the lithium ions "rock" back and forth between the two electrodes on each discharge–recharge cycle, and for this reason these cells are called *rocking-chair batteries*.

The first rocking-chair battery was demonstrated in 1980, but only in the 1990s has it been developed into a technologically useful and commercially viable device. In 1992 the Sony Corporation in Japan began marketing rechargeable lithium batteries that could tolerate as many as twelve hundred recharging cycles (fig. 6.5). Other battery companies—Duracell in the United States and Varta in Germany—are now working to develop their own versions. The Sony cell has electrodes of graphite and the lithium cobalt oxide $LiCoO_2$, from which lithium ions can come and go between CoO_2 layers; the electrolyte is a lithium salt in an organic solvent. But if such batteries are to achieve the market penetration that is hoped for, the cobalt compound may have to be replaced by another material, as cobalt is both rather toxic and expensive. It seems likely that similar compounds containing manganese in place of cobalt will be preferred in the long run. Commercial rechargeable lithium batteries now offer specific energy densities of over 100 watt hours per kilogram, and they are replacing other batteries in cellular phones and laptop computers.

Safety is far less a problem than it was for the old-style lithium-metal cells, but there have still been setbacks. In 1995, short circuits in a batch of the new rechargeable lithium batteries caused a fire in a manufacturing plant in Birmingham in the U.K.; the cells "exploded like missiles" as the fire reached them, according to the chief fire officer at the scene. And the Apple computer company was forced to withdraw a new range of laptop computers in 1995 because the lithium batteries that powered them were catching fire.

For use in electric vehicles, lithium batteries have to be scaled up to provide larger specific energy densities, and that brings with it new hazards. Mitsubishi has developed a prototype electric car called the Chariot which is powered by an array of rechargeable lithium cells. But in January 1996 one of these caught fire at the Mitsubishi headquarters in Cypress, California, destroying the expensive vehicle. Clearly, there is still plenty of scope for improvement in these devices; but the possibilities for making better and safer lithium batteries by finding the right electrode materials are by no means exhausted.

The Right Mix

Intercalation of ions into the electrodes is also the principal behind another new and apparently strong contender for the electric-vehicle battery. This device, called the *nickel–metal hydride battery* (plate 4), was developed in the early 1980s by the Ovonic Battery Company, a subsidiary of Energy Conversion Devices Inc. in Troy, Michigan. The positive electrode is made of nickel hydroxide; but it is the negative electrode that is the magic ingredient. This consists of a metal hydride alloy, typically based on a mixture of vanadium, titanium, zirconium, and nickel but also containing other elements (transition metals and so-called rare earth metals). At the positive electrode the reaction is essentially the same as that in a nickel–cadmium cell. During the charging cycle, nickel hydroxide ($Ni(OH)_2$) reacts with hydroxide ions in the electrolyte (a solution of potassium hydroxide) to form nickel oxyhydroxide ($NiOOH$) and water. This process produces a "spare" electron, which flows into the charging circuit. During discharge, the process is reversed. The process at the negative electrode, meanwhile, can be most simply described as the uptake of a positive hydrogen ion by the metal hydride alloy during charging, and its release during discharge: the hydrogen ions are intercalated into the metal alloy lattice. So there is very little change in the structure of the alloy electrode during each cycle—it merely acts like a hydrogen sponge (fig. 6.6).

The storage capacity of this cell depends on the capacity of the hydrogen sponge—the amount of hydrogen that the metal hydride alloy can take up during charging. It is important that the hydrogen stored within the alloy lattice is able to find its way out again during the discharge process: the hydrogen must be able to diffuse easily through the metal lattice. In addition, the metal hydride alloy must not react with the concentrated hydroxide electrolyte, which is highly alkaline. Under these conditions, many metals will react to form oxides and hydroxides. So the demands on the negative electrode material are rather stringent. Some metals will, by themselves, satisfy some of these criteria, but none will satisfy all. For example, metals such as magnesium, titanium, and vanadium offer good hydrogen-storage capacity. Others, such as vanadium, manganese, and zirconium, offer the right kind of bond strengths with hydrogen to allow the process of uptake to be reversible. Others, such as chromium and molybdenum, provide corrosion resistance.

So the researchers at Ovonic, headed by Stanford Ovshinsky, have worked to develop a suitable material simply by mixing several of these metals together in

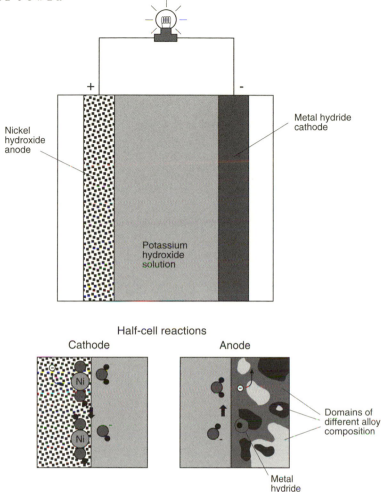

FIGURE 6.6 The reaction at the cathode of a nickel-metal hydride cell is much the same as that in a nickel-cadmium cell: the compound NiOOH acquires an electron from the circuit and a hydrogen ion from a water molecule to form nickel hydroxide, $Ni(OH)_2$. But the anode reaction is very different. A hydrogen atom incorporated into the complex alloy electrode combines with a hydroxide ion to form a molecule of water, liberating an electron. The availability of hydrogen for this reaction depends on fine-tuning of the electrode's hydrogen-uptake properties by careful selection of the mix of metals in the alloy. The alloy is disordered on the atomic scale, and also inhomogeneous at larger scales, being separated into domains between which the mix of metals differs. Crucially, uptake and release of hydrogen is fully reversible.

an alloy. But it is not quite so simple: they found that in order to exploit the desirable properties of each component without suffering from the less desirable ones, they had to introduce disorder into the structure of the alloy in a carefully controlled manner. Disorder over distances of a few tenths of a micrometer helps to ensure good hydrogen uptake. This kind of disorder corresponds to a variation in the actual composition of the alloy, leaving it like a marble cake. Disorder over

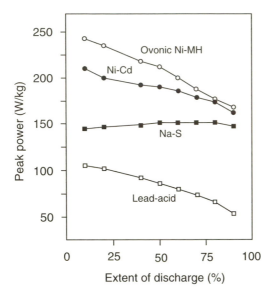

FIGURE 6.7 The Ovonic battery has a higher power output throughout its discharge cycle than any other kind of rechargeable cell on the market, including nickel-cadmium (Ni–Cd), sodium-sulfur (Na–S), and lead-acid batteries.

scales of about 10 to 100 nanometers, meanwhile, ensures that the electrochemical reactions at the interface with the electrolyte take place efficiently. Finally, disorder on the atomic scale provides a range of different kinds of sites for binding hydrogen—offering the hydrogen ions this range of choice also appears to enhance their uptake. The resulting material has a hierarchy of structures at different scales.

The result of this "alloy engineering" is a designer material that gives the nickel-metal hydride cell a higher energy density and power output than any other battery on the market (fig. 6.7). Small nickel-metal hydride rechargeable batteries are now being manufactured worldwide under license from Energy Conversion Devices for use in portable electronic devices. They have replaced the lithium batteries in the combusting Apple laptop computers mentioned earlier.

Scaling up this technology for electric-vehicle applications looks highly promising. The Ovonic battery can be fully discharged and recharged over one thousand times without impairing the performance, and if just 30 percent of its energy is discharged on each cycle it can withstand ten times this many cycles. When the lead-acid battery of General Motors' Impact electric vans is replaced with an Ovonic nickel-metal hydride battery, the range of the vehicle is increased from 190 to no less than 480 kilometers. The Ovonics team points out that, even making conservative estimates about the lifetime of the batteries, this would reduce the cost of the energy per kilometer by a factor of about six, relative to the cost for a conventional vehicle running on gasoline (this ignores the cost of the battery itself, however). Ovonics also claims that its battery components can be safely disposed of or economically recycled into new batteries or other metal components.

All of this makes nickel-metal hydride batteries a very attractive proposition for use in the so-called midterm electric vehicles envisioned by the USABC. These devices already meet the midterm goals set for energy storage density, power output, and speed of recharging. One problem that remains to be overcome is that the batteries discharge slowly when they are not being used; but even here, the self-discharge rate meets the USABC's midterm goal. In 1993 the Ovonic Battery Company was awarded an $18.5 million contract by the USABC to develop nickel–metal hydride batteries for commercial use in electric vehicles, and it has established a joint venture with General Motors (GM Ovonic LLC) whose intention is to manufacture two thousand batteries per year for use in electric vehicles.

HYDROGEN POWER

The smooth, clean release of power that takes place when a battery discharges, and the series of miniature explosions that occur in an internal combustion engine—these appear on the face of it to be very different processes. But fundamentally, they're not. As I indicated at the outset, just about any chemical process, these included, is a transaction in electrons. When gasoline burns in air, electrons are traded between hydrocarbon molecules and oxygen molecules: in a manner of speaking, oxygen atoms benefit at the expense of carbon and hydrogen. But the energy that is released as a result of this transaction is inefficiently harvested in the internal combustion engine. Is it possible to harness the energy of this kind of explosive process in an electrochemical cell?

This is precisely what the English barrister William Grove demonstrated to the Royal Society in London in 1839. He devised a cell that generated energy by burning hydrogen gas: that is, by reacting hydrogen gas with oxygen gas, to form water. This reaction releases a tremendous amount of energy, and for that reason great care has to be taken when using hydrogen gas in an aerated environment. The explosive nature of the reaction is familiar to many from the school experiment in which hydrogen gas is liberated from an acid by the reaction with a metal, collected in an inverted test tube, and ignited with a pop. More strikingly, the Hindenburg airship disaster of 1937 demonstrated what can happen when a mixture of the two gases catches a spark.

But Grove proposed to control this energetic process by carrying out the reaction electrochemically: in two separate half-cells, one containing hydrogen and the other oxygen. In the hydrogen half-cell, the hydrogen atoms in the H_2 molecule give up an electron each, to become hydrogen ions. The gas is bubbled through solution over a metal electrode; the hydrogen ions pass into solution and the electrode carries off the electrons. In the other half-cell, oxygen gas is bubbled over a second electrode; the oxygen atoms combine with water and with electrons provided by the electrode to form hydroxide ions (fig. 6.8). The result is that the two gases are converted into hydrogen and hydroxide ions—the ionized form of water—with the exchange of electrons between the electrodes.

This device, which converts the chemical energy of the combustion reaction into electrical energy in the circuit connecting the electrodes, is called a *fuel cell*. Unlike secondary batteries, it is not an energy storage device; rather, it can be regarded as an electrochemical internal combustion engine, in the sense that the

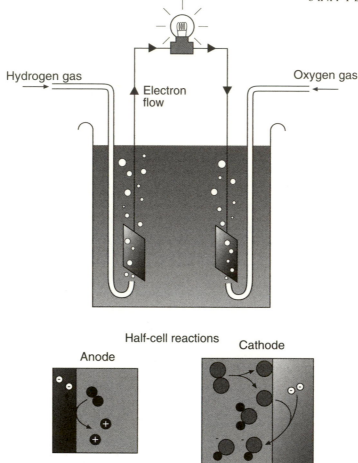

Hydrogen gas

Oxygen gas

Electron
flow

Half-cell reactions

Anode

Cathode

FIGURE 6.8 A fuel cell burns a combustible fuel such as hydrogen in a controlled fashion, by carrying out the reaction electrochemically in two half-cells. The simplest configuration of a hydrogen–oxygen fuel cell is shown here. At the anode, hydrogen gas bubbled over a metal electrode is converted to hydrogen ions, donating electrons into the circuit; at the cathode, oxygen atoms pick up these electrons and react with water to form hydroxide ions. The net result is the conversion of hydrogen and oxygen to hydrogen and hydroxide ions, the ionized form of water. In a fuel cell, the energy released in this reaction can be captured much more efficiently than in a combustion chamber.

power is evolved by feeding fuel into the cell. The hydrogen–oxygen fuel cell is, however, a "clean" combustion engine—it produces only water and electricity. But there is no reason in principle why *any* combustion process cannot be carried out electrochemically in a fuel cell, and indeed other such devices exist that "burn" hydrocarbon fuels such as methane instead of hydrogen. In this case, the process at the oxygen electrode is essentially the same, but at the other electrode the hydrocarbon is converted to carbon dioxide and water. This cell is then essentially a natural-gas burner, except that the energy of the combustion reaction is extracted much more efficiently, as electricity rather than as heat. Other kinds of fuel cell use methanol as the fuel—they are electrochemical alcohol burners.

William Grove's idea took around 120 years to mature, mainly because it was only in the 1960s that the materials requirements could be met for converting a cumbersome, wet electrochemical device into a compact, stable, and efficient device. As power sources, fuel cells do not exhibit particularly striking characteristics: the amount of power evolved per unit size of the cells is small. But in their favor they can be made very lightweight, and they make much more effective use of a fuel than would a combustion device. On an individual basis, they don't appear to be very attractive power supplies compared with batteries; but it can often be far more preferable to carry a large supply of fuel that will power a single fuel cell than to carry a supply of batteries that, one by one, will run out and have to be discarded.

This is especially true when power is required for long journeys, and so it was that fuel cells came into their own for the longest journey ever undertaken—to the Moon and back. In the 1960s, fuel cells were developed for powering instruments on board the Apollo spacecraft; they could make use of the same fuel—oxygen and hydrogen—that propelled the rockets through space. These cells were advanced versions of Grove's device, called *alkaline fuel cells* because they used a strongly alkaline electrolyte (a solution of potassium hydroxide). The efficiency of the electrochemical reactions at the electrodes was maximized by using highly porous metals, such as an agglomerate of small nickel particles, to increase the surface area over which the electron-donating and electron-consuming half-cell reactions took place. The nickel metal acts as a catalyst to promote the rate of the electrochemical reactions. A major problem with alkaline fuel cells is that their performance is greatly degraded by the small amounts of carbon dioxide that are commonly present as an impurity in the feed gases—this reacts with the alkaline electrolyte.

This problem can be avoided by using an acidic electrolyte, which is what is done in the phosphoric acid fuel cell. Here the electrolyte is phosphoric acid, which is held within a porous separator matrix between the electrodes, typically made of silicon carbide powder bound together with PTFE. The electrodes themselves consist of small clusters of a metal catalyst (for example, platinum) supported on carbon particles mixed with PTFE, all bound to an electrically conducting carbon cloth (fig. 6.9). These composite electrodes combine high porosity, good electrocatalytic activity, and good electrical conductivity. Phosphoric acid fuel cells can run on very impure feed gases, such as untreated air instead of pure oxygen, and hydrocarbons containing up to 30 percent carbon dioxide. Like most fuel cells, these devices are usually used in stacks, in which adjacent cells are separated by a grooved plate that gives the feed gases access to the electrodes while preventing them from coming into direct contact (fig. 6.10). A stack of two hundred phosphoric acid fuel cells can generate up to 210 kilowatts of power and a voltage of 140 volts.

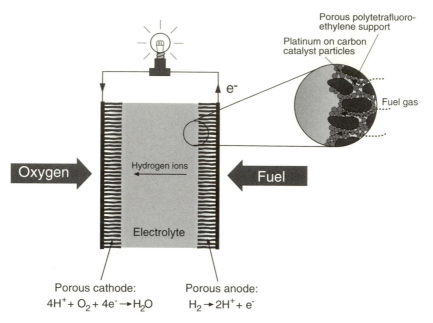

$$4H^+ + O_2 + 4e^- \rightarrow H_2O$$

FIGURE 6.9 The phosphoric acid fuel cell is a more sophisticated version of the cell in figure 6.8. The electrode reactions are the same, but the electrodes consist of a porous form of the polymer polytetrafluoroethylene, within which small particles of platinum are dispersed. These particles catalyze the electrode reactions, as the gases circulate between them. The electrolyte is acidic, so that water molecules (rather than hydroxide ions) are formed at the cathode. Hydrocarbon fuel can be used in place of pure hydrogen; in that case, carbon dioxide is also formed at the anode.

It would be nice to do away with the corrosive wet electrolytes in these cells altogether. That is now possible, thanks to the development of a class of polymers that can transport ions in the solid state. Foremost among these is Nafion, a sulfonated fluorocarbon introduced by Du Pont in the late 1960s. Nafion can be regarded as a kind of PTFE with charged sulfonate (SO_3^{2-}) groups attached to the polymer chains. The chains arrange themselves into tubular structures with the sulfonate groups on the inside, creating channels down which ions can pass. When saturated with water, Nafion will conduct hydrogen ions; as water is the only liquid present, the saturated polymer electrolyte is noncorrosive. Nafion and related ion-conducting polymers are now used in prototype hydrogen–oxygen fuel cells with platinum electrodes that are under investigation for use in hydrogen-powered electric vehicles.

Alkaline, phosphoric acid, and polymer-membrane fuel cells all run on hydrogen at relatively low temperatures (between 60 and 200 degrees Celsius). Low-temperature operation is an advantage, but the use of hydrogen fuel is a drawback. Unlike natural hydrocarbon gases, hydrogen gas cannot simply be pumped out of the ground; it must be made from hydrogen-containing compounds (generally from hydrocarbons), and this costs energy. Some hydrogen-powered fuel cells

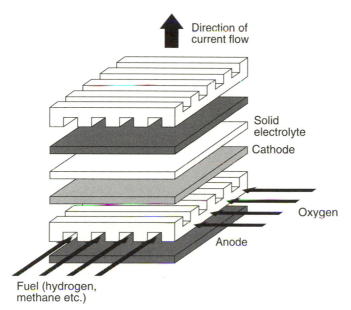

Direction of current flow

Solid electrolyte

Cathode

Oxygen

Anode

Fuel (hydrogen, methane etc.)

FIGURE 6.10 Fuel cells are commonly operated in stacks. Grooved plates between each cell provide access for the gases while preventing them from mixing (note that here I show an "exploded" stack for clarity—the layers are married together during operation). The electrolyte in these stacked cells is commonly a solid material—either a porous solid impregnated with a liquid electrolyte, or an ion-conducting polymer membrane such as Nafion (see text).

use methanol as the raw fuel, which is converted to hydrogen "on board" the device in a catalytic chemical reaction. (Because this reaction generates carbon dioxide, a greenhouse gas, as a by-product, methanol fuel cells are not as locally "clean" as those that run on pure hydrogen.) The energy cost of making hydrogen, whether in an industrial plant or within the device itself, restricts the overall energy efficiency of hydrogen-powered fuel cells: a typical phosphoric acid fuel cell is less than 40 percent efficient in this sense. (That's still better than an internal combustion engine; but to be competitive, fuel cells will have to be a *lot* better.) For this reason, the economic viability of these low-temperature cells is a matter of debate. Phosphoric acid cells provide power at the considerable cost of around $3,000 per kilowatt, for instance—double the cost from a conventional power generation plant.

But if all this sounds discouraging, jump aboard a city bus in Vancouver. If you're lucky, your journey will be pollution-free, on board the hydrogen-powered vehicle by Ballard Power Systems that has plied the city's streets since 1993. Researchers at Ballard have focused on polymer-membrane hydrogen–oxygen cells containing a proton-conducting polymer electrolyte sandwiched between porous electrodes impregnated with platinum. In 1991 they incorporated these "proton-exchange membrane" (PEM) cells into a prototype bus that delivered 125 horsepower, capable of carrying twenty passengers for 100 miles. When this vehicle was introduced to the Vancouver streets, other transport authorities began to take interest. A commercial prototype bus demonstrated in 1995 (plate 5), which more than doubles all of the figures for the earlier vehicle, is to be put into service in 1996 by the Chicago Transit Authority, which hopes ultimately to replace all of its fleet of two thousand with these zero-emission buses. The hydrogen fuel is carried on the roof of the bus in lightweight cylinders of a tough graphite/polymer composite material. Ballard intends to commence full commercial production of these vehicles in 1998. The hydrogen-powered vehicle is on its way!

Adapting this technology from buses to cars presents challenges, however. In particular, cars have less room to store the fuel. For partly this reason, Ballard has focused on methanol fuel cells for use in cars: methanol has a higher energy density than hydrogen, so a car would need to carry less of it to get the same refueling range. In addition, methanol costs about as much as gasoline, and can be handled in much the same way. The drawback, as indicated above, is that its use in fuel cells generates carbon dioxide—but less than is emitted by conventional engines and without the other noxious oxides they produce. The methanol is converted to hydrogen in a catalytic conversion device, and this is then "burned" in a PEM cell. Ballard has teamed up with the German automobile company Daimler Benz to create a methanol-powered vehicle, and in April 1994 this joint venture unveiled the Necar, a prototype methanol-powered van. Meanwhile, Ballard is developing its PEM hydrogen cells for use in industrial plants, submarines, laptop computers, and other niche applications.

Hot Shots

Fuel cells that burn organic fuels such as hydrocarbons, methanol, and carbon monoxide operate at higher temperatures—typically 450 to 1,000 degrees Celsius—but are nevertheless more energy-efficient overall because they can use unprocessed fuels. Two such devices are in general use: the molten carbonate and solid-oxide fuel cells. The first have been under development since the 1970s but continue to face problems that hinder widespread commercial use. Some molten carbonate cells do in fact run on hydrogen, but others use carbon monoxide and oxygen as the fuel, the overall reaction being the combustion of carbon monoxide to carbon dioxide. The electrolyte consists of mixture of molten potassium and lithium carbonate salts, held within a porous ceramic matrix of lithium aluminate ($LiAlO_4$) at a temperature of 650 degrees. At the negative electrode, carbon monoxide reacts with carbonate ions in the molten electrolyte to generate carbon dioxide, liberating electrons into the circuit; at the positive electrode, the reaction of oxygen with carbon dioxide, driven by an injection of electrons, regenerates carbonate ions. These cells give voltages of around 0.9 volts and power outputs approaching 10 kilowatts are now being obtained. But the molten salt electrolyte is corrosive and tends to dissolve the nickel oxide cathodes. This and other performance problems will have to be overcome before molten carbonate fuel cells become useful commercial devices.

Solid-oxide fuel cells are the hottest of the lot—operating temperatures of 700 to 1,000 degrees are typical. They burn hydrogen, hydrocarbons, or mixtures of hydrogen and carbon monoxide over metal oxide electrodes separated by a solid-oxide ceramic electrolyte. The latter eliminates many of the corrosion problems encountered with "wet" electrolytes, but the development of materials that conduct ions in the solid state has posed quite a challenge to materials scientists. The best solid ion conductor so far seems to be zirconia (ZrO_2) doped with yttria (Y_2O_3). Yttria introduces gaps in the zirconia lattice where oxygen ions should be, and this allows the oxygen ions to move about through the solid. The Westinghouse corporation has constructed fuel cells based on this material that offer power outputs of 25 kilowatts, which it hopes to increase to 100 kilowatts in the near future. This kind of power output will make the cells very attractive for some applications, while smaller ceramic fuel cells, running at lower power and lower temperatures, are being investigated for use in electric vehicles.

WHO WILL BUY AN ELECTRIC CAR?

Any scientist who works on battery technology for electric vehicles will surely come to realize that in the commercial world, scientific excellence in product development is only a small part of the story. Even if there were now available a battery that outstripped the internal combustion engine in terms of power, cost, and environmental friendliness, the motor-vehicle industry would be unlikely to ditch gasoline power overnight. The key commercial question is: Will it sell? The answer to that depends as much on the psychology of the buyer and the cunning of the advertiser as on the merit of the product. Can a population addicted to the roar of the internal combustion engine be persuaded that a silent, non-polluting battery will serve them just as well? Electric-vehicle enthusiasts say that once you've driven one you'll never look back; but here they are probably considering the car as a transport device, not as a status symbol. And in any case, will car users want, or be able, to simply ditch their present vehicles, with maybe a good twenty years of life still in them, in order to snap up the newest thing on the market? Furthermore, will the petroleum companies find a way to fight back? (You bet they will!) Will car manufacturers be prepared to change their assembly lines?

And it is still far from clear whether batteries *will* be able to compete with the internal combustion engine on all levels, at least while oil reserves last. The specific energy density of a fuel tank full of gasoline is around 10,000 watt hours per kilogram, whereas for a battery a specific energy of around 80 watt hours per kilogram is considered good, and even the most optimistic predictions do not take this beyond 200 in the near future. The reason for this is that most of the battery's mass consists of the casing and electrodes—the components—whereas the mass of a fuel tank is mainly the mass of the fuel itself, energy in a highly concentrated form.

And when you buy a car, you expect to get the engine included; the running cost of a gasoline engine is then (at least in the United States) relatively cheap.

But as we saw above, some of the batteries currently in development for electric vehicles cost as much as, or more than, a typical family car. Costs are coming down, but it is not clear that they can be made sufficiently low to provide an all-inclusive car-and-battery package at a competitive price, with the additional threat of a massive capital outlay once the car has covered the 120,000 miles or so of the battery's lifetime. There is therefore some debate among electric-vehicle developers as to whether the customer should actually buy the battery with the car, or whether instead the battery should be rented out by the manufacturer or vendor.

Estimates of costs vary widely. Mark DeLuchi of the University of California at Davis has estimated that an electric vehicle with attributes similar to those of a gasoline-driven Ford Taurus, priced at $17,300 for the year 2000, would cost $22,500 if powered by a methanol fuel cell, $24,000 for a hydrogen fuel cell, and $27,000 for battery power. Others (including a study for the U.S. Department of Energy) have concluded that fuel-cell vehicles need cost no more than $1,000 above equivalent conventional cars. And some automotive engineers believe that, if made in sufficient quantities, even battery-powered vehicles *with the battery included* can be produced at a similarly competitive price.

Then there is the question of refueling, or in this case, recharging. At present, this takes typically a few hours for battery-driven electric vehicles—that's a long stop at the "gas" station. On the other hand, fuel-cell cars can simply load up with fuel in the same way as conventional cars, especially if the fuel is a liquid like methanol.

With these questions in mind, perhaps only legislation like that imposed by California will be the way to a cleaner and less polluting transport system in the near future. All the same, some battery engineers predict that the primary market for electric vehicles will be from two-car families, who will use a gasoline-powered vehicle for long journeys and a battery-powered car for short about-town trips. If this means that our cities will no longer be overshadowed by a whiskey-brown haze of poisonous gases, it is surely worth looking forward to.

POWER FROM THE SUN

Everyone who takes a realistic view of the situation has to admit that the world needs energy. A lot of it, and undoubtedly more and more in the coming decades. This insatiable demand leads to talk, from time to time, of an impending energy crisis.

But there will never, in the foreseeable future, be an energy crisis. We will never be short of energy, since it is constantly supplied to us free of charge in overwhelming abundance. Over ten thousand times more energy than we need to keep the whole world running is delivered to us unceasingly, every day. The potential crisis, however, is one of energy *conversion*: we do not yet know how to make efficient use of this energy source, the origin of which is the Sun, our planet's star. So poorly do we fail in this energy-conversion problem that we largely ignore this tremendous resource, and instead look inward—into the bowels of the Earth—for the fuels that drive civilization along its shaky tracks. We dig up fossil fuels, in which the sunlight of millions of years ago is bound up in concentrated form, and use that old energy instead. Or we mine the planet for uranium, and let the heat liberated by its nuclear fission drive our steam turbines. These energy sources are immediate, and are on the whole relatively easy to exploit. But there is only so much oil, coal, and gas in the world—reservoirs that took millions of years to form can be exhausted in decades. And as we burn up these fuels, we transform the world bit by bit into a gigantic greenhouse, whose carbon-dioxide-laden atmosphere lets the Sun's heat in but is ever more reluctant to let it back out, so that the world warms up. Nuclear power does not have these environmental costs—but it has others, which may turn out to be far less severe but which very understandably frighten people far more, and which accumulate inexorably for future generations.

Nature is almost entirely solar powered. Plants have their own solar cells, called chloroplasts, which convert solar energy into chemical energy, in the form of high-calorie carbohydrate compounds which act as energy stores. These cells are not particularly efficient—they convert less than one percent of the light that falls on them—but they do not have to be, for sunlight is so abundant. Ultimately, all animals make use of these vegetable energy stores for their own metabolism. And solar energy drives the great geological processes of the planet—the circulation of warm and cool water in the seas, and of warm and cool air in the skies. The energy content in these circulation systems is very diffuse, and so we can easily overlook the enormity of it. But there is enough energy in the winds of North and South Dakota alone to power most of the United States, if it could be harnessed, and already one percent of California's power consumption is met by wind power.

Arguably the most attractive ways to make use of solar energy involve its direct capture and conversion into a useful form. Solar thermal technologies, such as solar furnaces, capture the heat of the Sun's rays for warming water or driving heat engines. Large-scale installations commonly deploy arrays of mirrors to focus the heat onto a central furnace; these systems can achieve efficiencies of 10 to 30 percent, meaning that between a tenth and a third of the incident solar energy is converted into usable form. Domestic solar heating systems consist of flat panels of absorbing substances that warm up a building's water supply. But I want to focus for the remainder of this chapter on the technology of solar cells, which capture sunlight and transform it directly into electricity. This technology has experienced a tremendous growth in the past three decades, largely as a result of the development of new materials and new processing methods that have turned solar cells from expensive, relatively inefficient devices for highly specialized applications (in the aerospace industry, for instance) to commercial systems that can be found powering anything from an Indonesian village (fig. 6.11) to a pocket calculator. Large solar-cell power-generation plants and smaller urban systems can now be connected to national power grids for domestic and industrial supply.

FIGURE 6.11 Solar panels can provide clean, locally controlled power sources for remote villages in developing countries. Here they are being used in Indonesia. (Photograph courtesy of R&S Renewable Energy Systems BV, The Netherlands.)

Crystal Power

All of the solar cells now available commercially, and the many that exist as laboratory prototypes, make use of the *photovoltaic* effect. I described this phenomenon in chapter 1, as it provides the operational principle of the photodetectors used in optoelectronic technology. Solar cells are in fact nothing more than large-area photodetectors, which generate electric power when irradiated with light. Almost all solar cells are built from solid-state semiconducting materials, and in the vast majority of these, including all commercial devices, the semiconductor is silicon. There are several good reasons for this choice: silicon is abundant in the Earth, and a silicon-based solar-cell industry can ride on the back of the sophisticated silicon-processing methods developed by the microelectronics industry (it can, moreover, make use of the latter's castoffs, as we shall see). From a more fundamental perspective, silicon is able to absorb sunlight at those wavelengths at which it is most intense—from the near-infrared region (wavelengths of around 1,200 nanometers) to the violet (around 350 nanometers). Silicon is not without its drawbacks, however; but there are now many options for overcoming or minimizing these.

The photovoltaic effect involves the generation of mobile charge carriers—electrons and their complementary partners, positively charged holes—by the absorption of a photon of light. To recap on the earlier discussion, this pair of charge carriers is produced when an electron in the highest filled electronic band of a semiconductor (the valence band) absorbs a photon of sufficient energy to pro-

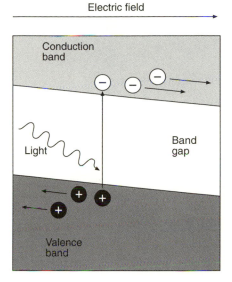

Electric field

Conduction band

Band gap

Light

Valence band

FIGURE 6.12 The photovoltaic effect. Absorption of a photon by a semiconducting material excites an electron from the valence to the conduction band, creating an electron–hole pair. Under the influence of an electric field, these photoexcited charge carriers move in opposite directions, setting up an electric current.

mote it into the overlying empty energy band (the conduction band). This excitation process can be induced only by photons with an energy larger (and thus a wavelength shorter) than a certain threshold, corresponding to the width of the energy gap that separates the valence and conduction bands (fig. 6.12).

The creation of an electron–hole pair can be converted into the generation of an electrical current in a semiconductor junction device, wherein a layer of semiconducting material lies back to back with a layer of either a different semiconductor or a metal. In most photovoltaic cells, the junction is a p–n junction, in which a p-doped and an n-doped semiconductor are married together. At the interface of the two, the predominance of positively charged carriers (holes) in the p-doped material and of negatively charged carriers (electrons) in the n-doped material sets up an electric field, which falls off to either side of the junction across a "space-charge" region. When absorption of a photon in this region generates an electron–hole pair, these charge carriers are driven in opposite directions by the electric field—that is, both are propelled away from the interface and toward the top and bottom of the two-layer structure. Charge carriers generated outside the space charge region will diffuse through the crystal at random, but they will have a certain probability of wandering into the electric-field region, whereupon they too will be propelled toward the top and bottom faces of the device. Metal electrodes on these faces collect the current so generated. The electrode on the lower face is called the back electrode; on the top face (through which light is absorbed), the electrode is divided into strips so as not to obscure the semiconducting layers below, and the exposed area is covered with an antireflective coating to reduce the amount of light lost by reflection (fig. 6.13).

In the most widely used commercial solar cells, the p-doped and n-doped semiconducting layers are formed within a monolithic piece of crystalline silicon. The

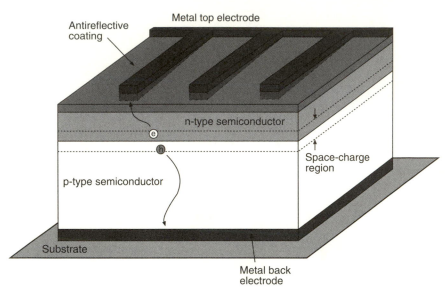

FIGURE 6.13 A solar cell made from a p–n junction of doped crystalline silicon. An electric field is set up within the space-charge region at the interface of the n-doped and p-doped material. Photoexcited charge carriers (electrons **e** and holes **h**), either created in this region by photon absorption or that find their way here by diffusion following absorption elsewhere, are propelled by the field in opposite directions, toward the two electrodes at the top and bottom. The top of the device, which is exposed to the illumination, is given an antireflective coating to reduce light losses by reflection.

first efficient silicon solar cell of this type was made in 1954 and had an electrical energy conversion efficiency of around 6 percent—the other 94 percent of the incident energy was squandered in other ways. That might seem rather wasteful, but it is considerably more efficient than nature's photosynthetic solar cells. And by the early 1960s this efficiency had been improved to around 15 percent, which was sufficient to justify the use of such solar cells to power devices on spacecraft and satellites.

To reduce losses from reflection, most commercial crystalline silicon solar cells are chemically etched on their top faces to create a rough surface—a rugged vista of tiny pyramids which suppresses reflection like the pitted surface of frosted glass. But such tricks will never raise the efficiency of crystalline silicon solar cells above a fundamental limit of about 29 percent: some wastage of the incident photons is inevitable. For a start, silicon cannot absorb all of the photons in sunlight; it has an absorption window that opens at wavelengths of about 350 nanometers (just inside the violet part of the spectrum) and closes again for wavelengths greater than about 1.1 micrometers (in the near-infrared), because photons of longer wavelength have insufficient energy to excite an electron across the band gap. Although most of the energy in sunlight lies within this window, there is also an appreciable amount of energy both in the ultraviolet and at wavelengths out to around two and a half micrometers; but these ultraviolet and infra-red photons cannot be used by a silicon cell. Another major source of loss comes from recombination and thermalization processes, whereby a fraction of the photoexcited charge carriers lose some or all of their energy before they can make their way to the electrodes. When an electron is excited into the conduction band by absorption of a photon with more energy than is strictly required for the jump across the band gap, the excess energy is very quickly lost as heat—this is called thermalization. And unless they are separated very rapidly, photoexcited elec-trons and holes will have a tendency to recombine by the electron falling back into the hole and emitting its energy either as heat or as a photon.

One of the major obstacles in making efficient solar cells from crystalline sili-con is that it is an "indirect band-gap" material (see chapter 1), which means that it can absorb photons (to create an electron–hole pair) only if the momentum of the photoexcited electron is altered. (The reverse process of electron–hole recom-bination also has a momentum "cost".) Photons can change an electron's energy, but not its momentum—the latter must be induced by an interaction between the electron and the vibrations of the atomic lattice of the crystal. In other words, absorption of a photon requires a chance coincidence of certain lattice vibrations to make it possible. This means that crystalline silicon absorbs light rather weakly, so to make efficient use of the incident light it is necessary to make the p–n double layer fairly thick, typically a few hundred micrometers. This imposes a materials cost—more material means greater expense. It also constrains the type of materials-processing methods used for fabrication—crystalline silicon slabs this thick must be sliced from large ingots.

These slices of crystalline silicon are no different from those used as the basis of microelectronic silicon chips, and the solar-cell industry uses the same technol-ogy for making them. In fact, because efficient silicon solar cells can be made from silicon wafers that have many more defects and imperfections than the mi-croelectronics industry can tolerate, companies that produce silicon solar cells commercially often buy up the defective castoffs of the latter. These silicon wafers are usually produced by a crystal-growth technique developed in the 1940s, called the Czochralski method, in which silicon that has been extracted and purified from its natural oxide quartz is melted in a crucible and slowly drawn out into cylindrical rods. These are then sawn like cucumbers into fine slices. In 1975 the German company Wacker developed a new way of manufacturing sili-con wafers, which involved pouring molten silicon into graphite molds before crystallizing it. This and other, more recent casting methods generate a lower grade of crystalline silicon called multicrystalline silicon, which is riddled with defects. The material is cheaper to produce than Czochralski-grown silicon, but would be uselessly imperfect for microelectronics. It is, however, perfectly ade-quate for making wafers for solar cells, and multicrystalline cells with efficiencies of around 18 percent have now been reported.

But although silicon solar cells can use the rejects from microelectronics, they are still a very expensive source of power, since to capture enough light they need to have a large surface area. Improvements in light-capturing efficiency and in processing methods allowed the price of these devices to drop by a factor of ten between the 1970s, when their main use was in high-tech applications such as space vehicles, and the late 1980s, when domestic applications started to become feasible. All the same, the high cost remains an obstacle to the widespread development of large-scale solar-energy plants that can make an appreciable contribution to national power requirements.

Less Is More

One of the most promising routes to lower-cost solar cells makes use of a different form of silicon, in which the material does not have a regular, crystalline atomic structure but is instead amorphous, without any long-ranged atomic order. The amorphous silicon used for solar cells also contains some hydrogen—it is called hydrogenated amorphous silicon, denoted a-Si:H. The key attribute of this material for photovoltaic technology is that it behaves as a direct-band-gap semiconductor, meaning that electrons can absorb the energy of photons without having to undergo a simultaneous change in momentum. This means that absorption happens much more readily, so instead of having to use silicon wafers hundreds of micrometers thick to ensure that light is captured efficiently, highly absorbing solar cells can be made from a-Si:H films only about half a micrometer thick. This saving in materials consumption lowers the price considerably. The use of thin films also means that different kinds of technology can be used to fabricate the cells. Instead of having to slice up large slabs of crystalline silicon grown by slow crystallization processes, atomic deposition methods such as chemical vapor deposition or ion-beam deposition can be used to make large-area films.

The top electrode in these devices is generally made from a conducting transparent material such as the inorganic substance indium tin oxide. Most of these cells do not, strictly speaking, contain a p–n junction, but have a wide undoped ("intrinsic") region between thin n- and p-doped regions at top and bottom. This is because the photoexcited charge carriers undergo rather rapid recombination processes in the doped materials, so it is necessary to keep these doped regions narrow. So thin-film amorphous silicon cells are in fact p–i–n devices of the sort found in photodetectors (chapter 1): charge carriers excited in the intrinsic region get swept up toward the electrodes by the electric fields at the i–n and i–p junctions (fig. 6.14a).

The efficiencies of these a-Si:H cells are generally lower than those of crystalline devices, commonly in the range of 10 to 12 percent. But higher efficiencies can be attained in cells in which the p-doped and n-doped materials are different alloys—efficiencies greater than 13 percent have been achieved for a-SiC/a-Si:H structures (fig. 6.14b). These structures, in which the p–n (or p–i–n) junction occurs between different materials, are known as *heterojunctions* (*hetero* meaning "different" in Greek).

Since the 1980s, chemical vapor deposition methods have been used to create a range of *alloys* of amorphous silicon with other elements, such as carbon and germanium. These materials also exhibit the photovoltaic effect, but their opti-

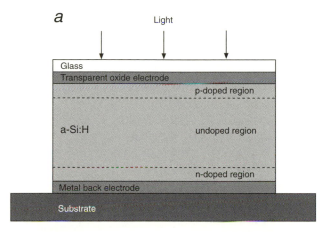

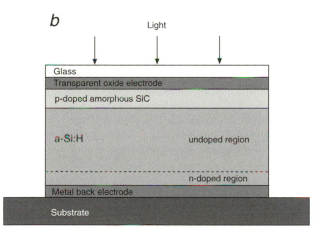

FIGURE 6.14 (*a*), Amorphous silicon solar cells contain a wide undoped (intrinsic) layer between the p-doped and n-doped layers to minimize recombination of photoexcited charge carriers (this process takes place readily in doped direct-band-gap semiconductors, of which amorphous silicon is an example). The top electrode is commonly a conducting transparent metal oxide such as indium tin oxide, to which metal leads are attached. *b*, The efficiency of amorphous silicon solar cells can be enhanced by using different materials for the various layers: here the p-doped region is amorphous silicon carbide, while the intrinsic and n-doped regions are a-Si:H.

mum absorbing wavelength depends on the precise composition—some might absorb most strongly in the blue part of the spectrum, some in the red, and so forth. This has made it possible to fabricate solar cells with different spectral characteristics, and by stacking them on top of each other one can create composite devices that make good use of all the wavelengths in sunlight. United Solar

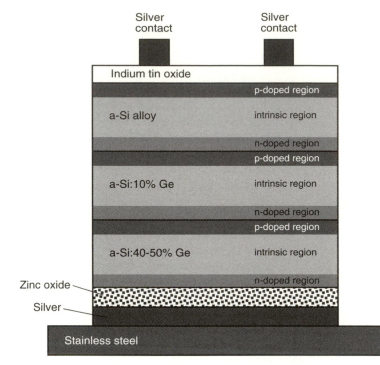

Silver contact   Silver contact

Indium tin oxide

a-Si alloy   p-doped region

intrinsic region

n-doped region

a-Si:10% Ge   p-doped region

intrinsic region

n-doped region

a-Si:40-50% Ge   p-doped region

intrinsic region

n-doped region

Zinc oxide

Silver

Stainless steel

FIGURE 6.15 The light-harvesting efficiency of amorphous silicon solar cells can be enhanced by stacking several different cells that capture light in different wavelength bands. A triple stack developed by the United Solar Systems Corporation contains p–i–n cells sensitive to blue light (top), green light (middle), and red and infrared light (lower). The bilayer of zinc oxide and silver is a reflector which sends light that is not absorbed on the first pass back through the device. The whole structure can be fabricated on a continuous stainless steel strip that rolls through successive deposition chambers. United Solar put these cells into full-scale commercial manufacture in 1996.

Systems Corporation, a joint venture between Energy Conversion Devices in Michigan and Canon Inc., has used this idea to develop a triple-junction device in which three amorphous silicon alloys capture blue, green, and red/infrared light, respectively (fig. 6.15). Each of the three layers is itself a p–i–n device. In the top layer, the intrinsic region is an amorphous silicon alloy with a band gap suited to absorption of blue light; the middle layer has an intrinsic region of an amorphous silicon/germanium alloy containing 10 percent of the latter, which absorbs green photons; and the lower layer contains the same alloy but with 40 to 50 percent germanium, giving it a smaller band gap so that it absorbs at red and infrared wavelengths. Light that is not absorbed in any of these layers on its first pass is reflected back by a silver/zinc oxide reflector, which is textured so as to maximize the capture of light within the device by internal reflections. Prototypes of this device have an efficiency of 11.8 percent, and United Solar has developed a continuous processing method in which the materials are deposited on a stainless steel sheet 14 inches wide and up to half a mile long as it rolls through successive deposition chambers. The company begun full-scale commercialization of the cells in 1996.

In commercial amorphous-silicon products, several cells are hardwired together in the factory into a so-called photovoltaic module, which can have light-collecting surfaces up to four square feet in area. The inevitable compromises involved in developing suitable manufacturing processes have the result that the efficiencies of these modules are typically around 8 to 9 percent; the United Solar triple-junction devices described above will be available in modules with greater than 10 percent efficiency. Despite having lower efficiencies than crystalline silicon solar cells, thin-film a-Si:H modules are viable in the marketplace because of their lower cost. They are already used in rooftop panels for domestic power supplies, an application in which appearances matter as much as performance. Thus the technologies have had to be adapted to suit the aesthetics of the market: in Japan, the Sanyo Electric Company has produced traditional Japanese roofing tiles that incorporate built-in amorphous silicon cells (plate 6), while United Solar is introducing flexible photovoltaic panels that resemble conventional asphalt shingles in appearance, function, and installation. Amorphous silicon devices are also used in vehicle technology, for example being integrated into the sunroofs of automobiles. In 1990 an airplane powered by amorphous silicon cells flew across the United States from San Diego in California to Kitty Hawk in North Carolina—a flight that, unlike Icarus's, welcomed the Sun.

Elegant Variation

Silicon dominates the solar-cell business just as much as it does the microelectronics business; but it does not have the field entirely to itself. It has at least two old rivals, and several new ones. So-called chalcopyrite semiconductors, in particular the compound copper indium diselenide ($CuInSe_2$), have been considered for solar cells since 1974, when a group at AT&T Bell Laboratories described a crystalline $CuInSe_2$ cell with a respectable efficiency of 12 percent. Because this material has a direct band gap, it is highly absorbing and so can be used in the form of thin films less than a micrometer thick. These films can be made by depositing the compound from an atomic vapor, by electrochemical means, or as a finely ground polycrystalline powder in a volatile organic solvent. The $CuInSe_2$ is deposited as the p-doped layer on a metal back electrode, typically of molybdenum. The n-doped material is a different substance—cadmium selenide was used in the earlier devices, but zinc oxide is now preferred. In between these two materials at the p–n heterojunction, a very thin "buffer layer" must be inserted to prevent a chemical reaction from occurring. This buffer layer is typically a 10- to 50-nanometer layer of cadmium sulfide.

In the laboratory, the efficiencies of these cells are generally around 14 to 15 percent, and researchers at the U.S. National Renewable Energy Laboratories in Golden, Colorado, have achieved a performance of 18 percent. Again, the efficiencies of $CuInSe_2$-based modules marketed commercially are lower, closer to

Light

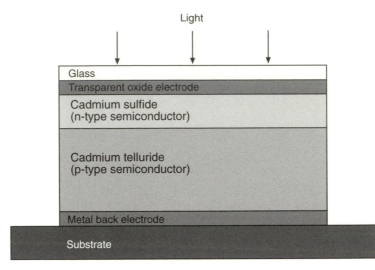

FIGURE 6.16 A cadmium telluride solar cell is a thin-film device in which light is absorbed and charge carriers are created in a cadmium telluride layer that acts as a p-doped semiconductor.

10 percent. Siemens has produced a module with an area of nearly half a square meter and an output power of over 40 watts.

Another competitive material for thin-film solar cells is cadmium telluride (CdTe). This material is again a direct-band-gap semiconductor and absorbs light strongly in the wavelength range from the ultraviolet part of the spectrum (below 350 nanometers) to 850 nanometers (in the near infrared). Cadmium telluride can be used in the undoped form, as this already contains a lot of positive charge carriers (holes) and so acts as a p-type semiconductor. It is used in a heterojunction device with cadmium sulfide as the n-type material, on top of which sits the top electrode of a transparent conducting oxide (fig. 6.16). The cadmium sulfide layer absorbs light in the blue part of the spectrum and so prevents it from reaching the cadmium telluride layer—so these cells have a wavelength cutoff at the wavelength of green light unless very thin cadmium sulfide layers are used. The efficiencies tell a similar story to other thin-film devices—around 16 percent in the laboratory, more like 7 to 8 percent in commercial modules. But questions about the use of cadmium, a highly toxic metal, hang over the long-term prospects of these cells.

Related to both of these materials is the mineral pyrite, otherwise known as fool's gold, whose chemical identity is iron disulfide (FeS_2). This substance has a long history as a semiconductor, and was used as a rectifier (a kind of diode) in early radio receivers in the 1920s. Its potential as a photovoltaic material began to be exploited in earnest only in 1975, however, when researchers at Bell Laboratories in Holmdel, New Jersey, built heterojunction solar cells from natural pyrite and cadmium sulfide. Iron disulfide absorbs all visible light strongly, and so in principle could be used in solar cells in layers less than a micrometer thick. But it has a rather reactive surface, which makes it difficult to form stable, uniform heterojunctions with other solid-state materials. Recent attempts to use this material in solar cells have therefore focused on developing *photoelectrochemical cells*, in which the pyrite sits in contact with a liquid electrolyte. The process that occurs at the interface is essentially the same as that at a p–n junction, although it sounds very different. Electron–hole pairs are generated by absorption of photons in the pyrite (which can act either as an n-type or a p-type semiconductor, depending on the conditions of its formation). Because there is an internal electric field set up at the interface with the electrolyte, photoexcited electrons are propelled away from the interface toward a back electrode in contact with the pyrite, while the holes go in the other direction, into the electrolyte. What this means, however, is that rather than traveling as mobile charge carriers through the electrolyte (as they do through a solid-state semiconductor), the holes take part in an electrochemical reaction with ions in the electrolyte. For example, when the solution contains doubly charged vanadium ions (V^{2+}), these can be oxidized to triply charged ions as the ions "pick up" an additional hole (that is to say, lose an additional electron) at the surface of the pyrite. In this way, a flow of charge is set up, and a platinum electrode immersed in the electrolyte can carry this charge into an external circuit (fig. 6.17). Photoelectrochemical cells of this sort have been

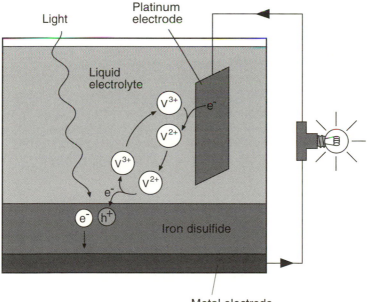

FIGURE 6.17 Photoelectrochemical solar cells can be constructed from the semiconductor iron disulfide. Light absorption in this material creates electron–hole pairs. The electrons are collected by a metal back electrode, while the holes are annihilated by electrons donated by vanadium ions (V^{2+}) in the liquid electrolyte above. The vanadium ions then replenish their electrons from a platinum anode. In effect, it is as if the photoexcited electrons and holes travel to opposite electrodes as in a solid-state solar cell.

pioneered by Helmut Tributsch and colleagues at the Hahn-Meitner-Institut in Berlin, Germany, who have achieved efficiencies of around one percent. Yes, this is very low compared with solid-state devices; but the cells could conceivably be extremely cheap to make, relying on "bucket chemistry" rather than sophisticated semiconductor technology.

A related "wet" solar cell has been developed by Michael Grätzel's group at the Swiss Federal Institute of Technology in Lausanne, Switzerland. His devices mimic some of the characteristics of nature's solar cells, the chloroplasts that allow plants to carry out photosynthesis. In the chloroplast, the various elements of light-to-electricity energy conversion are performed by separate components. Light absorption is carried out by chlorophyll molecules, which absorb most strongly in the red and blue parts of the spectrum (leaving the green light to be reflected). The absorbed photon's energy is then converted into electricity—the motion of electrons—by a series of events involving other molecules. The energy is sufficient to knock an electron from the excited chlorophyll onto another, similar molecule called a pheophytin, from where it is shuttled onto other molecules in the membrane of the chloroplast, taking it ever farther from the chlorophyll molecule and thus ever less likely to recombine with the hole that it has left behind. In this way, charge flows across the membrane. In contrast, in a solid-state solar cell the light-sensitive semiconductor is responsible both for producing the charge carriers and for conducting them; the result is that the chance of wasteful recombination processes is quite high unless the device structure is carefully engineered.

In Grätzel's photoelectrochemical cells, the tasks of charge-carrier generation and transport are assigned to different species. His devices consist of an array of nanometer-sized crystallites of the semiconductor titanium dioxide, welded together and coated with light-sensitive molecules that can transfer electrons to the semiconductor particles when they absorb photons. The light-sensitive molecule is called the sensitizer, and it plays a role equivalent to chlorophyll in photosynthesis. Grätzel's sensitizer is a ruthenium ion bound to organic bipyridine molecules, which are themselves bound to the titanium dioxide surface (fig. 6.18). These ruthenium complexes absorb light strongly in the visible range, and when a photoexcited complex transfers an electron to the titanium dioxide nanocrystals, the ruthenium ion's positive charge increases from two to three.

The titanium dioxide plays the role of the electron relay mechanism in the chloroplast membrane, carrying the electrons that it receives rapidly away from the electron donor. To sustain this light-induced charge transport, the sensitizer molecules must get back an electron so that they can absorb another photon. So the particle/sensitizer assembly is immersed in a liquid electrolyte containing molecular species that can pick up an electron from an electrode immersed in the solution and ferry it to the sensitizer. Dissolved iodine molecules (I_2) will do this job—they will pick up electrons to form iodide ions (I^-), and then donate these to the ruthenium ions in the surface-bound complex.

These wet cells can convert sunlight to electricity with an efficiency of up to 10 percent in direct sunlight, and they are even more efficient (around 15 percent) in the kind of diffuse daylight that prevails on cloudy days. So cells of this sort

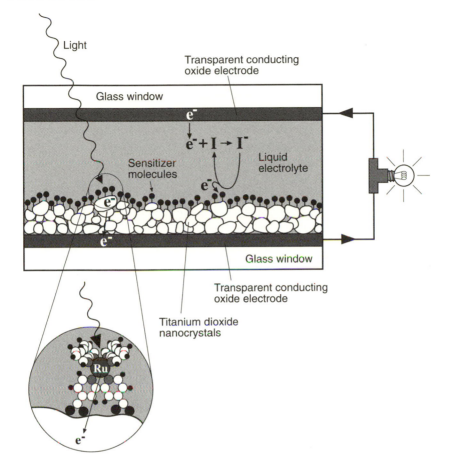

FIGURE 6.18 Michael Grätzel's nanocrystal-based photoelectrochemical cell mimics the photosynthetic process of plants. A light-sensitive complex containing a ruthenium (Ru) ion is chemically bound to the surface of semiconducting, nanometer-sized titanium dioxide crystals. Absorption of light induces the transfer of an electron from the ruthenium ion to the titanium dioxide particles, which bear it to the electrode on which they are deposited. Thus the molecular species responsible for generating the charge carriers and the material responsible for their transport are distinct entities, which reduces the chance of recombination. The ruthenium ion regains an electron from an iodide ion in the liquid electrolyte, which in turn picks up an electron from the other electrode. The iodide ions therefore act as an electron relay, like the vanadium ions in figure 6.17.

look particularly attractive for use in countries (such as mine) that do not enjoy a great deal of bright sunlight. Moreover, titanium dioxide is a very cheap material: it occurs as the natural mineral ilmenite, and is used in powdered form as the white pigment in paint. About 10 grams is sufficient to make a film of area one square meter in Grätzel's cells, at a cost of about one cent. Grätzel estimates that they could be produced at a commercial cost some five times lower than that of conventional silicon devices.

In chapter 1 we saw how silicon in microelectronics is coming increasingly under threat from photonic materials such as gallium arsenide. It should therefore be of little surprise that these materials are finding their feet in the solar-cell field too. Gallium arsenide and other III–V semiconductors are direct-band-gap materials, many of which can absorb light across the visible spectrum, and so these materials can be used in thin-film solar cells. These devices have the same kind of construction as silicon-based cells: a p–n junction sandwiched between a metal back electrode and strip-type metal top electrodes. Because their absorption is so high, and the probability of recombination processes is relatively low, these III–V cells can achieve very high efficiencies: a single-junction gallium arsenide cell has been made with an energy conversion efficiency of 28.7 percent, while a "tandem" device with two junctions stacked back to back (one with a gallium arsenide p–n junction and one with a gallium antimonide p–n junction), made by researchers at the Boeing Corporation, holds the world record for conversion efficiency at around 33 percent. The Applied Solar Energy Corporation began to sell aluminum gallium arsenide/gallium arsenide solar cells in the 1980s for powering instruments on space satellites, and the record-breaking Boeing devices are also now being tested for this application. While this shows the promise of III–V cells, the specific application is telling: the devices are still very costly, and so are limited so far to uses where high power density, rather than cost, is at a premium.

A Bright Future?

It might seem strange to be splitting hairs over the extra percent or so of energy-conversion efficiency that makes for a record-breaking solar cell, while a good four-fifths of the incident light energy continues to be squandered in all commercial devices. But solar cells do not have to make use of every last photon in order to provide a viable alternative source of energy (and that is just as well, because the limitations that prevent them from doing so are fundamental). Rather, they merely have to reach an operating threshold at which they can provide power as cheaply as traditional fossil-fuel sources. In these terms, the difference between 10 percent efficiency (which is typical of the best commercial modules) and 15 percent (which is a commonly quoted target figure) might tip the balance—it would mean that the price of each watt of power would be just two-thirds of its current price. The fact that the greater part of the Sun's energy would still go unused is neither here nor there—there is plenty of it (around 1,000 watts per square meter of the Earth's surface), and it is free. But the real challenge that solar-cell researchers face is not to improve efficiency but to lower manufacturing costs without sacrificing the performance of present-day devices. After all, halving the cost of a device lowers the price of the power it generates just as effectively as does doubling the efficiency.

Moreover, as Wim Sinke of the Netherlands Energy Research Foundation has pointed out, a fairer comparison between the costs of photovoltaic and fossil-fuel or nuclear power sources would take into account more than just the bills for constructing and running the devices. The latter energy sources have considerable societal costs—they are all in some sense polluting technologies, with considerable clean-up premiums, and their detrimental effects on the quality of our environment might be hard to quantify but are still harder to deny.

For small-scale power generation, where it is an advantage to have a self-contained unit that does not need continual refueling or recharging, solar cells are already proving their worth, and I have mentioned several examples of these. One of the most promising arenas for the future is in providing energy for developing countries: a self-contained solar power source for a small, remote village would be far preferable to, and potentially far cheaper than, a link-up with some national power grid. The World Bank is sponsoring initiatives to develop power sources of this sort. It is conceivable that developing countries might be able to leap a generation in energy production, bypassing the need for polluting fossil-fuel plants (for whose operation these countries might be largely dependent on other, richer nations) and moving straight to localized, clean solar power.

But can solar energy meet the large-scale requirements of an industrialized nation? In densely populated urban areas in Europe, the United States, and Japan, distributed power generation using photovoltaic devices arrayed on rooftops and building facades to supplement power from the national electricity grid is becoming an attractive option. Each module in such an array typically generates 1 to 100 kilowatts, and when linked together these distributed systems can now reliably produce enough power to be hooked up to the grid. In Heerhugowaard in the Netherlands, for instance, photovoltaic roof systems on some family houses operated by the PEN Utility Company deliver one and a half kilowatts under normal daylight illumination and are connected to the Dutch 220-volt grid.

Several intermediate and large-sized photovoltaic power plants have now been built in countries throughout the world, including the United States, Germany, Japan, and Italy. These deploy arrays of crystalline silicon solar panels over large areas to generate many kilowatts of power. Some incorporate banks of "concentrators"—lenses or reflectors—to focus sunlight from a large area onto solar cells. In this way, a lot of sunlight can be harvested without the expense of covering the whole of the collecting area with solar panels, so more efficient (and thus more expensive) cells can be used. The disadvantage is that the focusing of sunlight is possible only when it is direct, not when it comes in the form of the diffuse light of a cloudy day (in the same way, you can't use a microscope to burn a hole in paper on an overcast day). Other solar power plants use rotating panels that will track the path of the Sun to optimize light utilization.

These installations engender optimism about the future of solar technology. But there is some distance still to go in improving efficiencies and reducing costs before we can stop living off the Earth and live off the Sun instead. As the world goes on getting warmer with the incinerated breath of coal and oil, this is a race we should all be watching.

Tunnel Vision

POROUS MATERIALS

> It could be a jaw-bone
>
> or a rib or a portion cut
>
> from something sturdier:
>
> anyhow, a small outline
>
> was incised, a cage
>
> or trellis to conjure in.

—Seamus Heaney, "Viking Dublin, Trial Pieces"

Materials with finely sculpted interiors are now being built by molecular architects. Far too small for the eye to see, the channels and cavities of these materials have shapes and sizes that can be precisely defined and controlled. These porous materials serve as molecular-scale filters, industrial catalysts and miniature laboratories for conducting chemistry just a few molecules at a time.

SOME MATERIALS are like music: the gaps are as important as the filled-in bits. The presence of holes in a material can convey upon it all sorts of useful bulk properties that the equivalent dense material would not possess. Take bone, for example: its open, porous structure provides spaces for capillaries and soft tissues to pass through, and ensures that the material, while rigid and strong, does not overburden us with its weight. Making holes in a material generally reduces its density at a much faster rate than it reduces its strength. And the gaps need not be regarded simply as voids—rather, a porous material can be seen as a kind of composite of solid and air, in which the air space plays an active role in determining the properties. We are all familiar with the thermal insulating properties of open, air-filled structures, from string vests to fiberglass roof insulation. An air-filled porous material can also provide acoustic insulation, absorbing sound where a dense material would merely transmit it. And materials laced with tiny channels can also act as *light* insulators, preventing light and other electromagnetic radiation to pass.

For reasons like these, there has always been an interest in making porous materials—foam rubber, for instance, and expanded polystyrene and cement. But in some of today's advanced porous materials, the pores take on a new significance. They are a structured internal world, a sculpted microenvironment in which one can conduct scientific experiments on the scale of nanometers. Nature provides us with natural minerals laced with pores whose width is no greater than the size of small molecules. These channels, all of which are identical, can be used as "molecular test tubes" for carrying out chemical processes only a few atoms at a time. And materials scientists are now becoming adept at making synthetic materials with pores this small, whose size and shape they can design at will. Manipulating matter in such small quantities is one of the hallmarks of the burgeoning science of *nanotechnology*, whose practitioners are attempting to conduct engineering at the scale of individual molecules.

MINERALS WITH HOLES

I don't know that anyone has yet succeeded in getting blood from a stone, but getting water is an easy trick—if you can lay your hands on the minerals known as zeolites. The name itself derives from the Greek for "boiling stone," and the Swedish mineralogist Baron Axel Cronstedt is credited with first observing that, when heated, these minerals are left bubbling and steaming. The water comes from within a labyrinth of channels and cavities just a few angstroms wide—about the width of a small molecule like benzene—which run throughout the crystals. Pores this small are called *micropores*. Unlike the much larger pores in pumice and sandstone, the micropores of zeolites are not simply the result of flaws, bubbles, or gaps between mineral grains but are instead intrinsic elements of the crystal structure. As such, these channels do not form a random network but are ordered in a regular, periodic way, like the elevator shafts and hallways of an office building.

Natural zeolites are aluminosilicates, meaning that they contain the elements aluminum, silicon, and oxygen interlinked into an extended network. The basic structural elements of this network are tetrahedral SiO_4 and AlO_4 units, in which the four oxygen atoms at the corners of a tetrahedron surround the central silicon or aluminum atom. These units are linked via the oxygen atoms at their corners into rings, which define the necks of pores and the faces of cavities into which the pores feed (fig. 7.1). The pores themselves typically have diameters of 4 to 7 angstroms, whereas the cavities (called supercages) they open onto may be 10 to 13 angstroms across. The aluminosilicate framework carries a negative charge, and this is balanced by positive metal ions, commonly sodium, which sit within the cavities.

As a result of this highly porous structure, zeolites have an internal world of seemingly miraculous dimensions. They show that a phenomenal amount of surface can be packed into a small volume. Indeed, they are virtually *all* surface. A crystal of a typical natural zeolite weighing just a gram can have an internal surface area of about 900 square meters—the size of two basketball pitches. It is

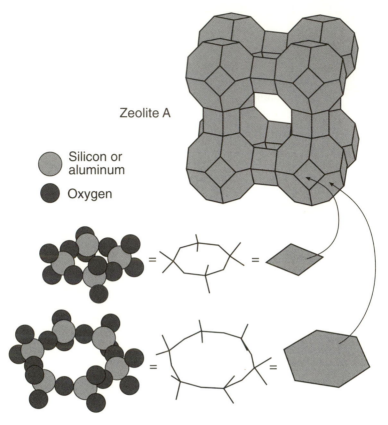

Zeolite A

Silicon or aluminum

Oxygen

FIGURE 7.1 Natural zeolites are aluminosilicate minerals whose crystal structure is composed of tetrahedral units, in which either silicon or aluminum atoms are surrounded by four oxygens. These units are linked together into rings that form an open network of cavities connected by pores. The width of the pores, typically 10–13 angstroms, is about the same as the size of small organic molecules like benzene. Zeolites can therefore act as "molecular sieves."

within this expansive microworld that natural zeolites pack the water to which they owe their name.

Because the metal ions in zeolites sit rather loosely in the porous framework, they can be exchanged relatively easily for others. The discovery of this capacity for ion exchange in the mid-nineteenth century led to the first use of natural zeolites—as water softeners. When water containing calcium or magnesium ions (which make water "hard") is passed through a sodium zeolite, the sodium ions in the crystal are replaced by the other metals. Ion exchange in zeolites is now used to remove radioactive heavy metals from nuclear waste.

During the 1950s, researchers in the petrochemical industry realized that zeolites could facilitate some very tricky chemical transformations of the hydrocarbons that make up crude oil. When hydrogen atoms from water become attached to the aluminosilicate framework, this becomes an extremely powerful acid, several million times more acidic than concentrated sulfuric acid. As a result, zeolites are supremely adept at attaching positive hydrogen ions (protons) to molecules held within the pores and cavities, including hydrocarbons—which normal acids leave untouched. Once protonated, hydrocarbons become highly reactive and have a tendency to rearrange their constituent atoms into new configurations—for example, by shedding small hydrocarbon fragments or shifting a side-branch appended to the main hydrocarbon chain. Ultimately the zeolite framework regains the proton, restoring its acidity, when the transformation of the organic molecule is concluded. So zeolites are able to act as catalysts for transforming hydrocarbons, assisting the reaction but emerging unchanged.

While some liquid "superacids," such as hydrofluoric acid dissolved in sulfuric acid, will also bring about these reactions, the attraction of zeolites is that they are not corrosive and environmentally hazardous chemicals but safe solids that can be held in the hand. But one of the critical reasons why zeolites make useful industrial catalysts is that the nature of the reactions they catalyze is determined by the size and shape of the pores, and so these reactions are highly selective rather than indiscriminate. For example, hydrocarbons that are too big to fit within the pores will not be transformed, so that a zeolite will selectively extract and rearrange only certain compounds from the complex hydrocarbon mixture of crude oil. The size of the cavities within which the rearrangements take place also determines the nature of the products—any that are too big to fit inside the cavities will not be formed. So these microporous catalysts acquire a catalytic specificity that is somewhat akin to that of nature's catalysts, enzymes.

Because of these properties, zeolites have become some of the most important catalysts in the chemicals industry. They are used, for example, to "crack" large hydrocarbons, breaking them down into smaller molecules such as ethylene and benzene, which are of more use as raw starting materials for chemical synthesis. They can also improve the fuel performance of gasoline by removing some of the straight-chain hydrocarbons while leaving alone the bulkier, branched ones—this boosts the fuel's octane rating. Not all of the reactions that zeolites catalyze involve hydrocarbons alone—they are also used in the production of synthetic gasoline from methanol (made from natural gas) or from "synthesis gas," which is itself prepared from natural gas or coal and water. In New Zealand, the zeolite-catalyzed generation of gasoline supplies about one-third of the nation's transportation fuel.

Because of their ability to absorb molecules selectively into the internal pore space, zeolites have been accorded the label "molecular sieves." Because the selectivity of both this absorption process and the subsequent reactions that absorbed molecules might undergo is determined by the size and shape of the pore mouths and the internal cavities, a lot of effort has gone toward synthesizing artificial zeolites with pore geometries different from those in the natural minerals: this would expand the repertoire of reactions that these materials can catalyze. There are forty or so known natural zeolites, but now these have been augmented by around one hundred artificial varieties (fig. 7.2).

FIGURE 7.2 Many non-natural zeolites have now been synthesized by chemical means, providing a varied gallery of pore sizes and shapes. Both natural and synthetic zeolites are used extensively as catalysts in the petrochemicals industry. Shown here is a single crystal of the synthetic zeolite Dodecasil 3C, over half a millimeter across. These "giant" zeolite crystals were made by Geoffrey Ozin's group in Toronto, who used a new synthetic method to create much larger crystals than those commonly generated by traditional approaches. They formed the material by linking together aluminate and silicate ions in an organic solvent, rather than in the conventional solvent of water. Large zeolite crystals like this are needed for some of the more sophisticated applications envisioned for zeolites, such as in optoelectronics. (Photograph courtesy of Geoffrey Ozin, University of Toronto.)

The first synthetic zeolites were made in the 1930s, largely through the research of Richard Barrer at Imperial College in London. Barrer made his artificial zeolites from dissolved silica and aluminum hydroxide under high temperatures and high pressures, mimicking the geological conditions under which natural zeolites form. In the late 1940s, researchers at Union Carbide found a route to synthetic zeolites that required less extreme conditions, and which gave materials with larger pores to boot. It was these materials that led to the first catalytic applications of zeolites in the 1950s. But the field of zeolite synthesis really took off at the beginning of the 1960s when a team at the Mobil Oil Corporation, led by George Kerr, discovered a way of controlling the pore size—something that previously had relied largely on trial and error.

Kerr's innovation was to include organic tetraalkylammonium ions in the reaction mixture. These became incorporated into the microporous crystal lattice in place of the metal ions. Tetraalkylammonium ions contain a central, positively

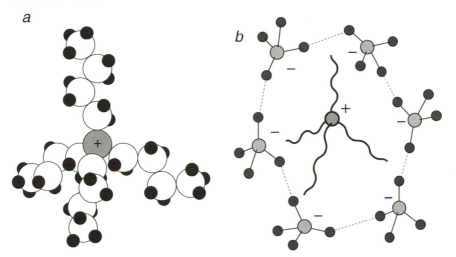

FIGURE 7.3 Tetraalkylammonium ions (*a*), in which a positively charged nitrogen atom sports four hydrocarbon appendages, can act as templates around which silicate and aluminate ions link together to form the walls of a zeolite's cavity (*b*). George Kerr and colleagues at the Mobil Oil Corporation found in the 1960s that this templating approach provided some control over the size and shape of a synthetic zeolite's pore network. They used tetraalkylammonium ions in which the hydrocarbon chains contained between three and seven carbon atoms.

charged nitrogen atom to which are attached four alkyl (hydrocarbon) groups (fig. 7.3*a*). They are very much larger than ions of sodium or calcium, and Kerr believed that this difference in size might generate different types of aluminosilicate framework. Zeolite chemists now believe that the organic ions act as a kind of template for the formation of the zeolite's pores: crudely speaking, one can imagine that the negatively charged silicate (SiO_4^{2-}) and aluminate (AlO_4^-) ions in solution come together around the positive alkylammonium templates and link into shells whose size is determined by that of the templating agent (fig. 7.3*b*). The pendant alkyl chains, meanwhile, might dictate the disposition of the pore openings. But the details of the process must be somewhat more complex than this, because the resulting pores are not simply arranged at random but form an ordered microporous framework. This ordering must involve some cooperative interactions between the different templated clusters, but it is still poorly understood.

The Mobil approach yielded a synthetic molecular sieve that is now one of the most important zeolitic catalysts, called zsm-5. This material, which is templated by alkylammonium ions with three carbon atoms in each chain, contains a series of parallel pores—the "elevator shafts"—interconnected by short channels, the "hallways" (fig. 7.4). The pore mouths contain rings of ten aluminum or silicon atoms linked via oxygens, giving the pores a diameter that is intermediate between those of two common natural zeolites, called (prosaically) zeolite A and zeolite Y. This size turns out to be just right for the selective catalytic conversion

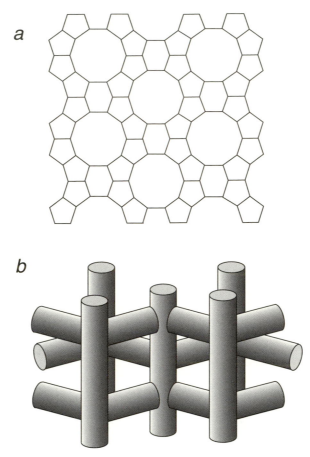

FIGURE 7.4 The Mobil templating approach afforded a new zeolite called ZSM-5, whose pore openings consist of rings of ten silicon or aluminum atoms (*a*). The pore network (*b*) is not like any found in natural zeolites. ZSM-5 is now an extremely important industrial catalyst, used for making synthetic gasoline from methanol.

of methanol to gasoline, and also of toluene and methanol to *para*-xylene, the molecular precursor to nylon.

The gallery of synthetic molecular sieves now contains some exotic figures, some of which have frameworks made up of atoms other than aluminum, silicon, and oxygen. In the early 1980s researchers at Union Carbide's laboratories in Tarrytown, New York, introduced a family of aluminophosphate molecular sieves (ALPOS), in which PO_4 tetrahedra play much the same structural role as SiO_4 tetrahedra. In 1988 researchers at the Virginia Polytechnic Institute made an aluminophosphate molecular sieve that they called VPI-5, whose unusually large pore mouths are made up of rings of eighteen aluminum or phosphorus atoms. And in 1991 French and Swiss researchers created a gallophosphate molecular

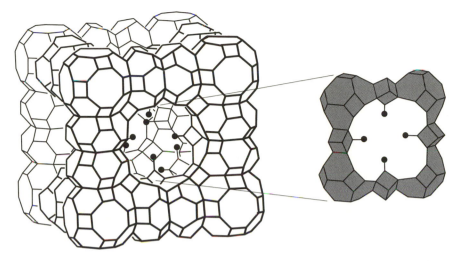

FIGURE 7.5 The synthetic gallophosphate molecular sieve cloverite, made by Swiss researchers in 1990, has pore openings 13.2 angstroms across whose shape (expanded in gray) is reminiscent of a four-leafed clover. Phosphorus and gallium atoms sit at the vertices of the framework, and the black circles are hydroxyl groups.

sieve, with gallium in place of aluminum, that they christened cloverite because its spacious pore openings had the shape of a four-leaved clover (fig. 7.5). This variety of shapes allows chemists ever more choice of holes in which to fit molecular pegs. But sadly, phosphate-based materials do not have the same practical potential as aluminosilicates because they are not very stable against heating (something that most catalytic reactions require). It has proved much harder to make aluminosilicate zeolites with large pores than to do the same for phosphate-based materials.

If we can do away with aluminum and silicon, why not oxygen? During the 1990s this question has been pursued with increasing vigor. An obvious candidate to take oxygen's place in microporous frameworks is sulfur, which sits below it in the periodic table. The idea is all the more tempting because sulfur forms semiconducting materials with many metals—the vision of microporous semiconductors raises all sorts of prospects, from high-surface-area electrodes for electrochemical cells to molecular-scale composite materials for microelectronic and sensor technologies. Although dramatic applications of this sort have resolutely refused to materialize, a very wide range of microporous sulfides is now known. These tend to look very different from zeolites—tetrahedral units with four sulfur atoms around a central metal atom do crop up, but so do all manner of other building blocks, including octahedra (six sulfur atoms around a metal), pyramidal units, and prism-shaped units. These are linked into a diverse selection of rings and channels. Sulfur's lower neighbor, selenium, provides another menagerie of curious microporous materials with a similar diversity of building blocks. They are made in every which way—in as mild a manner as precipitation from solution at room temperature, or as vigorous as melting metals and sulfur at 800 degrees Celsius.

And still more exotic inorganic microporous solids continue to appear, welded together by cyanide ions, by nitrogen atoms, by arsenic. The outsider might reasonably wonder what is the purpose of all this chemical tinkering. And in truth, that is precisely what it is—you can be sure that, whenever a potentially useful new material is discovered, journal pages will soon be filled with reports of minor variants in which one element or another has been replaced, just to see what happens. But that is the way much of materials science works—it is rare that a new material is discovered that cannot be improved upon by modifying the composition. Sometimes researchers can be guided by logic, or by intuition, in choosing those modifications, but equally there are times when they just have to try things out and trust to chance. Without such tinkering, it is doubtful that we would have seen the new high-temperature superconductors exhibit ever higher superconducting transition temperatures, leaping over one hundred degrees in under a decade. It is hard to say how many of the new microporous materials now known will ever prove useful—but one hit can be worth a hundred misses.

Between the Sheets

Gardeners who suffer from waterlogged clay soils are not accustomed to thinking of clay as porous. And indeed, relative to the fine, granular texture of a good loamy soil, clay is a dense and troublesome medium. But in reality clay is more like a sponge than you might imagine: the clay called montmorillonite, for example, will swell to many times its dry volume by absorbing water, and will absorb vast quantities of gases. Compared with dense, monolithic rocks like basalt, clays are indeed a sparse medium, containing a volume of empty pore space that can rival that of zeolites. For this reason they too have attracted much interest as potential catalysts and materials for separation processes, and today several important industrial reactions, such as the conversion of ethylene to ethanol, are carried out in clay catalysts.

Like most zeolites, clays are aluminosilicate materials. But the nature of their pores is rather different: whereas in zeolites the tetrahedral aluminate and silicate units are linked together in three-dimensional networks, in clays they form flat sheets that are stacked on top of one another. These sheets bear negative charges, which are counterbalanced by the positive charges of metal ions such as sodium and calcium sitting between the sheets (fig. 7.6). The pores are therefore slitlike: in one direction (between adjacent sheets) they have dimensions similar to those of small molecules, but in the other two directions (parallel to the sheets) they are more or less infinite. This means that long, thin molecules can wriggle between the sheets, but fat, stout molecules may be too big.

In the 1970s John Thomas and colleagues at the University of Aberystwyth in Wales found that the metal ions between aluminosilicate layers in some clays can be replaced by protonated water ions, molecules of H_2O with a proton (hydrogen ion) attached (H_3O^+). This converts the clays into solid-acid catalysts rather like

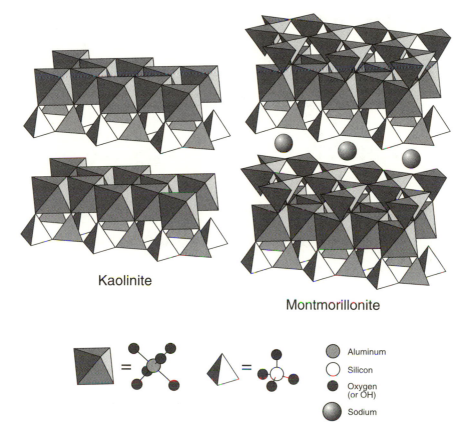

FIGURE 7.6 Clays are minerals with a sheetlike structure. Generally the sheets are composed of two kinds of layer—one in which tetrahedral SiO_4 units are linked together, and the other containing either aluminum or magnesium atoms surrounded by six oxygen atoms or hydroxyl groups in an octahedral arrangement. In most clays, each sheet is comprised of either a tetrahedral silicate layer back to back with a layer of aluminum- or magnesium-containing octahedra, or one of the latter sandwiched between two silicate layers. An example of the former is kaolinite (the main constituent of china clay), and of the latter is montmorillonite (the main constituent of Fuller's earth). Montmorillonite is an example of a so-called smectic clay, which can absorb large amounts of water—the water is incorporated between the sheets, forcing them apart and causing the material to swell.

zeolites, making them able to effect chemical transformations of hydrocarbons and other organic molecules. One particularly important reaction catalyzed by Thomas's acidic clay was the formation of methyl tertiary butyl ether from methanol and a hydrocarbon called 2-methylpropene: this product boosts the octane rating of lead-free gasoline, and also reduces the formation of environmentally malign gases as gasoline burns in an idling engine.

The diversity of clay catalysts has been greatly enhanced by the development of so-called pillaring reactions. These involve the introduction of small molecular

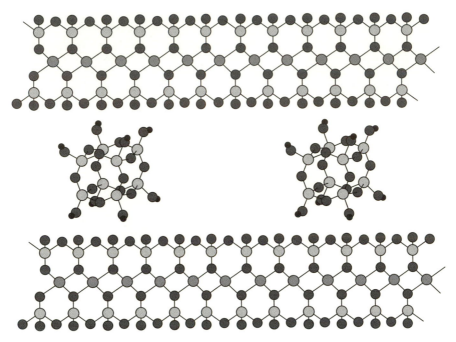

FIGURE 7.7 In pillared clays, the sheets are held apart by large atomic clusters inserted between them. Commonly these are so-called polynuclear cations, which contain several metal ions and are positively charged, such as "Keggin" ions like $[Al_{13}O_4(OH)_{24}(H_2O)_{12}]^{7+}$. Here I show pillaring by a cube-shaped, silicate-based cluster, $Si_8O_{12}(OH)_8$. Pillared clays contain open "galleries" between the aluminosilicate layers, and are highly porous. Some serve as catalysts for industrial organic reactions.

units between the sheets to prop them apart, thereby expanding the slitlike pores. Clays immersed in water will swell as the water molecules slip in between the sheets and push them apart; while in this state, the pillars can be introduced, so that when the water is removed the pillars stay in place and, like miniature Samsons, hold the ceiling from the floor (fig. 7.7). Richard Barrer showed in the 1970s that large, negatively charged many-atom ions called Keggin ions can be used as the pillars in a variety of clay materials. Because these ions act as acids themselves, they make the pillared clay a solid-acid catalyst, of a kind that is particularly useful for the catalytic production of important classes of organic compounds like esters and ethers.

SCALING UP THE SCAFFOLD

While synthetic molecular sieves and modified clays have greatly expanded the potential of microporous materials for selective molecular filtration and catalysis, materials scientists and chemical engineers would dearly love to be able to generate porous solids with any arbitrary pore size and shape—in other words, to produce molecular sieves to order with any kind of mesh. The problem, however, is that, while nature does not necessarily abhor a vacuum, she is not keen on holes. There is an energy cost to making a surface, so that the wider a pore becomes (and thus the greater its surface area), the greater the driving force for it to collapse into a more compact state. This has meant that the useful selectivity that zeolites show toward small molecules cannot easily be extended to larger molecules for lack of ways to create materials with regular networks of larger pores.

In 1992 that situation changed in a way that has sent repercussions throughout materials science. For while, on the one hand, the discovery of a means for synthesizing large-pore molecular sieves had the agreeable but scarcely revolutionary consequence of expanding the range of these materials, on the other it demonstrated the huge potential of what amounts to an emerging new philosophy in materials processing: exploiting the interplay between organic and inorganic substances.

In the early 1990s, a team from the Mobil Corporation's research laboratories in Paulsboro and Princeton, New Jersey, led by Charles Kresge and Jeffrey Beck, was exploring the use of surfactant molecules for the templated synthesis of molecular sieves—the approach developed by Kerr and colleagues in the 1960s. This method generally provides pores no larger than about 12 angstroms across. Kresge's team was using tetraalkylammonium ions as the surfactants, just like those that generate microporous materials such as ZSM-5, but with longer-chain alkyl groups, containing a dozen to twenty or so carbons rather than the three-carbon chains of the ZSM-5 template. Furthermore, they used rather higher concentrations of these surfactants. The Mobil researchers anticipated that these differences might generate stronger interactions between the organic molecules, perhaps enabling them to orchestrate new framework structures. But they found that their approach exceeded all expectations.

The reaction mixture yielded a rather disordered silica-rich material, much of which looked like a sheer mess. But Kresge asked a colleague to look at this unpromising stuff carefully with an electron microscope, just to check whether there was anything worthwhile in it. In a few isolated regions, the microscope revealed an ordered array of hexagonal pores up to one hundred angstroms across, about ten times as big as the pores of zeolites (fig. 7.8). The surfactant molecules just weren't big enough to imprint pores of this size, and at first the researchers thought they were seeing the structure of the grid used to support the sample in the microscope. Yet more persistent investigation confirmed that the porous material was indeed silica, shaped into an ordered array of nanometer-scale pores (these are called *mesopores*, in distinction from the micropores of zeolites). There was no precedent for this kind of material (which the researchers unmemorably christened MCM-41), and Kresge took some persuading before he would accept that this was not an instance of what the American Irving Langmuir in the 1920s had derisively called "pathological science"—a scientific unicorn.

Somehow the surfactant molecules were generating voids much larger than the size of the molecules themselves. Kresge and colleagues knew that surfactants have a tendency to cluster together in water to generate aggregates like micelles (see page 175), in which the hydrophobic tails are hidden from the water

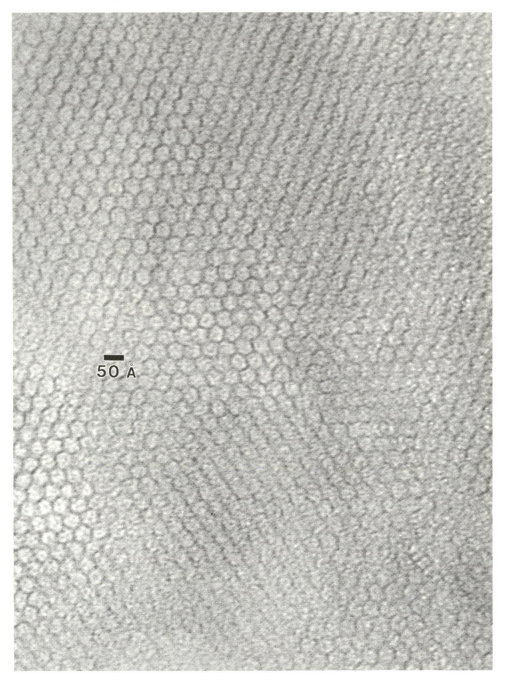

FIGURE 7.8 The silicate material MCM-41, made by Mobil researchers in 1992, contains an ordered network of pores fully ten times larger than those of conventional zeolites. Pores of this size are known as mesopores. It is believed that the surfactant molecules used in the synthesis of MCM-41 form large aggregates that pack together in a regular way, giving rise to an organic

molecules. The concentration of surfactant in their reaction mixture was rather greater than that at which micelles first form, and they recognized that under these conditions micelles can coalesce into cylindrical aggregates. These cylindrical micelles can line up side by side, like stacked logs, to form a hexagonal array (fig. 7.9). It was tempting to suppose that the surfactant molecules in their mixtures were clustering into this hexagonal phase, and that the positively charged surfaces of the cylinders were then acquiring a shell of silicate ions. As the inorganic ions join together into solid silica, the material would be left with long hexagonal channels formed around the templates of the organic aggregates.

This templating mechanism gained credibility from the Mobil group's observation that a similar procedure could generate a layered silica material, rather like a synthetic clay. They hypothesized that in this case the surfactants were forming bilayers, which stacked up into a lamellar phase (page 174). These bilayers would then act as templates for the formation of silica layers.

The approach was quickly taken up by other laboratories keen to explore the range of pore sizes and shapes that could be generated in this way. At the University of California in Santa Barbara, a group led by Galen Stucky, who had a wealth of experience in making new microporous solids, found evidence suggesting that the templating process might not be as simple as had been supposed by the Mobil team. For one thing, it was possible to create mesoporous materials using concentrations of surfactants that are much too low to form the ordered hexagonal phase. Stucky and colleagues were able to make the materials using a surfactant concentration of just one percent, under which conditions the surfactant molecules would simply cluster into isolated, spherical micelles. They suggested that the silicate anions and the surfactant molecules were somehow acting in a cooperative manner, leading to different behavior from that which each of these components would show on its own. It now seems likely that the detailed mechanism by which these mesoporous materials form depends on the exact composition of and conditions in the reaction mixture.

The most appealing aspect of these discoveries is their use of both organic and inorganic materials to create highly ornate patterns at the microscopic scale. We saw in chapter 4 that this is precisely how nature generates complex structures in inorganic materials such as the shells of marine organisms. Thus, from innocent and apparently unrelated beginnings, the search for new porous materials has led us to identify a way of mimicking nature. I described in chapter 4 some of the astonishing new directions that this idea of organic templating is now taking. But the Mobil team is of a more practical persuasion—its research program was, after all, launched to find new materials that would actually be useful for the petrochemicals industry. How successful was it?

One reason to welcome mesoporous materials is that they should act as sieves for separating molecules much larger than those that zeolites will sift. But we can also exploit this new range of size selectivity to carry out new kinds of selective

---

template that directs the linking-up of the silicate ions. The detailed mechanism of this process is not yet clear. (Photograph courtesy of Charles Kresge, Mobil Central Research Laboratories, Princeton, New Jersey.)

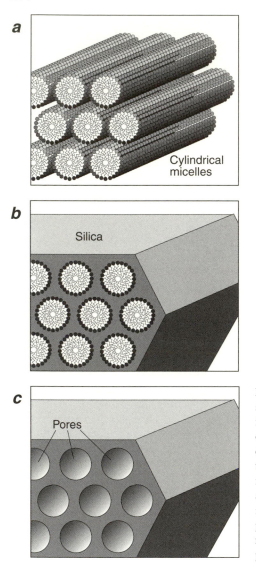

FIGURE 7.9 The creators of MCM-41 suggested that its hexagonal pore network is formed as a result of the self-assembly of the surfactant molecules into cylindrical micelles, which stack together like logs (*a*). The inorganic material then precipitates in the spaces around the cylinders, thereby becoming imprinted with a hexagonal honeycomb pattern (*b*). When the surfactants are removed by heating, large cylindrical pores are left behind (*c*).

catalysis. Silica itself is not a solid-acid catalyst—aluminum atoms are needed in the framework to give them zeolite-like acidity. Thomas Pinnavaia at the University of Michigan has taken a different route to making mesoporous catalysts, however: he incorporated titanium atoms into the framework in place of some of the silicon atoms. A *micro*porous version of such a material, called titanium silicalite I or TS-1, was developed in the 1980s by researchers at the Enichem company in the United States, and it proved to be a useful catalyst for carrying out shape-selective oxidation of unsaturated hydrocarbons to epoxides; the latter are precursors to alcohols, so this reaction converts components of crude oil into useful organic compounds. The titanium atoms in the framework of TS-1 play a crucial part in this process. Pinnavaia used a variation of the Mobil approach, with a titanium alkoxide compound in the reaction mixture, to prepare a mesoporous titanium silicate that could catalyze epoxide and alcohol formation from much larger hydrocarbons than those that TS-1 would transform, such as cyclohexene.

John Thomas, now at the Royal Institution in London, has shown that mesoporous silica can be turned into a selective catalyst in another way: by leaving the silica framework intact but attaching to it a molecule that is itself a catalyst. The catalytic molecules sit in the channels of the material and benefit from their geometric shape selectivity, so that the composite material is much more choosy than the individual catalyst molecules would be on their own. Thomas and colleagues grafted so-called titanocene complexes to the walls of MCM-41 via chemical bonds. These complexes contain a titanium ion attached to a five-atom carbon ring (fig. 7.10). Once affixed to the silicate pore walls, the complexes were heated

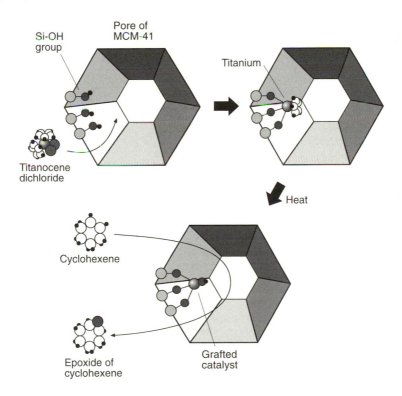

FIGURE 7.10 MCM-41 can be turned into a useful catalyst by anchoring catalytic groups to the silica framework. A titanocene complex attached in this way will catalyze the epoxidation of unsaturated hydrocarbons (alkenes), a reaction that provides the first step in the conversion of these components of crude oil to alcohols and ketones. By grafting the catalytic group onto the porous framework, one can exploit the molecular sieving action of the pores to create a catalyst that converts molecules selectively according to their size and shape.

in oxygen, causing them to shed their carbon ring and replace it with a hydroxyl group (-OH). The grafted titanium ions are then in much the same form as those in the network of TS-1 and are able to perform the same kind of selective catalysis as Pinnavaia's mesoporous titanium silicate.

The Nanoscale Laboratory

One of the fundamental tenets of the sciences that deal with the behavior of matter is that this behavior does not depend on the amount of stuff that you are playing with. A kilogram of sugar has the same physical and chemical properties as a milligram. The conductivity of copper remains the same whether it is fashioned into a wire half a millimeter across or a cable as thick as your wrist. But it has been long known that this is not a precise principle; in particular, when matter is confined to very small dimensions (by which I mean something of the order of less than a micrometer), we might start to see deviations from the behavior of a lump that you can see and hold. Since the beginning of the century, scientists have known that the boiling point of a liquid alters when it is held within the tiny channels of microporous materials, and that the change in boiling point gets larger as the pores get smaller. The effects on the properties of a substance of limiting its extent to a microscopic scale in one or more dimensions are now central to many fields of research. We saw in chapter 1, for instance, that silicon becomes luminescent when etched into pillars just a few nanometers in width, and many new microelectronic devices exploit the consequences of confining electrons to conducting layers of a comparable width.

In these latter cases the effect of confinement can be explained by quantum mechanics, which tells us that the energy states of a system are dependent on its size (although this dependence usually becomes noticeable only at the nanometer scale). Other effects of confining matter, such as the shift in boiling point of liquids in pores, are "classical," being in this case a result of the fact that the edges of the confined substance can be "felt" at more or less any point throughout it. The consequences of confinement are fascinating from a fundamental point of view, in part because they confound our everyday experience, and they are also of potential practical value.

In the past two decades, researchers have begun to recognize the tremendous value of zeolites and other porous materials for research on ultrasmall systems. Not only do they have pores small enough to reveal the influence of confinement on substances held within them, but those pores also have very well defined sizes and shapes, so that we can investigate confinement effects in a systematic and controlled way rather than having to unscramble the influences of a range of different confining geometries. Materials scientists are now accustomed to regarding molecular-sieve materials as nanometer-scale laboratories in which they can study matter in handfuls of just a few atoms or molecules.

In chapter 4 I explained how micelles are used as miniature casts for making particles of semiconductors just a few nanometers across—the light-emitting properties of these particles might be useful for making new kinds of optoelectronic devices. Researchers have also used the cavities of zeolites to mold such particles. At the Du Pont company in Delaware, Norman Herron and colleagues have fashioned cadmium sulfide particles, containing just four atoms of cadmium and four of sulfur, within the cavities of zeolite Y. The particles remain trapped in the aluminosilicate network because they are too big to escape through the pore mouths. Thomas Bein at Purdue University in Indiana has attempted to create single-molecule wires from polymers that conduct electricity (see chapter 9) by synthesizing them within a zeolite's channels; the idea was that the molecular-scale channels would house isolated strands of the polymer. Bein worked with polyaniline, one of the best-studied conducting polymers. He first filled a zeolite's pores with the monomer (aniline) molecules, and then added a compound that caused the monomers to link up into a polymer. Bein's initial attempts at making wires did not succeed—the monomers linked up, but the polymer was not conducting. This is apparently because wires just one molecule thick (which is all the zeolite's channels could accommodate) are too narrow to allow a current to pass along them. By performing the same operation in the much wider pores of the Mobil group's mesoporous materials, however, Bein was able to make polymer wires thick enough to conduct.

The mesoporous solids have indeed expanded the scale and scope of these miniature laboratories. Geoffrey Ozin has used them as a mold for making silicon "nanowires" in the pores. Like those generated by etching away silicon into a highly porous form (chapter 1), Ozin's nanometer-scale silicon wires are luminescent in the visible part of the spectrum. But because they are encased in a porous support, rather than forming a free-standing spongy material, they are more robust.

As materials chemists become ever more adept at fashioning new materials with a variety of pore sizes and geometries, we can expect to see a convergence of "bucket" chemistry with the highly advanced technologies of microelectronics and photonics, whose delicately patterned materials have previously been obtained only by painstaking lithographic sculpture or atomic-vapor deposition under highly controlled conditions with expensive instruments. The moral could serve for this book as a whole: materials chemistry is invariably cleverer than we might think!

HOLES BY DESIGN

The Mobil approach to making mesoporous solids is very much that of the chemical engineer, involving a carefully controlled precipitation process. But very recently, chemists have started to explore a different approach to the synthesis of porous solids in which the channel size is dictated at the outset by the nature of the molecular building blocks from which the material is made. This approach has huge potential, because it affords much more control over the geometry of the pores. In effect, the products are "designer" porous media, for which the chemist can first sit down and decide in a rational way exactly what kind of pores she would like to construct. She might, for example, want to make pores that will admit benzene but not toluene; or that have catalytic metal ions located at particular points around their circumference; or that conduct electricity along their length. She will then go to the drawing board to design the appropriate molecular building blocks, and set about synthesizing them in the laboratory. Most often, these constituents will be organic—that is, carbon-based—because it is for crafting such molecules that the chemist's skills are most finely honed.

An example of this rational design of porous materials was described in 1993 by M. Reza Ghadiri and colleagues at the Scripps Research Institute in La Jolla, California. The building blocks of their pores are peptide molecules, which have the same constituent parts as proteins: amino acids, linked together via peptide bonds. Chains of amino acids can adopt a bewildering variety of conformations, which are determined by the specific sequence of amino acids and the interactions among them. This has led many researchers to regard synthetic peptides as potentially very versatile materials for creating new molecular architectures.

The architecture pursued by Ghadiri's group was that of a hoop. They reasoned that hooplike peptides might stack one atop the other to form tubular channels. The trick, however, was to make peptides that would adopt and retain this hoop shape. This might seem easy enough—can one not just link up the ends of a chainlike peptide? But the reality is not so simple, because the bonds between the amino acids are rather flexible, so a cyclic peptide is likely to be more like a rubber band than a wedding ring, liable to fold up on itself.

Most peptide molecules do have a tendency to crumple up—this, indeed, is an essential step in enabling protein molecules to achieve their biologically active form. Ghadiri's team worked with peptides containing chains of just eight or so amino acids, linked together in a ring (fig. 7.11a). As we saw in chapter 4, every amino acid (with the exception of the simplest, glycine) comes in left- and right-handed varieties. Ghadiri and colleagues reasoned that if they alternated left- and right-handed amino acids around the ring, it would take on a puckered shape that would be relatively resistant to collapse. They chose to use alanine as the right-handed amino acid, which would alternate with a left-handed one (alternately glutamine and glutamic acid) around the ring.

Each peptide bond between amino acid units contains an amide group—a carbon atom joined by a double bond to an oxygen (-CO), and linked to a nitrogen–hydrogen unit (-NH). These two groups provide the complementary halves of a hydrogen bond, in which the oxygen forms a weak bond with the hydrogen. The puckered structure of Ghadiri's peptide rings leaves the oxygens and hydrogens pointing upward and downward more or less perpendicular to the plane of the ring—exactly where one wants them in order to fix the rings on top of one another by hydrogen bonds.

Ghadiri and colleagues precipitated long, needlelike crystals from a solution of their cyclic peptides. Under the electron microscope these crystals showed striations along their long axis which could be the signature of stacked tubes lying side by side. Painstaking attempts to make larger crystals finally produced ones big enough to use for analyzing the structure by X-ray diffraction (see chapter 10),

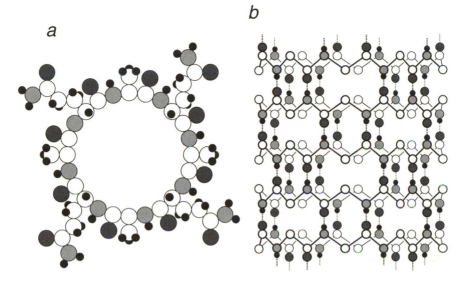

FIGURE 7.11 The cyclic peptide molecules synthesized by Reza Ghadiri and colleagues (*a*) will stack up into cylindrical channels of hoops, held together by hydrogen bonds between adjacent rings (*b*). Crystals of these peptide tubes can be thought of as organic analogs of zeolites, with regular pores of molecular dimensions.

which shows up the precise positions of the atoms. This confirmed that indeed the rings were stacked into tubular channels (fig. 7.11*b* and plate 7).

This crystalline assembly can be thought of as a kind of organic analog of a zeolite, with well-defined, molecular-sized pores. Ghadiri's team has now produced channels of different diameters using larger peptide-ring building blocks. As peptides comprise the fabric of nature's own catalysts, the prospects for performing catalysis in these organic systems look very encouraging. (For all that one sometimes hears such porous molecular solids described as "organic zeolites," however, it is as well to remember that the true inorganic zeolites used as catalysts by the chemical industry commonly do their job immersed in boiling oil, under which conditions these organic systems wouldn't last a minute. We will have to find more genteel catalytic tasks for them.)

A Sticky End

Hydrogen bonds—the glue that binds the peptide tubes—are ideal for sticking together molecular building blocks, because they make them sticky and able to self-assemble without leaving them so reactive that they are hard to control. Jim Wuest at the University of Montreal has used hydrogen bonds to link up small molecules to form a rather complex three-dimensional porous network (fig. 7.12). Here there are no preformed pore-forming units; instead, the pores are generated as several of the building blocks come together. But nevertheless the geometries

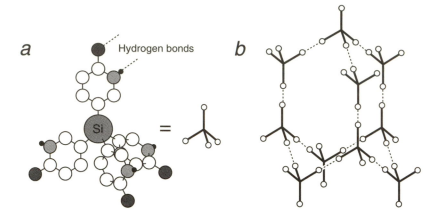

FIGURE 7.12 Tetrahedral molecular building blocks ("tectons") made by Jim Wuest and colleagues (*a*) will crystallize into an open, porous network (*b*) glued together by hydrogen bonds. The pores of the crystal are filled with solvent molecules.

of the pores can be predicted from the shape of the building blocks: Wuest's tetrahedral blocks (he calls them *tectons,* from the Greek word for "to build") have sticky extremities with hydrogen-bonding capability, which enable them to assemble into the same kind of tetrahedral network as that formed by carbon atoms in diamond (fig. 8.1), with a tecton instead of a carbon atom at each vertex. Rather than straight channels, this solid has a labyrinthine network of interconnected voids.

This is precisely the structure that Wuest had anticipated when he chose to use tetrahedral building blocks. But nature countered his ingenuity with a riposte: the crystal formed from these tectons turned out to contain two independent tetrahedrally connected networks, one weaving in and out of the other (plate 8). This finding illustrates a common problem faced by those who wish to take this building-block approach to making porous materials: as I indicated earlier, nature does not like leaving gaps within a crystal structure, but prefers instead to pack the molecular constituents as closely as possible. The tetrahedral shape of Wuest's tectons frustrates this desire on one level, because it constrains the molecular units to stick together in an array that inevitably leaves big voids. Nature responds by filling up some of this space with a second interconnected network, intricately woven into the first.

But if this is the case, you might wonder how is it that natural microporous materials, such as zeolites and clays, form at all. After all, the restrictions on packing of the (ionic) components in these materials are far less severe than those in the building-block materials just described; and indeed, nonporous silicate and aluminosilicate minerals, such as pyroxenes and perovskites, are very common in nature. The fact is, however, that zeolites are not actually as full of empty space as you might think, despite the tremendous surface areas of their pores. The most revealing way of classifying the openness of a porous material is in terms of its packing density: the ratio of the volume of space filled by atoms to the volume of empty space. A material that contains no empty space at all has a packing density of one, and a "material" that is just empty space has a packing density of zero. Clearly, all the possible options must fall somewhere in between.

It turns out that just about all materials, whether considered porous or not, occupy a rather small range in packing density—from about 0.75 to about 0.5. Since all materials are essentially assemblies of roughly spherical atoms, the theoretical upper limit to the packing density is less than one, because the closest possible packing of spheres still leaves some empty space; this much is clear from a stack of oranges on a market stall. The densest packing possible for spheres of equal size corresponds to so-called hexagonal close packing (because in every layer of spheres each is surrounded by a hexagon of six others), which has a packing density of 0.7405. Many pure metals adopt this crystal structure, but subtle aspects of the bonding between atoms can lead to slightly less dense packings, such as the body-centered cubic structure, with a packing density of 0.6802.

All of these packings are still pretty compact—there is no definable pore space between the spheres. Surprisingly, though, the porous structures of zeolites do not sacrifice very much at all in terms of packing density: the zeolite ZSM-5 has a packing density of about 0.6, and the aluminophosphate molecular sieve ALPO-8 clocks in at around 0.5. So the sacrifice in making a porous structure of this sort is rather little, and is more than compensated by other factors. The characteristic that distinguishes porous materials like these from a compact one is not so much the packing density but the way in which the empty space is distributed: in zeolites, the empty space (around 40 percent of the total volume) is merged into continuous channels rather than being distributed more evenly between atoms.

(I should mention that it is of course possible to make materials with a much higher ratio of empty space to filled space than this, by carving out voids. This is, after all, how the porous silicon in chapter 1 was made. But this porosity is on a larger scale than that of zeolites, and is not dictated by the constraints of packing atoms together. The solid material in porous silicon—the silicon wires—is packed very densely, being essentially close-packed crystalline silicon.)

A single diamondlike network of Wuest's tecton crystals would have a packing density of just 0.235—considerably lower than that of zeolites. This is too much for nature to feel comfortable with, so she crams another network inside to double the packing density to 0.47. That still leaves the crystal with a very open structure, however, with large square channels, about the width of those in a common zeolite such as zeolite A, that extend throughout the material. So this building-block approach can produce materials that are highly porous on the atomic scale, even if the pore space may be reduced by interpenetration.

Metal Welding on the Molecular Building Site

Metal ions are another valuable glue for the building-block approach to designed pores. Like hydrogen-bonding units, they form bonds that are highly directional (that is, they point in particular relative directions in space) and are thus able to steer the organization of the network away from close-packed structures; and the bonding propensity is selective, so that the sticky ends of the building blocks do not just stick together any old way. For the molecular construction engineer, one of the most attractive features of metal-ion linkages is that their geometry is often very strongly influenced by the chemical nature of the metal and by its charge. For example, ions of copper that bear a single positive charge tend to form structures in which four other molecular entities (called ligands) are attached to the ion at the corners of a tetrahedron. If the ion is doubly charged, on the other hand, an array of four ligands at the corners of a square is preferred.

Metal ions have been used for stitching together open networks of organic molecules. The approach is like a microscopic version of Tinkertoy construction: the organic molecules (the ligands) are the rods, and the metal ions are the vertex connectors. Jeffrey Moore and coworkers at the University of Illinois have ex-

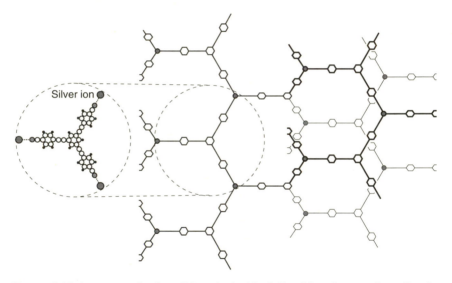

FIGURE 7.13 A porous molecular solid synthesized by Jeffrey Moore's group from silver ions and organic molecules with silver-loving ends has a complex "hinged" structure, which might give the solid unusual mechanical properties when it is stretched: pulling in one direction might cause the material to *expand* (rather than contract) in a perpendicular direction. These properties are predicted, but the researchers have not yet been able to measure them for the real material—the crystals aren't big or stable enough.

ploited this approach to make the network shown in figure 7.13. Their building blocks are the rather complex organic molecules shown in the figure, which contain cyano (-CN) "sticky ends"; these form strong complexes with silver ions. This is a very open structure, with a lot of pore space. But again nature's abhorrence of empty space intervened, and Moore found that the crystal contained no fewer than six independent, interpenetrating networks (plate 9). Even so, the material still possessed open channels 15 by 22 angstroms across.

Richard Robson and colleagues at the University of Melbourne in Australia

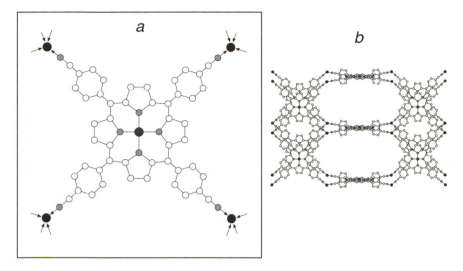

FIGURE 7.14 A square porphyrin building block (*a*) can be linked together via copper ions at its corners into a crystal with wide channels (*b*). (Note that I have not shown any of the hydrogen atoms in (*a*) for clarity.) The cyano (CN) groups at the corners of the building blocks will bind to singly charged copper(I) ions (black circles) in a roughly tetrahedral geometry; in (*a*) I indicate with arrows how each copper(I) ion is bound to three other porphyrin blocks as well as that shown here. These blocks lie face-on at the sides of the channels in (*b*), and edge-on at the top and bottom of the channels.

have made several molecular networks based on metal-ion coordination chemistry. In some of these the building blocks contain the porphyrin group (a central component of the light-harvesting molecule chlorophyll), to the periphery of which they attach nitrogen-containing chemical groups that can form bonds with copper ions (fig. 7.14*a*). The porphyrin units of this assembly represent "square" junctions, where four arms of the network meet in a square arrangement; and the copper ions provide tetrahedral junctions. These units assemble from solution to form an open network with straight pores about 20 angstroms wide (fig. 7.14*b*). Again the packing density of the crystals is increased by intertwining two networks; but these sit more or less on top of each other, leaving the large pores unobstructed. The use of porphyrin units here is particularly appealing, because in nature such groups are found not only in natural molecules that interact with light, such as chlorophyll and light-harvesting proteins, but also in biological catalysts. Robson's hope is that by using building blocks that have known chemical functions, it might be possible to make organic zeolites with similar light-induced or catalytic functions.

The interpenetration problem, which makes it hard to predict the details of a particular crystal structure, is just one of the difficulties faced by those trying to construct these and other designed porous materials. Another critical problem is that, unlike zeolites, the materials are not very stable unless the pores are filled with something to prop them up. All of the molecular materials described so far are made by precipitating the crystals from a solution of the components, and the pores in the resulting materials are invariably filled with solvent molecules. But if one tries to remove these molecules, by heating or simply drying, the pore system usually collapses. (Nevertheless, the solvent molecules can often be exchanged for other small "guest" molecules, for example by soaking the crystals in another liquid, provided that this exchange is gradual and does not involve emptying the pores before refilling them.) It seems that these molecular pores are not generally strong enough to withstand the strains imposed by extracting the solvent molecules. Some groups are now seeking ways to address this problem: Wuest, for instance, has speculated about incorporating chemical groups near the ends of his sticky-ended tectons which can react together to form robust covalent bonds. His idea is then that, once the tectons have assembled themselves into the network structure, one would trigger reactions between these chemical groups so as to form permanent cross-links between the building blocks.

FROZEN SMOKE

How far can we increase the porosity of a material before there is simply nothing left but empty space? Zeolites, as we have seen, are extremely porous materials, having an internal surface area many hundreds of times larger than the external area of their crystal faces. But even so, they are only about 40 to 50 percent empty space, and to the eye they look like dense, solid minerals. But a kind of material first devised in the 1930s by S. S. Kistler at Stanford University in California can now be synthesized with a pore fraction of 0.99—that is to say, it is 99 percent air. These materials, whose fabric may consist of a range of materials from silica to metals to rubber, are called *aerogels*. As one would expect from their extreme porosity, aerogels are remarkably lightweight: while a cubic-meter block of everyday silica glass weighs about 2,000 kilograms—about the mass of an automobile—a silica aerogel block of the same volume might weigh just 20 kilograms, or about 44 pounds. But for all their tenuousness, aerogels remain relatively robust.

Aerogels are sometimes dubbed "frozen smoke," and it is not hard to see why: a lump of a silica aerogel has a smoky, translucent appearance (plate 10). But the analogy goes further: the structure of an aerogel really is like that of smoke. Smoke particles are composed of many tiny fragments of carbonized material that aggregate in the air to form tenuous, branched aggregates. Aerogels are simply large-scale versions of such filamentary aggregates. Silica aerogels, for example, consist of tiny, dense silica particles about one nanometer across, clustered together into porous aggregates that are themselves then joined into the chainlike branches of a tenuous skeleton (fig. 7.15).

Many suspensions of small, solid particles can be induced to come together in solution to form such microscopic porous aggregates. Silica aggregates can be formed by acidifying a solution of a silicate salt, which neutralizes the negative charges on the silica ions by capping them with positive hydrogen ions and

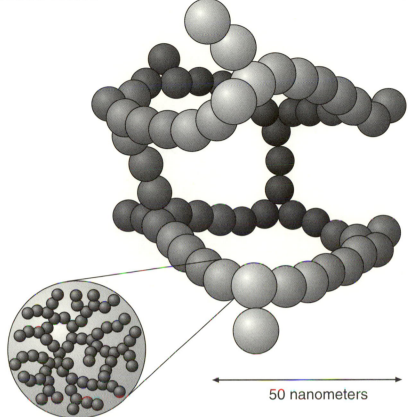

50 nanometers

FIGURE 7.15 Silica aerogels are highly tenuous, hierarchical structures. Their fundamental units are dense silica particles about one nanometer across, which polymerize into highly branched aggregates. These aggregates then themselves link up into a highly porous network of branches.

allows them to link together into chains and clusters. This was the first step in Kistler's original synthesis of silica aerogels: he added hydrochloric acid to a solution of sodium silicate, commonly known as "water glass." The discrete silicate aggregates so formed then slowly link up to produce a gel—a viscous substance consisting of a tenuous network interspersed with solvent molecules.

These gels collapse into a dense mass called a *xerogel* (meaning a "dry gel") when the solvent is removed by evaporation. This is not, as might be supposed, because the aggregated network is too flimsy to support its own weight; rather, it is because the drying process itself creates forces that pull the network in on itself. As the liquid-saturated porous material dries, a meniscus develops between the liquid and the air. The surface tension of the meniscus pulls the sides of the pore inward, causing the pore network to collapse.

Kistler realized that if the gels could be dried while avoiding this collapse, highly porous materials would result. This meant finding a way to get rid of the

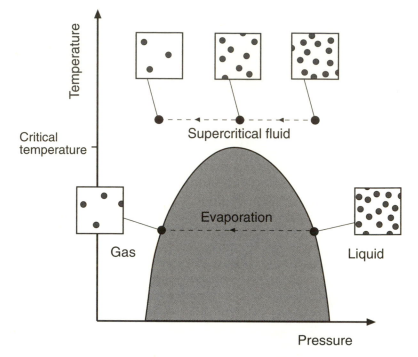

FIGURE 7.16 A supercritical fluid is one that does not have distinct gas and liquid states. Rather than undergoing abrupt changes of state (condensation to liquid or evaporation to gas), it can pass continuously from a dense liquidlike state to a rarefied gaseous state as the pressure is changed. This supercritical state is attained above a critical temperature, which denotes the point at which the distinction between liquid and gas disappears. For water, the critical temperature is 374 degrees Celsius.

meniscus. But how do you dry a liquid-filled pore network without setting up an interface between the liquid and the air into which it is evaporating? Kistler's answer involved doing something rather extreme to the liquid: subjecting it to high pressure and temperature, under which conditions the difference between a liquid and a gas disappears. All fluids possess a so-called critical temperature, above which they can pass from a dense, liquidlike form to a rarefied, gaslike form without the abrupt change in state that accompanies normal evaporation. In this regime the fluid is said to be supercritical and is no longer strictly a liquid or a gas at all but simply a fluid whose density varies smoothly as the pressure is altered (fig. 7.16). By subjecting the liquid inside a gel to supercritical temperatures and pressures, it can be extracted without setting up a meniscus, because there are no longer distinct liquid and gas states.

Kistler found that by using this supercritical drying technique he could make aerogels from a whole range of substances—not only silica but metals and organic materials like rubber and the protein albumin. His procedure was rather complex, however, involving first replacing the water solvent with alcohol (this was necessary because water dissolves the gel at high temperatures). More recently, several techniques have been developed for making silica aerogels that avoid this need for a tedious change of solvent, making it possible to prepare aerogels in a matter of hours rather than weeks. These methods use a different starting material from Kistler's water glass: a colorless liquid called tetramethoxysilane (TMOS) which, when mixed with water in the presence of an acid or an alkali, generates clusters of silicic acid that form the fabric of the gel's spongelike structure. An even better starting material is tetraethoxysilane (TEOS), since unlike TMOS it is nontoxic. In these compounds the silicon and oxygen building blocks of the final silica material are attached to four methyl (CH_3) or ethyl (C_2H_5) groups, respectively.

These approaches are examples of sol-gel processing, a technique that is now very widely used for making all kinds of ceramic and inorganic materials, both porous and dense. Sol-gel processing of a sort was practiced in the nineteenth century and seems to have been used for making transparent silica coatings on mirrors used in tanks in the Second World War. But as a general approach for materials synthesis, its use was pioneered by Rustum Roy of Pennsylvania State University, who employed it from the 1950s to make new kinds of dense, mineral-like materials. While still in the sol phase, the solution can be coated onto a surface, so that drying then produces a very thin film of the material; this makes the sol-gel process very useful for making thin coatings. Such coatings, which are generally transparent, are finding many applications, for example as antireflective films for solar cells and computer screens.

Aerogels Spring Back

Although sol-gel processing makes aerogel synthesis very much easier than Kistler's method, the need for supercritical drying is a significant drawback from the commercial point of view: it involves high temperatures and pressures, which makes the process hazardous, expensive, and incompatible with the needs of some potential applications (these extreme conditions might damage the support on which an aerogel film is deposited, for example). If one could sidestep the need for supercritical removal of the solvent, the possible uses of these materials would be greatly expanded.

In a normal silica gel, the walls of the pores are coated with hydroxyl groups, because the oxygen atoms of the silicon–oxygen network are capped with hydrogen atoms. When two such groups on opposite faces are brought close together by the pull of the meniscus during normal drying, they can react, ejecting a molecule of water (which contains the two hydrogen atoms and one of the oxygens) and leaving an oxygen atom bridging the two silicon atoms (fig. 7.17a). So the pore walls get permanently cross-linked into a dense, collapsed state—a xerogel.

C. Jeffrey Brinker and colleagues at Sandia National Laboratories in Albuquerque, New Mexico, have found a way to make gels whose pore surfaces are modified to prevent these bridging reactions. They first made a standard silica gel from TEOS, and then reacted this with trimethylchlorosilane (TMCS), a compound in

a

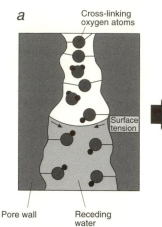

Cross-linking
oxygen atoms

Surface
tension

Pore wall   Receding
                   water

Collapsed pore

b

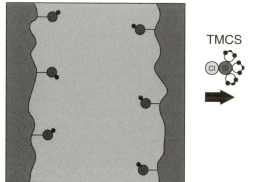

TMCS

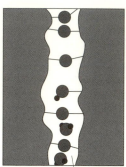

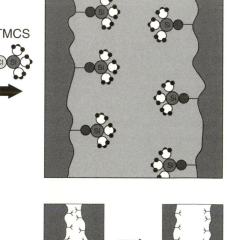

Drying   Spring-back

FIGURE 7.17 As the delicate gel precursor of an aerogel dries, the pore walls are pulled together by the surface tension of the liquid meniscus as the solvent retreats down the pores. This can lead to the formation of cross-links between surface groups on the pore walls. Shown in (*a*) is the condensation reaction that takes place between hydroxyl (-OH) groups on the pore walls of a silica gel. These cross-linking reactions cause the gel to collapse into a dense mass during drying, and the porosity is lost. In conventional aerogel synthesis, this collapse is avoided by heating the solvent to supercritical temperatures before extracting it, so that there are no distinct liquid and gas states with a meniscus between them. Jeffrey Brinker and colleagues have made aerogels that can be dried without having to make the solvent supercritical (*b*). They reacted the precursor gel with trimethylchlorosilane (TMCS), which gives the water-filled pore walls a coating of methyl (CH_3) groups. As these do not form cross-links with each other, the pores spring back after the retreating liquid meniscus has passed during drying.

which the silicon atoms are attached to three CH_3 groups and one chlorine atom. The silicon atoms in TMCS became incorporated into the surface of the silica gel, losing their chlorine atoms in the process (fig. 7.17*b*), and the result is a silica surface coated with CH_3 groups rather than with hydroxyl groups. As the former have no propensity to react with each other to form cross-links, the gel acquires a springiness: during drying, the pores shrink under the usual capillary forces but then expand again once the meniscus has passed by (fig. 7.17*c*). In this way, Brinker's team was able to prepare silica aerogels by subcritical drying with more than 98 percent empty space—comparable, in other words, to the kinds of material previously available only by supercritical drying.

Tenuous Promises

This development should greatly expand the possibilities for putting aerogels to practical use, since the difficulty of making them has so far limited their potential. The fact is that aerogels have several interesting physical properties that could in principle be exploited. One of the most significant of these is their low thermal conductivity—they are very poor conductors of heat (about a hundred times poorer than normal glass), which means that they could act as an excellent thermal-insulation material. Because thin sheets of a silica aerogel are also fairly transparent, this raises the possibility of a kind of super-double-glazing. In normal double glazing, a layer of air trapped between two glass panes provides a low-thermal-conductivity barrier to reduce the loss of heat from a warm room. Heat loss could be reduced still further by removing the air from between the panes, creating a vacuum—this, after all, is what is done in thermos flasks. But a vacuum between two large window panes is not an easy thing to sustain; the air pressure on the outer faces of the panes would push them together. An aerogel layer between the two panes would, however, allow one to set up what amounts to a "solid" vacuum—the aerogel is mostly empty space, and so by removing the air from the pores one would end up with a gap of perhaps 99 percent vacuum supported by a robust, low-thermal-conductivity solid. It is estimated that such an arrangement could reduce by a factor of three the heat losses experienced by the most advanced window-insulation systems currently available (in which argon gas replaces air in the cavity). Aerogel-double-glazed windows would have a slightly "frosted" appearance, since even a thin slab of aerogel scatters some light; but in many situations the heat savings would surely be worth the reduction in visibility.

A similar glass/aerogel sandwich might also prove useful for harvesting solar energy. Passive solar panels are basically flat plates that absorb the Sun's heat. As they warm up, some of the heat they collect is inevitably radiated back out again. But a translucent, low-thermal-conductivity sandwich of glass and aerogel might provide a one-way barrier to the re-radiation of heat from an absorbing surface. For example, the wall of a house or of an external water tank could be coated with black paint to improve the efficiency with which it absorbs the Sun's heat; and a glass/aerogel screen in front of the black surface would hold in the heat so that most of it is passed on to the house or the body of water.

One of the most surprising properties of aerogels is that sound travels through them at extremely low speeds, close to the speed of sound in air. In most solid materials the speed of sound is over ten times greater than that in air, because the medium is so much more dense and rigid; sound speeds as low as those in aerogels are otherwise observed only in soft, rubbery materials. Because of this, aerogels might be able to improve the performance of devices that convert electrical signals into sound waves, called piezoelectric transducers (chapter 3). These devices are commonly used in sonar systems for gauging distances by sensing the reflection of ultrasonic waves. Because the acoustic properties of piezoelectric materials and air are so drastically different, a considerable part of the energy emitted by the sonar device is wasted in ultrasonic waves that are reflected back from the sharp interface between the piezoelectric medium and the air. A coating of an aerogel would reduce these losses by providing a buffer layer with acoustic properties intermediate between those of the piezoelectric and those of the air, making the change in acoustic behavior across the interface less abrupt.

Aerogels show that there is scarcely any limit to the amount of perforation we can make in a material. Research on microporous organic and inorganic crystals, meanwhile, illustrates how finely it is now possible to design the perforations. In the face of advances like these, we can expect to see tiny holes playing an ever more important role in our society's technology, from housing construction to biomedicine.

CHAPTER EIGHT

Hard Work

DIAMOND AND HARD MATERIALS

> Does the law perhaps punish "fabricators and vendors of fake
> diamonds"? Do there exist fake diamonds?
>
> —Primo Levi, *Order on the Cheap*

Diamond is the hardest material known. The diamonds that feature in today's cutting and grinding tools are not dug from the Earth, but are synthesized in hot presses. Diamond films can now be grown at low pressures from carbon-containing vapors, and they promise to usher in ultrahard protective coatings and new kinds of electrical and optical devices.

ON A GOOD DAY in the Geophysical Laboratory in Washington, D.C., it is like being halfway to the center of the Earth. The pressure can reach about two million times the atmospheric pressure, while the temperature rises to over a thousand degrees Celsius. These conditions do not prevail throughout the laboratory, you understand, but merely in the all-important cubic millimeter or so of it that lies between the beveled teeth of a diamond anvil cell. This instrument, invented in 1959, is capable of squeezing materials to pressures comparable to those deep within the bowels of the Earth. In 1975, researchers at the Geophysical Laboratory first pushed the instrument beyond the million-atmosphere mark, a pressure equivalent to being buried under 2,000 miles of rock.

The diamond anvil cell has allowed researchers to explore the nature of the deep Earth from within the laboratory. They can watch rocks squeezed within the apparatus undergo the pressure-induced changes in mineral structure and composition that are thought to take place at great depths. Such studies not only provide a better understanding of the constitution of our planet; they also allow us to investigate the origin of geophysical processes such as deep earthquakes like those that constantly shake Japan. And high-pressure experiments on more exotic substances, such as solid hydrogen, promise to cast light on the internal structure of the giant planets like Jupiter, whose interiors are thought to consist primarily of hydrogen under great compression.

All of this is made possible by the tremendous strength of diamond, from which are fashioned the teeth of a vise in which the samples are squeezed. By making their tips very small (typically less than a millimeter in diameter), one can concentrate the force with which the teeth are brought together over a very small area, creating extremely high localized pressures at the turn of a screw. But the teeth themselves have to be able to withstand these pressures without cracking.

Since diamond is actually formed at great pressures within the Earth (below about 150 kilometers depth, where the pressure reaches several thousand atmospheres), it is not surprising that it can withstand high pressures in the laboratory. But unlike most high-pressure states of minerals, diamond remains stable when the pressure is removed. Diamonds carried rapidly to the surface of the Earth by water vapor and other gases discharging along deep fissures in the crust sit there until a lucky prospector chances upon them. So stable and robust do they seem that we are told how a diamond is "forever," although we will see later how contingent that longevity is.

Diamond is the hardest material known. The practical value of this was evident to the ancient Greeks, who, according to Pliny the Elder, used splinters of natural diamonds (produced with a hammer, anvil, and much effort) in tools for engraving other gemstones. Diamond powder has been widely used since the sixteenth century as an abrasive for shaping and polishing jewels, and John Donne mentions the use of diamond splinters for engraving of glass. Diamond-studded drill bits for boring into rocks date at least from the eighteenth century, and their use for mining was widely practiced by the late nineteenth. Today the uses of diamond as an abrasive and in cutting tools remain its major industrial applications: a diamond-studded saw will slice through almost anything. In addition, protective coatings of diamond can prevent wear of moving parts in mechanical devices, and much effort is today devoted to finding ways of coating bearings and engine parts with thin films of diamond to reduce their tendency to become ground away.

Yet the value of diamond as an engineering material is by no means defined purely by its hardness and strength. Diamond does most things to excess. It is an excellent electrical insulator, and has a high dielectric constant, meaning that it is extremely good at stopping a charge from leaking away through it. This makes diamond valuable to the electronics industry, for example as the insulating medium between the plates of capacitors. Yet diamond can be doped with other elements to make it a semiconductor, in which guise it turns out to have properties that render it a better prospect for some electronic applications than the current mainstay, silicon.

High-quality diamond is highly transparent to light from the visible to the infrared parts of the spectrum, which, in combination with its hardness and consequent resistance to scratching, makes it a good material for windows that are subjected to extreme temperatures and pressures. The refractive index of diamond (a measure of the extent to which light that passes through it is slowed down) is greater than any other clear material, and one consequence of this is the tendency of a faceted diamond to "disperse" strongly any light that enters it—to break the light down into the various spectral components. This high dispersion gives diamond gems their brilliance and sparkle. And diamond is a superlative conductor of heat, a property that makes it a useful heat sink in electronics engineering.

But we all know very well that diamond does not come cheap. If we must rely on the diamond that we can dig out of the ground, most of these applications become highly, perhaps prohibitively, expensive—one thinks twice before investing in a cutting tool at the price of several diamond rings. Their technological potential, their monetary worth, and their sheer beauty—all of these things have inspired scientists to attempt to create synthetic diamonds for over a century. These attempts have been as hazardous as they have been ingenious, and have not been lacking in color and controversy. The first synthetic diamonds were made in the 1950s, and this success has now spawned a mature technology for making diamond synthetically. How that came about is a major topic of this chapter. In recent decades the conventional route to diamond synthesis has found itself under challenge from a new method that does not have to mimic the fearsome conditions of natural diamond formation. This new synthetic diamond is without doubt one of the brightest advanced materials of the present age, and so I shall have much to say about it.

Moreover, there now exist several materials that rival diamond in their hardness, and others that make up in toughness what they lack in hardness. These are primarily ceramic materials, and the ease with which they can be fashioned into molded parts makes possible the vision of hard-wearing, all-ceramic engines and turbines. Advanced ceramics really deserve a chapter of their own, but here I can only provide a sketch that gives some context to the status of diamond as a hard, strong engineering material. There is now a lot of excitement at the prospect of devising new ceramic materials that are even harder than diamond—such materials have been predicted theoretically, and the race is on to try to create them.

Why Diamonds Aren't Forever

The richness in natural resources of the southern African countries has not done them many favors—rather, it has left them a highly desirable prize for colonial powers, and the conflicts that have followed from colonial domination and disputes are all too familiar. Gold from Witwatersrand was one of the magnets that drew the European powers to the land, there to wrangle and wage war; and diamonds from Kimberley were another. Legend has it that the rich diamond veins in Kimberley were discovered in 1866 when the children of a Boer farmer brought him back a shiny pebble that they had found while playing on the bank of the Orange River—this proved to be a 21-carat diamond. During the late nineteenth century this region represented the main source of diamonds for the world at large, and the matrix rock in which diamonds form, which is rich in fluids containing carbon dioxide, is given the generic name kimberlite. Diamonds are believed to crystallize from molten kimberlite within the Earth's mantle, and to be blasted up fissures, along with the host rock, in a kimberlite eruption when the pressure from the carbon dioxide fluid gets too high. The diamond vein discovered in Kimberley represents the ancient remains of such a volcanic outburst.

Diamond is a crystalline form of pure carbon, and is one of two such forms that have been long recognized. The other is graphite, and it would be hard to imagine two materials less alike—diamond is the beautiful sibling, crystal-clear and harder than any other known natural material, whereas graphite is the dirty one, soft and black. Whereas diamond is an excellent electrical insulator, graphite has sufficient conductivity to qualify it as a semimetal or "almost metal." But despite these profound differences, both materials consist only of carbon atoms linked together into vast networks. At the tremendous pressures and white-hot temperatures of the deep Earth, diamond is the most stable of these two forms of carbon. But at atmospheric pressure and room temperature, graphite is more stable. So in principle, diamond ought to revert to graphite once it is released from the great pressures under which it is formed.

That this does not happen spontaneously is a consequence of the fact that the transformation from diamond to graphite requires a great deal of energy to get it underway. It involves the breaking of bonds between carbon atoms, and these are very strong. So even though the end result of the process would be a lowering of energy—the transformation is a "downhill" one—the energetic cost of setting the ball rolling is too large to enable the transformation to proceed spontaneously. In technical terms, diamond is said to be metastable, which means provisionally stable, relative to graphite. To effect the transformation, diamond must be heated to around 1,200 degrees Celsius. So for most intents and purposes, the metastability of diamond can be ignored, and diamond can be regarded as a stable material. The high activation energy for the conversion to graphite is crucial, however, to all practical applications of diamond—if it were not so, our diamond-toothed saws would slowly revert back to soft graphite (all the more rapidly because of the frictional heating they encounter).

The marked differences in properties of these two carbon materials are a consequence of the differences in the way in which their constituent atoms are linked together. In diamond, each carbon atom is joined by chemical bonds to four others, which are arrayed around it at the corners of a tetrahedron (fig. 8.1). The

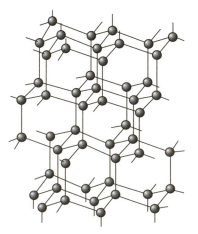

FIGURE 8.1 Diamond is pure carbon (although all natural diamonds are laced with impurities such as nitrogen). The carbon atoms are arranged on a crystal lattice in which each is surrounded by a tetrahedron of four others. Diamond has the greatest atomic density of all solids.

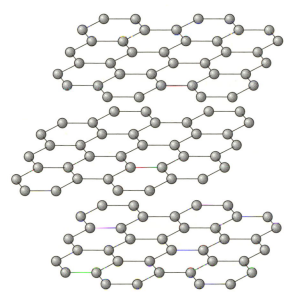

FIGURE 8.2 Graphite is another form of crystalline carbon, in which the atoms have only three nearest neighbors instead of diamond's four. They form sheets of six-atom hexagonal rings, which stack on top of one another, held in place by relatively weak forces. Carbon atoms have four bonding electrons each, and the "extra" electrons on each atom create a network of delocalized bonding that runs across the entire sheet. This delocalized network gives rise to graphite's appreciable electrical conductivity and to its strong light absorption, making the material black in contrast to diamond's transparent brilliance.

bonds between carbon atoms are short and strong, giving the crystal lattice both high strength and high rigidity in all three dimensions. In graphite, on the other hand, each carbon atom has three rather than four nearest neighbors. These are arrayed around the central atom in a triangular manner, confining all four atoms to a flat plane. The atoms are linked together in vast sheets, in which the triangular bonding arrangement creates hexagonal rings that form an extended, two-dimensional network reminiscent of chicken wire (fig. 8.2). As carbon atoms like to form four, rather than three, bonds, the triangular arrangement leaves each carbon atom with a residual bond-forming electron. These "spare" electrons get smeared out in a delocalized bonding network that extends over the entire sheet. The delocalized electron states give rise to their own energy bands in the solid: a filled valence band separated from an empty conduction band by a narrow band gap (see chapter 1). Electrons in the valence band can readily make the transition to the empty conduction band by absorbing photons of visible light; this strong light absorption throughout the visible range gives graphite its black appearance, and the excited electrons give it an appreciable electrical conductivity. Some geologists believe that the surprisingly high electrical conductivity of rocks in the Earth's crust results from the presence of thin films of natural graphite on the surfaces of their grains.

The atoms within the graphite sheets are linked together no less strongly than those in diamond, but the sheets themselves lie far apart, bound to one another only by a much weaker interaction called the van der Waals force, which results primarily from the sloppiness of the delocalized electrons. As a result, the sheets can slide over each other relatively easily, and the material as a whole is rather soft, and is a good lubricant. (I should emphasize, however, that the individual sheets are not at all weak—indeed, carbon fibers, which are essentially pure graphite, are among the strongest and stiffest synthetic fibers. Graphite becomes a weakling only when the sheets can slide over one another.)

Although the French chemist Antoine Lavoisier seems to have been the first person to deduce that diamond burns to form only carbon dioxide, the intimate relation to graphite seems to have been identified at a somewhat later date by the English chemist Smithson Tennant, who showed in 1797 that both produce the same gas when they are burned. This information, implying that diamond is nothing more than a denser form of graphite, led to speculations in the nineteenth century that the latter might be converted to the former by squeezing. Since diamond is about one and a half times as dense as graphite, it is reasonable to suppose that carbon will crystallize in this form under high pressure.

A Pressing Problem

In 1897 the eminent French chemist Henri Moissan tried many times to make diamond by squeezing other forms of carbon. In early experiments he added burnt sugar to molten iron in a furnace. Moissan's studies of natural diamonds had revealed iron in the source material, and he suspected that the metal might be a catalyst for diamond formation. But he came to realize that heat alone would not suffice—pressure was needed too. By quenching molten iron saturated with carbon in water or in molten lead, Moissan hoped that high pressures would arise in the still-molten core of a solidifying droplet. Whether or not he met with any success is unclear—he claimed to find within the quenched metal traces of a transparent substance with optical properties similar to diamond, and which gave off carbon dioxide when burned, but this was not unequivocal evidence of diamond formation. The transparent crystals were most probably silicon carbide, now a useful material in its own right but then a kind of "fool's gold" for the unwary diamond maker.

The American J. Willard Hershey claimed success when he repeated these dangerous experiments in 1940, but these claims are widely held to be suspect. More hazardous still were the experiments of James Ballantyre Hannay in 1880, who heated organic, carbon-rich mixtures in sealed iron tubes, causing many to explode. Again, whether any of these attempts produced diamonds has been the subject of much debate. During the early part of this century, the British chemist Sir William Crookes used explosives to generate high temperatures and pressures in steel tubes, and he believed that he had found diamonds in the blasted remains; but the hard, transparent crystals were again most probably silicon carbide. At the same time, the Irishman Charles Parsons made many attempts to synthesize diamonds at high pressure, including duplications of both Moissan's and Hannay's experiments. Initially claiming success, he later discovered that what he thought was diamond was in fact another impostor, a transparent, inert mineral called spinel. He concluded that neither his own attempts nor those of his predecessors had made diamond. But more significantly, these experiments set the agenda for much of the future work by highlighting the importance of high pressure.

From 1935 onward, Percy Bridgman of Harvard University began to attempt diamond synthesis from carbon materials under high pressures and temperatures. Bridgman is one of the pioneers of high-pressure research, and his specially designed apparatus completely rewrote the rule book, achieving well-calibrated pressures so much higher than previous studies that he felt it necessary to humbly defend their credibility. Yet even this was not enough: his instruments reached pressures of around 400,000 atmospheres and temperatures a little under 3,000 degrees Celsius, but were unable to withstand these extreme conditions for more than a few seconds, which was not enough to achieve diamond formation.

By 1950, diamond making was a tale of cracked presses, burst vessels, and even violent deaths. It took unusual confidence to believe that the transformation was possible. But the field was never short of true believers, and in the early 1950s they were finally vindicated. A team at the Allmänna Svenska Elektriska Aktiebolaget (ASEA) in Sweden began a diamond-making project in 1942, using a complicated and cumbersome assembly of interlocking metal anvils to squeeze a sample between their teeth. By the early 1950s, when the project was being led by Erik Lundblad, they were able to attain pressures of over 80,000 atmospheres, which they could sustain for up to an hour. In February 1953 the ASEA team compressed a mixture of iron carbide and graphite along with a seed of natural diamond, all embedded in a cube of thermite, an explosive that generated temperatures of several thousand degrees when ignited. Among the usual debris from this furious treatment they found transparent particles no bigger than grains of sand. To their joy, their crystallographers identified these grains unambiguously as diamond. The transformation that took place deep in the Earth had finally been brought into the laboratory.

The ASEA team repeated their success twice more in 1953 (it was not an easy experiment to repeat, because the anvil assembly was left in such a mess after each run). But then, inexplicably, they made no announcement of their achievement until seven years later—instead, they labored in silence to perfect their difficult, costly technique. It seems that in 1953 they believed no one else was close to making diamond, so they could afford to take their time and get things right before telling the world. But they were wrong.

In February 1955, a team at the General Electric Company in Schenectady, New York, called a press conference to announce that they had made diamond. Unaware of the secret ASEA results, the GE team believed they were the first. In the same year their results were published in the journal *Nature*.

The GE group were Francis Bundy, Herbert Strong, Tracy Hall, Robert Wentorf, Harold Bovenkerk, and technician James Cheney. They were a collection of individuals, each developing their own ideas about how diamond might be synthesized. Beginning in 1951 they explored diamond synthesis using a variety of approaches, favoring the "brute force" approach that Bridgman had championed: bringing a carbon-rich sample to sufficiently high temperatures and pressures that

FIGURE 8.3 The first reliable report of the synthesis of diamond came from a team working at the General Electric Company's research laboratories in Schenectady, New York, in 1955. The GE team used two presses to achieve these tremendous pressures—a large thousand-ton press (center), and an older, leaky, four-hundred-ton hydraulic press (left center). Ironically, it was in the latter that a reliable synthesis procedure was developed. Several members of the GE team are shown here attempting diamond synthesis in 1954. (Photograph courtesy of General Electric.)

diamond would be the most stable form. Drawing on the experience gained in Bridgman's work, the GE team developed high-pressure high-temperature apparatus that could sustain conditions of up to 100,000 atmospheres and around 2,000 degrees Celsius for hours of continuous operation. In these devices a tiny sample constrained by a gasket ring was squeezed between anvils or pistons of an ultrahard material called Carboloy, a type of tungsten carbide which Bridgman had also used. The squeezing was done by two presses—an old 400-ton hydraulic press dating from GE's early days, and a huge new 1,000-ton press acquired specially for this project (fig. 8.3).

Success depended not only on clever engineering—devising a piston arrangement that would achieve the required pressures—but on finding the right chemistry. The researchers believed that, rather than simply trying to compress graphite into diamond, they should include metals with the carbon source, which might assist the transformation by, for instance, first dissolving the carbon. Iron in particular—the element favored by Moissan—was thought to be a potential candidate. In December 1954, Herbert Strong ran an experiment in which he maintained the pressure for an unusually long time, right through the night. Strong packed carbon powder inside iron foil, along with two small natural-diamond seed crystals, and squeezed them to 50,000 atmospheres at around 1,250 degrees for 16 hours.

At first the run seemed to be another failure—the diamond seeds were recovered unaltered. But during analysis of a melted part of the iron foil, two hard, bright crystals came to light in the fused metal. Analysis of their structure confirmed Strong's suspicions: here, it seemed, were synthetic diamonds.

The result inspired Tracy Hall to run experiments the next day in the piston assembly of his own design, in which a doughnut-shaped belt confined the sample between two Carboloy teeth. Hall added iron sulfide to the carbon sample, and compressed it to fully 100,000 atmospheres at 1,600 degrees in the leaky old 400-ton press. Dozens of tiny diamonds were formed. Hall and Wentorf found that they could reproduce this result again and again—at last, diamonds could be made to order.

The GE team's *Nature* paper in 1955 described many of these synthetic diamonds (fig. 8.4), which they characterized by their crystal structure, their chemical composition, and their hardness (their ability to scratch natural diamond). Modern readers of that paper would notice a conspicuous shortcoming, however: the researchers never specified exactly how the diamonds were made, so that no one could have repeated the experiments from the account given. Today this would be regarded as very bad practice, but the GE team had no choice, as their work was classified under a U.S. Department of Defense secrecy order. It was not until this was lifted in 1959 that the team was able to provide a full account of their work. In the meantime, GE was able to exploit their success without fear of competition, since the crucial details of their technique remained secret. Efforts to turn the discovery to commercial advantage began immediately, and by late 1957 General Electric began selling Man-made diamonds, the name given to their product, as an abrasive. They had to work hard to persuade potential buyers that these synthetic diamonds were every bit as good as the natural ones sold by the

FIGURE 8.4 The GE diamonds, up to one millimeter in size, were made by subjecting graphite to pressures of around 100,000 atmospheres and temperatures of up to 2,000 degrees Celsius in the presence of iron. (Photograph courtesy of General Electric.)

South African cartel De Beers, which until then had enjoyed a near monopoly in the world's diamond market. But gradually the word spread, and by 1959, when the secrecy order was lifted and full details of the GE method were finally made available, the company had captured 10 percent of the U.S. market.

The publication of the GE paper showed the world that diamond could be manufactured. During the subsequent few decades, diamond formed by the high-pressure/ high-temperature (HPHT) techniques pioneered at GE gained rapidly in commercial status, and was soon being produced by several companies around the world. Even De Beers had to make room for the new order, and set up their own synthetic-diamond manufacturing operation under license from GE's patent (though they did not concede GE's priority to the HPHT process without a legal struggle).

It seems that the molten metals used in this process somehow assist the formation of diamond in preference to graphite; but the metal's precise role has been the matter of some debate. One idea is that the metal simply acts as a solvent to break the graphite apart, but others suggest that it acts as a catalyst, reducing the energy needed to initiate the transformation. There is some evidence to suggest that the metal fulfills both roles, acting as a solvent for the carbon atoms but also as a catalyst to convert them into a form that is more conducive to diamond formation. It appears that diamond forms only in the presence of metals that tend to give the dissolved carbon atoms a positive charge; if they are neutral or negatively charged, graphite is deposited even though in theory diamond should be more stable. The idea that positively charged carbon ions might form diamond was supported by experiments conducted in 1971 by S. Aisenberg and R. Chabot, who reported the formation of diamondlike carbon films directly from beams of vaporized carbon ions. Paul Schmidt and colleagues showed in 1976 that this approach could generate genuine diamond films.

In general, the diamonds formed in HPHT syntheses are tiny—a hundredth of an inch or so in size—and so the product comes in the form of a sandlike grit (fig. 8.5), which is separated out into various grades (depending on the particle size) for industrial use as abrasives. The precise shape and size of the diamond crystals can be controlled to some extent by varying the conditions of their synthesis. For instance, at 60,000 atmospheres the GE method gives cubic crystals when grown at 1,400 degrees but octahedral crystals at 1,600 degrees. This control over the properties of the diamond grit means that it can be tailored to the requirements of particular uses, such as cutting or abrasion. This gave the GE method a decided commercial advantage over the natural diamond grits supplied by De Beers.

FIGURE 8.5 Industrial diamonds synthesized by the high-pressure/high-temperature method developed at General Electric come in the form of a grit, with each diamond typically no bigger than a grain of sand. The diamond grit is used in cutting, drilling, and abrading tools, where it is embedded in a tough ceramic matrix. (Photograph courtesy of De Beers Industrial Diamond Division.)

The grits are incorporated into drill bits, saws, and cutting wheels, either by individual emplacement of larger diamond pieces or by bonding a powder into a metal or ceramic support. You would now be hard pushed to identify a manufacturing industry that does not use such components. During the 1960s Robert Wentorf and Herb Strong found a way to grow HPHT diamond on a small seed crystal so as to produce larger gemstonelike synthetic diamonds, up to two carats in size. These larger diamonds find use as heat sinks in electronic devices, where their high thermal conductivity allows them to transfer heat rapidly away from an area that is in danger of overheating. They are also used as windows for sensing devices on board space satellites; as diamond is transparent in the infrared whereas glass is not, diamond windows are needed for infrared sensors. But the large stones are very expensive to grow, so their application is limited to very specialized situations.

Explosions dogged the early days of diamond making (and are not unknown even in the modern high-pressure industrial facilities). Conducted purposely, however, their effects can be as much constructive as destructive. An explosion may create immense pressures and temperatures—fleetingly, but for long enough to bring about the transformation of graphite to diamond. The synthesis of diamond by compression using an explosive shock wave was first described in 1961 by Paul deCarli and John Jamieson of the Stanford Research Institute in California. This rather extreme approach was adopted on a commercial scale by the Du Pont company in the late 1960s and early 1970s, and in the late 1970s Nobuo Setaka and colleagues in Japan showed that it would work using carbon compounds, rather than pure carbon, as the starting material. Nature is apparently wise to this ruse too: the small diamond particles that have been identified in some carbon-rich meteorites may be formed by shock waves resulting from the impact when the meteorite falls to Earth.

A LIGHTER LOAD

Today the GE method is used to make many tens of tons of synthetic diamond a year, for example at GE's Superabrasives plant in Worthington, Ohio. But although the market for synthetic diamonds continues to be dominated by those made at high pressure and temperatures, this technology has since its inception developed in parallel with a competitor that provides diamond at *low* pressures, without the need for the cumbersome and expensive (not to say dangerous) equipment that a high-pressure process requires. So far, these two alternatives have proved to be complementary rather than competitive, because the nature of their products is rather different. Whereas the HPHT techniques are good for making small single crystals of diamond, for use primarily as abrasives and in cutting tools, the low-pressure approaches produce thin films composed of many microscopic diamond crystals welded together in different orientations—a polycrystalline film. These diamond films suggest themselves for rather different applications, particularly in microelectronics and as hard surface coatings. In many ways, however, HPHT diamonds represent an "old" technology, which was well established by the 1960s and has changed little since. The low-pressure approach, on the other hand, began to flourish only in the 1980s, and the diamond materials that it produces are only just beginning to find applications. By 1990 the synthesis of diamonds at high pressures and temperatures served an industry worth around $800 million, whereas diamonds made at low pressures were scarcely an industrial concern at all. But many industrial analysts now believe that these newer diamond materials will eventually come to dominate the market.

The story of how low-pressure diamond came into being is as curious and tangled a tale as that behind high-pressure diamond, in large part because much of the history is not documented in the regular scientific literature but lodged in patents, company reports, and outright hearsay. Between them, these threads weave an intriguing warp of controversy, serendipity, and innovation. The beginning of the field can perhaps be traced to 1911, when the German W. von Bolton reported the formation of diamond from acetylene mixed with mercury vapor and heated to 100 degrees Celsius. Acetylene is almost a gaseous form of pure carbon—it consists of molecules of two carbon atoms linked together by a triple chemical bond and capped with a hydrogen atom at each end. Von Bolton's claim was that the triple bonds had burst open, linking the carbon atoms into the tetrahedral framework of the diamond lattice. It is now well established that diamond crystals *can* be formed by pyrolysis (gas-phase heating) of acetylene, and indeed Benno Lux of the University of Vienna suggests that there is evidence that such a process was carried out by German researchers during the 1940s. Certainly there exists a report in the East German literature from one H. Z. Schmellenmeier in 1956 of the formation of solid films of some form of poorly characterized diamondlike carbon by decomposition of acetylene in a discharge apparatus. But there seems to be no indication that these early attempts ever met with less ambiguous success. In 1957, the German Hans Meincke reported rather more persuasive evidence of diamond formation from the pyrolysis of graphite by an electrical discharge; but his success went largely unnoticed, because by that time the high-pressure synthesis by the GE team had stolen the limelight.

The formation of diamond from acetylene is actually a form of chemical vapor deposition (CVD), a technique discussed in chapter 1. It involves the introduction of the component atoms of a solid material in the form of a volatile gaseous compound, and the breaking apart of that compound by heat to allow the atoms to condense on a substrate. This same principle was employed in the first successful synthesis of diamond at low pressures by William G. Eversole of the Linde Division of the Union Carbide Corporation in 1952. This date, you might observe, comes before the successful high-pressure syntheses of the GE team and even the unreported Swedish work at ASEA. It is now clear that Eversole's was in fact the first reliable, reproducible synthesis of diamond by *any* method. That Eversole was first past the post became evident only in retrospect, however: his first successful experiments, conducted in November 1952, were recorded only in

1956 in an internal company report written by A. D. Kiffer of the Linde Division. Eversole himself never published the work in scientific journals, and his own account appeared only in patents granted in 1958 and 1962.

Despite the prevailing view, propagated by Percy Bridgman and others, that high pressures were needed to stabilize diamond relative to graphite, Eversole chose to attempt diamond synthesis at low pressures—typically between 50 and 300 atmospheres. (This might not seem so "low" at all—but it was far below the several thousand atmospheres used by Bridgman and in the GE process, and is easily contained in standard apparatus without the need for extensive strengthening.) In late 1952 Eversole undertook a series of experiments that involved heating diamond seed crystals to between 820 and 1,000 degrees Celsius in a low-pressure atmosphere of carbon monoxide. He found that new diamond was deposited on the seeds.

As Eversole did not make public news of his success, it did nothing to alleviate the general belief that diamond synthesis at low pressure was impossible. (Curiously, this view was not shared by Percy Bridgman, who was happy to concede that the metastability of diamond at low pressures might not be an obstacle.) Yet in 1956, a group at the Institute of Physical Chemistry of the Soviet Academy of Sciences in Moscow, led by Boris V. Deryaguin, began independently to attempt diamond synthesis at low pressures. One of Deryaguin's colleagues, Boris Spitsyn, proposed that diamond might be deposited on seed crystals from hot vapors of volatile carbon compounds such as carbon tetraiodide (CI_4), at extremely low pressures of the gas (far less than one atmosphere). Spitsyn filed this idea as a patent in 1956, but the patent was not granted until 1980.

Then in 1959, with the Soviet work still unknown in the West and Eversole's results still all but invisible, a graduate student at the University of Michigan named John Angus independently hit on the idea that low-pressure growth of metastable diamond should be possible. He described this idea in a company report for the 3M Corporation in 1961, but it did not attract much interest at the time, because the prevailing view remained that at low pressures it would be impossible to avoid the formation of graphite from a carbon-rich gas.

In 1962 Eversole finally acquired a patent on his low-pressure process for diamond synthesis from carbon monoxide or methane. In the latter case the carbon atoms shed four hydrogen atoms apiece as they join the solid phase, and these hydrogen atoms combine as molecular hydrogen, H_2. It turns out that hydrogen plays a crucial role in low-pressure diamond synthesis.

During the early 1960s, John Angus and others began to speculate that if hydrogen gas was added separately to a carbon-containing gas, it might suppress graphite formation and so favor diamond. Toward the end of the decade, Angus (who had moved to Case Western Reserve University in Ohio) found that, although H_2 would slow down the nucleation of graphite from hydrocarbons, graphite crystals would nevertheless inevitably nucleate sooner or later. Once that happened, graphite, rather than diamond, would subsequently grow from the carbon vapor. Something more was needed to tip the balance in diamond's favor. To address this problem, Angus followed up a suggestion from his colleague Nelson

Gardner to introduce *atomic* hydrogen—individual hydrogen atoms, formed by passing H_2 over a hot tungsten filament. A lone hydrogen atom is a free radical—it possesses an electron that is not paired up in a chemical bond, and so is very reactive.

In choosing to use this voracious chemical species, Angus and Gardner were guided by several considerations. First, they knew that atomic hydrogen tends to "saturate" carbon-containing compounds—to encourage carbon atoms to form the maximum number of single bonds that it can (four, disported in a tetrahedral arrangement). Since diamond contains saturated carbon atoms (each joined to four others) while in graphite the carbon atoms are unsaturated (joined to three others), Angus and Gardner reasoned that a reactive hydrogen-rich environment might be good for diamond growth. In addition, they knew that hydrogen can stabilize the surface of diamond. The surface atoms are left with "dangling" bonds; to eliminate these, the surface has a tendency to rearrange its atomic constituents into a graphitelike structure, and that is not conducive to further diamond growth. But in 1963, James Lander and Jim Morrison of Bell Laboratories reported that the integrity of the diamond surface is retained when hydrogen atoms cap the dangling bonds.

It all seemed to make good sense. But the Case Western researchers also found a further benefit to atomic hydrogen: it can etch graphite away. Atomic hydrogen does not just stick to carbon surfaces, but reacts with them, stripping the carbon atoms away in the form of hydrocarbons. This reaction takes place at both graphite and diamond surfaces; but it happens faster for graphite than for diamond, and so removes the former preferentially. The Case Western group developed an approach that involved stopping the growth process periodically and introducing atomic hydrogen to etch away any graphite formed, before then starting the process up again. By 1971 Angus's team had achieved considerable progress in growing diamond films with this stop-start approach, and they described their success at an international meeting in Kiev.

This location is significant because it is questionable whether otherwise the news would have filtered back to the Soviet team of Deryaguin and Spitsyn. They had been working hard on the low-pressure approach throughout the 1960s, and in 1969 had reported at a meeting in Novosibirsk that they had reproduced Eversole's successes using pure methane at low pressure to deposit diamond on seed crystals. To remove the graphite that would inevitably form in this process, the team used oxygen; like the hydrogen in Angus's experiments, this strips away graphite faster than it does diamond. But this kind of stop-start process was not the way to grow diamond rapidly. When one of the team, Dmitri Fedoseev, attended the Kiev meeting and recounted the Case Western work to his colleagues, they began to introduce hydrogen into their own experiments. Rather than using atomic hydrogen only in a separate "clean-up" stage, however, Valentin Varnin from Deryaguin's team suggested that it might be introduced during the diamond growth process itself.

During this period, at the height of the Cold War, the flow of information between West and East was extremely limited, and the members of Deryaguin's group were forbidden by the Soviet authorities from discussing their results openly. This meant that there was much duplication of effort between the U.S. and Soviet groups, often on the basis of information passed on secondhand or by rumor rather than through the usual channel of scientific publications. It also means that reconstructing the evolution of the Soviet experiments is a patchy business that has to rely to a great degree on the recollections of the researchers themselves—now a somewhat dwindling resource (Deryaguin himself died in 1992). But when in 1976 Deryaguin and his colleagues were able to report (in a Russian journal) the growth of diamond films on substrates other than diamond seed crystals, such as copper, gold, and silicon, they made no mention of the use of atomic hydrogen. So it is not clear how far, by this stage, they had gotten with the approach that Varnin had proposed.

The 1976 report was significant all the same, since it raised the prospect of making diamond-coated, wear-resistant materials or of using diamond as an insulating material for silicon-based microelectronics. The Soviet team suggested that diamond growth on foreign substrates takes place first by formation of a thin layer of a compound containing both the substrate material and carbon—a carbide—and that the diamond films subsequently grow on this.

Hot-Wired for Speed

Varnin and other Soviet researchers did make reference to the probable consequences of using atomic hydrogen during the growth stage in a Russian paper in 1978, but it was not until 1981 that Deryaguin's group published the results of an experimental study that used atomic hydrogen to enhance diamond growth. This provided the crucial clue that led a Japanese team to revolutionize diamond synthesis.

Until that time, all of the low-pressure CVD techniques grew diamond films only very slowly—the film thickness increased at a rate of typically less than a micrometer per hour, so that growing a film just a millimeter thick would take over a month of continuous growth. This limitation stymied potential industrial applications. Japan's entry into the field began in earnest only in 1974. During the 1960s, several patents were filed by Japanese researchers for methods of diamond synthesis, but few led to publications or had a strong impact on the field. The first synthesis of diamond in Japan seems to have been by H. Honda and coworkers in 1964, who used spark discharges between metal electrodes to break apart carbon-rich kerosene at high pressure. Their solid product seemed to be diamondlike but was clearly not a very pure material. But the Japanese effort to grow diamond from pyrolysis of gaseous carbon compounds at low pressure began in 1974 at the National Institute for Research in Inorganic Materials (NIRIM) in Tsukuba, stimulated by the successes of the groups of Angus and Deryaguin. The NIRIM group, led by Nobuo Setaka (who is now at the Showa Denko company in Tokyo), was one of the first to foresee the potential of diamond as an electronic and optical material as well as purely a superhard material.

At the end of 1981, the NIRIM group announced at a meeting in Japan that they had developed a method for growing diamond films rapidly using a mixture of methane and hydrogen that was "activated" by a hot filament. The Japanese group cites the inspiration of an article in an obscure popular magazine by one W. B. Wilson, who had suggested in 1973 that the growth of metastable diamond might be enhanced by creating carbon atoms in an excited electronic state, in which form they would have a greater propensity to forge the chemical bonds that hold together the diamond lattice. Wilson proposed several methods for producing excited-state atoms; but he also pointed out that it would be necessary to find some way of extending the time that the atoms spent in the excited state before decaying back to the ground state, as this was typically very short. That Wilson presented his ideas in a forum where it was almost guaranteed to remain overlooked is symptomatic of the peculiarities of early diamond research.

When in 1981 Deryaguin's group cast more light on the role of atomic hydrogen in assisting diamond growth, the Japanese workers guessed that this was the key to realizing Wilson's idea. They deduced that the reaction between methane and hydrogen gas in a hot feed gas might produce activated hydrocarbon species that would generate excited carbon atoms when they fell apart. Specifically, the reaction with hydrogen would strip methane molecules of one or more hydrogen atoms, leaving hydrocarbon free radicals (CH_3, CH_2, and CH). These species, which would be generated only at much higher temperatures by fragmentation of methane on its own, are highly reactive because they possess dangling bonds. The NIRIM team used a heated substrate to nucleate diamond: the activated hydrocarbons fell apart on the substrate to yield carbon atoms in an excited state (which formed diamond) and hydrogen atoms (which reentered the gas phase). Any graphite that was formed instead of diamond was quickly etched away by the atomic hydrogen. Moreover, this reaction regenerated activated hydrocarbon radicals, ensuring a continual supply of these species and thereby of excited-state carbon (fig. 8.6).

The key to this process is the activation of the methane/hydrogen mixture to generate hydrocarbon radicals. The Japanese team used a hot filament of tungsten for this activation step, which split the molecular hydrogen into atoms; these then pulled other hydrogen atoms off methane molecules. Their technique became known as hot-filament-assisted CVD (fig. 8.7*a*). After the announcement in 1981, the NIRIM researchers published several papers in which they reported diamond-film growth rates of several micrometers per hour. In addition to the hot-filament method, the group developed a method of activating the feed gas mixture using microwaves, which generated a plasma of free radicals and atomic species. This was called microwave-plasma CVD (fig. 8.7*b*). These advances showed at last that diamond-film growth could be a commercially viable technology, and they spawned a new burst of enthusiasm for low-pressure diamond synthesis, which has led to its current status as a nascent industrial concern.

The results of the NIRIM team vindicated the efforts of Robert deVries at General Electric in Schenectady to reinstate work on low-pressure synthesis during the 1970s, and by the mid-1980s GE researchers, led by Tom Anthony, were routinely growing diamond films at low pressure. Rustum Roy from Pennsylvania State University visited the NIRIM laboratories in 1984, and as a result was able to win U.S. government support for research into low-pressure synthesis at

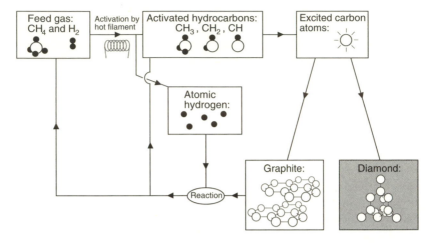

FIGURE 8.6 Research into diamond films grown by CVD blossomed into a commercially viable prospect when a group from the National Institute for Research in Inorganic Materials in Japan, led by Nobuo Setaka, found a way to accelerate the rate at which the films grew. They used a hot tungsten filament to "activate" the methane used as the carbon source, converting it to highly reactive hydrocarbon free radicals which then fell apart on the surface of a substrate to form diamond. The process is self-sustaining: any graphite formed is etched away by hydrogen atoms, regenerating activated hydrocarbon free radicals.

Pennsylvania and elsewhere. Gradually, U.S. industrial companies began to show interest, while in the same year Michael Pinneo founded Crystallume in California, the first U.S. company devoted solely to making diamond films.

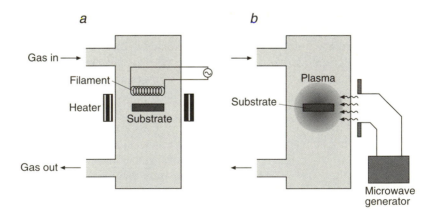

FIGURE 8.7 Diamond films can be grown rapidly by CVD methods in which either a hot filament (*a*) or a microwave discharge (*b*) is used to activate the feed gas of methane and hydrogen (sometimes containing a small amount of oxygen gas too). The latter approach forms a plasma of activated free radicals, and is called microwave-plasma-assisted CVD.

A Growing Concern

The diamonds that are grown by CVD are hardly jewels. The films are conglomerates of many tiny crystals, each typically a few micrometers across (fig. 8.8). Diamond crystals are nucleated at many surface sites simultaneously, and each crystal grows independently until it encounters its neighbors. Each of the tiny crystals is by itself of essentially gem quality, but the film as a whole is like a compaction of many tiny diamond chips that have fused together.

CVD methods have now been refined to a degree that allows film growth rates to reach almost one millimeter per hour, using a plasma activation method akin to the NIRIM technique. (This speed is unusual, however; rates of a few micrometers per hour are more typical.) But other methods of low-pressure synthesis are still being explored, with varying degrees of success. Foremost among these is the combustion of acetylene, which harks back to von Bolton's experiments in 1911. Yoichi Hirose of the Nippon Institute of Technology showed in 1988 that under carefully controlled conditions acetylene can indeed be burned in oxygen,

FIGURE 8.8 Diamond can be made without having to mimic the extreme conditions under which it is formed in the Earth. At lower pressures, graphite is more stable than diamond; but researchers have been able to trick carbon into forming diamond under these conditions by using the technique of chemical vapor deposition and mixing atomic hydrogen into the carbon vapor. In this way, diamond thin films can be deposited on surfaces. The films are polycrystalline—they consist of many microscopic crystallites of diamond that have grown together. Diamond films grown by CVD are now finding a range of technological applications. (Photograph courtesy of De Beers Industrial Diamond Division.)

in what amounts to an oxyacetylene blowtorch, to generate diamond. In this process the feed gas contains a little more acetylene than oxygen, so it is not all simply burned up into carbon dioxide and water. This is the simplest method of all (although not the most efficient); indeed, it seems likely that welders employing oxyacetylene torches constantly generate minute quantities of diamond as they work.

As with the high-pressure and shock-wave approaches, nature probably got there first with CVD diamond synthesis. A process akin to CVD may account for the formation of tiny diamonds in interstellar space. When a star explodes in a supernova, hydrogen and carbon gases are thrown out at high energy, which condense into diamond crystals as the gases cool. In addition, a certain class of stars, called red giants, hold large quantities of carbon and hydrogen in their outer atmospheres under conditions similar to those used in CVD, and these conditions can give rise to diamond crystals which find their way into interstellar space. Some of the microscopic diamonds found in meteorites are believed to have such origins. In 1995 researchers in England reported evidence that meteoritic diamonds collected from around the Ries meteoritic impact crater in southern Germany were formed by a CVD process in the hot plume of material ejected by the impact.

THE DIAMOND FILM INDUSTRY

In his novel *2061—Odyssey Three*, Arthur C. Clarke predicts that six decades hence buildings will be constructed from diamond-coated materials, making them impervious to the ravages of time. Perceptively, he picks out the key attribute of the new diamond technology: that it provides us with diamond films and coatings, rather than the diamond powder and grit of the older HPHT technology. The cost of these coatings will have to fall rather substantially before it becomes routine to use them on building materials, but CVD diamond films already look set to provide a new generation of hard, tough cutting and grinding tools. In themselves, diamond films are extremely hard but also quite brittle, so that a blade made of pure diamond would not last beyond the first microscopic crack it develops. But by depositing diamond films onto tools made from tough materials such as silicon carbide (discussed later in this chapter), one can combine hardness and toughness in a single component. Prototypes of such diamond-coated tools are already in production.

The wear resistance of diamond films also makes them ideal as coatings for bearings and engine parts, although at present the cost of making such coatings is too great to allow its widespread use. CVD diamond films tend to have a rather rough surface because of their polycrystalline nature, but chemical etching methods have been developed for polishing them to a smooth state. These use molten metals to dissolve the diamond away, rounding off the sharp edges of the tiny crystals. Diamond films derive their excellent wear resistance not only from their hardness but also from their high thermal conductivity, which allows them to dissipate the heat generated by friction before it can damage the film.

Toward the Carbon Chip

The communications industry is pushing the speed of silicon microelectronics to its limits, and we saw in chapter 1 that photonics now threatens to replace electronics in this demand for speed. But diamond films give electronics the chance to strike back by supplanting silicon with its sibling, carbon. According to K. V. Ravi of the Lockheed Company in Palo Alto, California, the promise that CVD-grown diamond holds in electronics makes this "potentially the most revolutionary of the applications of this material."

You might ask why diamond, the paradigm of an insulating material, should be considered for use in semiconductor electronics at all. Pure diamond has a band gap five times larger than that of silicon, so its valence electrons cannot be thermally excited to the conduction band at room temperature. This is why pure diamond is an excellent insulator (fig. 8.9a).

But just as the conductivity of silicon can be enhanced by adding dopant atoms that provide more charge carriers, so too can diamond be made semiconducting by doping. Boron atoms, for instance, will supply holes to the valence band (p-type doping). Boron is adjacent to carbon in the periodic table, and so the two

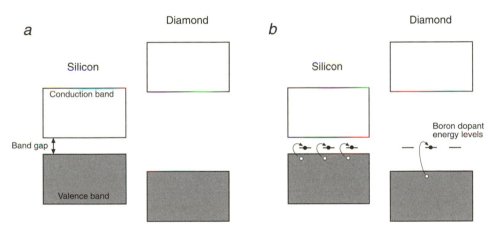

FIGURE 8.9 (a), Diamond is an insulator, whereas silicon, with the same electronic band structure, is a semiconductor. This is because the band gap—the band of "forbidden" energies between the electron-filled valence band and the empty conduction band—is five times greater for diamond than for silicon, so valence electrons in the former cannot acquire enough thermal energy, at room temperature, to make the jump to the conduction band. (b), Diamond can be made semiconducting by doping it with boron. But a high level of doping is needed, because the empty energy levels created by the boron dopants lie a fair way above the top of the valence band, so that not many valence electrons can reach them. Because doping with boron introduces positive charge carriers (holes), it is an example of p-type doping. In principle, n-type doping with phosphorus or nitrogen is also possible. But in practice it is difficult to incorporate high concentrations of phosphorus into the diamond lattice, whereas nitrogen suffers from the same problem as boron: its energy levels are too far below the conduction band to provide significant numbers of charge carriers, unless very high levels of doping are attained.

atoms have very similar sizes, meaning that a boron atom fits comfortably into the diamond lattice. In theory, doping with phosphorus, which supplies electrons to the conduction band (n-type doping), should also be possible, as it is for silicon; but phosphorus atoms are much larger than carbon atoms, so it is extremely hard to get any significant quantity of phosphorus into the diamond lattice. Other, smaller n-type dopants have been tried, such as nitrogen and lithium, but with limited success—so most studies of semiconducting diamond have focused on p-type doping with boron.

Usually the dopant atoms are incorporated into the CVD diamond films as they are grown, by including boron in the carbon-rich vapor. It is also possible to inject boron ions into preformed films by firing them at the material at high velocity, a method called ion implantation—but this brute-force approach can damage the film. High dopant concentrations are necessary to achieve a useful conductivity, because fewer charge carriers (holes) are created per boron atom in doped diamond than in doped silicon. This is because the energy levels of the dopant atoms lie further above the top of the valence band in diamond than in silicon (fig. 8.9*b*), so valence electrons need more energy to reach them: at room temperature, the valence electrons can access fewer than one percent of the dopant energy levels in boron-doped diamond. On the other hand, this means that diamond works better at higher temperatures: it remains a good semiconductor up to at least 500 degrees Celsius, whereas the conductivity of silicon can no longer be controlled under such hot conditions. So diamond presents a much more attractive prospect for electronic devices that are required to work at high temperatures—in contrast to silicon-based devices, which cease to be effective at around 150 degrees Celsius.

Devices made from semiconducting diamond have the potential to work at greater speeds than silicon. This is firstly because the particles that carry charge around in solid-state materials—electrons and holes—move much faster in diamond than in silicon, making diamond more responsive to high-frequency electronic signals. Second, diamond can support higher voltages (that is, higher electric fields) than silicon without breaking down by supporting a sparklike discharge. In an electronic device this kind of breakdown can cause the device to fail and perhaps even to be irreparably damaged. The breakdown threshold for diamond occurs at electric field strengths about thirty times greater than that of silicon, so that diamond devices could operate at higher voltages. Equivalently (because the electric field across a material is given by the voltage applied divided by the thickness of material), diamond devices could be much smaller (and thus faster) than silicon devices operating at the same voltage.

The potential of diamond in electronics has been called legendary; but realizing that potential is another matter. The main obstacle to a diamond-based microelectronics is the difficulty of making sufficiently high-quality thin films—perfect single crystals free from defects. Because defects scatter charge carriers, they degrade the electrical conductivity. So the polycrystalline films produced by standard CVD methods have rather poor electrical properties even when doped.

One approach for creating large single-crystal diamonds is to use small single crystals of natural diamonds as seeds, and simply to increase their size by careful deposition of a synthetic coating. Because the deposited film has a crystal structure that exactly matches that of the substrate, it does not matter if nucleation occurs simultaneously at several places—the growing films will match up precisely, and in the same orientation, where they meet. In effect, the seed acts as a template to guide the orientation and structure of the deposited film. This approach, called *homoepitaxial growth*, was pioneered by Deryaguin's group in the early 1960s. To make large-area single-crystal films this way, however, one faces a chicken-and-egg dilemma: a small crystal can grow in all directions into a larger one, but a flat diamond-film template must have the same area as the film one wants to grow. One solution is to use as the template a nondiamond substrate with a crystalline interatomic spacing similar to that of diamond. This is called *heteroepitaxial growth*. Crystals of nickel and copper, and also of the ceramic materials silicon carbide, beryllium oxide, and boron nitride have interatomic spacings that match closely that of diamond, and all have been used to template diamond-film growth. But even a small mismatch in atomic spacing will eventually lead to the formation of defects, and this is indeed what happens in most cases—the heteroepitaxial films are too full of defects to be used for electronics.

Michael Geis of MIT's Lincoln Laboratory and coworkers have produced a kind of "cheat" version of homoepitaxial film growth in which one tries to mimic a perfect single-crystal template film of diamond by an array of small seed crystals aligned so that they all have their crystal planes facing in the same direction. The method is labor intensive, because it requires that a collection of conventional HPHT diamond crystallites first be sorted to collect a batch that is almost uniform in size and shape. These crystallites are then inserted into tiny etched pits in the surface of a silicon wafer, by allowing them to settle there from a slurry of crystallites and liquid (fig. 8.10). A CVD diamond film is then deposited onto this array of aligned seeds. These films are never perfect crystals, because it is not possible to ensure absolutely perfect alignment of the seeds, but the misalignment of

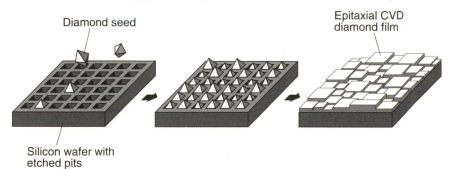

Diamond seed

Epitaxial CVD diamond film

Silicon wafer with etched pits

FIGURE 8.10 One approach to the growth of single-crystal diamond films uses an array of aligned diamond seed crystals as a template for growing a CVD film. The seed crystals, each of which is a tiny single crystal grown by the high-pressure/high-temperature method, are placed in pits etched into a silicon wafer. The shape of the pits matches the crystal facets of the seeds, and so ensures that their crystal planes all point in more or less the same direction. This process, developed by Michael Geis in 1992, is called MOSAIC, reflecting the character of the slightly imperfect diamond film grown on the seeds.

crystal planes in the films that grow out from each seed crystal is only slight, and does not interfere too much with the electrical properties.

Another difficulty in making diamond-based electronic devices is that of fashioning diamond films into device structures, because diamond is much more resistant to etching than silicon. New etching methods have therefore had to be developed to pattern diamond films, for example using highly reactive plasmas or the gas nitrogen dioxide as the etching agent.

So what has been achieved in diamond microelectronics? Most of the prototype devices fabricated so far have been transistors, in particular those known as metal–oxide–semiconductor field-effect transistors—MOSFETs, the most common type of switching device in modern computers. These devices are, as the same suggests, sandwiches in which a semiconducting layer lies beneath an insulating oxide layer, which itself lies beneath a metal contact called the gate. Doped diamond (usually boron (p)-doped) is used as the semiconductor, and the oxide layer is silicon dioxide. MOSFETs based on diamond can show better performance than those based on silicon, because in the latter case charge carriers tend to accumulate at the interface between the silicon and the silicon dioxide layer when a gate voltage is applied, and this shields the charge carriers within the semiconducting layer from the effects of the applied field. Diamond MOSFETs do not suffer from this drawback.

All of the diamond-based transistors made so far, however, have been strictly laboratory prototypes, because they have generally used doped CVD diamond films grown homoepitaxially on natural diamond substrates. This is utterly impractical for commercial applications, since it makes no sense to contemplate building a computer on a base of pure diamond. The problem of heteroepitaxy—of growing single-crystal films on nondiamond substrates—will have to be solved before diamond electronics becomes a reality. On the other hand, some researchers have concentrated on exploring just what can be achieved electronically using the polycrystalline films that standard CVD methods supply. These too can be doped to provide mobile charge carriers, but because of the many defects present, the carrier mobility is much smaller—typically by a factor of sixteen to twenty—than that in single-crystal films. Nonetheless, some devices have been created even from this relatively poor semiconducting material, including diodes and field-effect transistors.

Return of the Vacuum Tube

Diamond films may also find a rather different niche in electronics, one related more closely to the old-fashioned technology of vacuum tubes than to the modern technology of transistors. In an old-fashioned vacuum-tube diode, a metal filament in a vacuum chamber is heated by an electric current to temperatures of over a thousand degrees Celsius. At such temperatures the filament emits electrons. When the filament is given a negative charge (making it a cathode) and a separate "collector" plate placed nearby is given a positive charge (making it an anode), the electrons pass through the vacuum from the cathode to the anode. Thus an electric current can flow from the cathode emitter to the anode—but not vice versa. The device is a kind of one-way valve for electrons: a diode.

Diamond has the unusual property of being able to emit electrons into free space *without* being heated. This is because diamond has a "negative electron affinity," meaning that the bottom of the conduction band is higher in energy than the vacuum energy, the energy of a free electron in space (fig. 8.11*a*). So electrons placed in the conduction band (by n-type doping, for instance) can lower their energy simply by "stepping put" into space. Most metals have a positive electron affinity, which means that their conduction band lies below the vacuum energy;

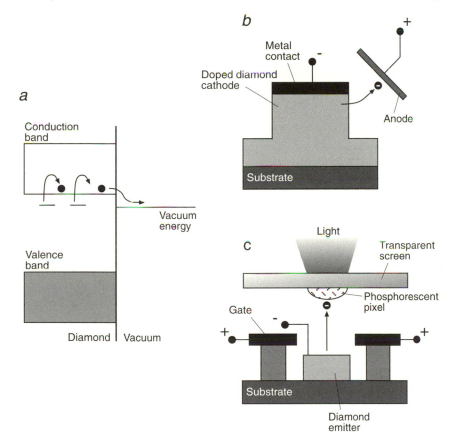

FIGURE 8.11 A new kind of miniaturized "vacuum-tube" electronic technology is promised by diamond devices called cold cathodes. These take advantage of the fact that semiconducting diamond can emit electrons spontaneously, without having to be heated like the cathodes of conventional vacuum tubes, because electrons in the conduction band have a higher energy than free electrons in the vacuum (*a*). When the emitted electrons are collected by a positive electrode, a current flows through the device and it becomes a kind of diode (*b*). These devices offer the prospect of higher speeds for information processing. The emitted electrons can be accelerated in an electric field into a beam that, when directed onto a phosphorescent material, provides the basis of a miniature cathode-ray tube for display devices (*c*).

then, conduction electrons can make the jump only when thermally excited up to the vacuum energy. Some semiconductors, such as silicon and gallium arsenide, can be given a negative electron affinity by coating them with cesium, but this is of limited practical use because cesium metal is so reactive that the tiniest amount of air or moisture destroys the performance rapidly.

Because diamond does not need to be heated to emit electrons, it raises the prospect of making "cold-cathode" diodes, analogous to vacuum tubes but (by virtue of modern microfabrication methods) on a much smaller scale (fig. 8.11*b*). Such devices might ultimately be incorporated into integrated circuits, where they could offer faster performance than solid-state devices because the charge carriers travel through a vacuum rather than through a solid material. Cold-cathode devices would also be more resistant to radiation-induced damage—this introduces defects into conventional solid-state structures, which scatter the charge carriers. The first diamond cold-cathode diode was made by Michael Geis in 1991, who used p-doped diamond. Geis envisages a new vacuum-tube microelectronics based on such devices, which could potentially offer higher-speed information processing than even diamond-based transistors.

Electron emission from cathodes can also be exploited in display devices—most famously in the cathode-ray tubes of televisions—by accelerating the electrons in an electric field and firing the resulting beam onto a phosphorescent material, which glows in response. Diamond cold cathodes are now being investigated for use in phosphor display devices (fig. 8.11*c*), which might one day compete with light-emitting diodes and liquid-crystal displays.

HARD CUSTOMERS

The only natural material that even approaches the hardness of diamond is the mineral corundum, a form of aluminum oxide (Al_2O_3). It is perhaps most familiar (although far more rare!) in the form of the gemstones sapphire and ruby, in which trace amounts of other metals turn the crystals blue and red, respectively. Like diamond, these gems too can be made synthetically using high-pressure methods, and artificial corundum is widely employed as an abrasive and in cutting tools. Where corundum falls short in comparison to diamond is not just in hardness but in toughness: like most solids in which the bonding is predominantly ionic, corundum is somewhat brittle. But today most of the superhard rivals to diamond are *covalently* bonded materials: so-called covalent ceramics, predominant among which are carbides and nitrides.

Some of these materials have a venerable history, since they proved easier to make than diamond. Henri Moissan, a pioneer of diamond synthesis, was also pivotal to the early development of hard carbide technology. As we saw earlier, he first synthesized silicon carbide in the late nineteenth century by accident, during his diamond-making attempts (and promptly mistook it for diamond). Industrial-scale synthesis of silicon carbide (SiC) was made possible by the experiments of the American A. G. Acheson in 1881—and Acheson too was trying at the time to synthesize diamond instead. He was hoping to recrystallize graphite into diamond within molten aluminum silicate, but instead the carbon reacted with the silicon, precipitating silicon carbide. Today the material is made by using sand (silica) in place of the silicate, and coke in place of graphite; but the process is still much the same as Acheson's. The product is obtained in a powdered form, which can then be pressed into molds before being heated to fuse the grains together into a dense, hard material. Because it remains hard and tough at temperatures of over 1,000 degrees Celsius, and because it resists oxidation and corrosion under these conditions too, silicon carbide finds uses in high-temperature engineering applications: as components for motor vehicle engines and gas turbines, for example. Bound into a softer matrix in the form of thin, hairlike crystalline whiskers, silicon carbide is used for reinforcement in composite materials. In powdered form, it is used as an abrasive.

Boron carbide is another venerable superhard covalent ceramic. It is made industrially in a manner similar to Acheson's process, by heating borax (boron oxide: B_2O_3) with coke. Moissan came across this material too in his high-pressure experiments on carbon, and he proposed (incorrectly) that it had the formula B_6C—this was corrected to B_4C in 1934. At room temperature, only diamond and boron nitride (see below) are harder than boron carbide, and boron carbide is superior even to diamond above 1,100 degrees Celsius, provided that it is not in an oxidizing environment (like air). So it is used for wear-resistant parts that are subjected to high temperatures, such as milling and grinding equipment.

Also deserving of mention here is tungsten carbide, the material that—under the trade name Carboloy—provided the optimal material for the anvils used by Bridgman and the GE team in their high-pressure diamond-making attempts. Carboloy was marketed as a hard material from the 1930s by General Electric, which devoted an entire plant in Detroit to its synthesis. Today tungsten carbide forms the basis of a material known as a *cemented carbide*, in which cobalt is used as a binding agent to cement the carbide grains together. The cobalt confers a degree of ductility to the material, making it tougher than the carbide alone. Cemented carbides (sometimes called *cermets*) are used in cutting tools and mining and drill bits.

Hard nitride ceramics have come into their own only since the 1950s, owing to the difficulties involved in their synthesis. Boron nitride (BN) can be made by the reaction of borax and ammonia (NH_3) at high temperature—but the product of this reaction is a relatively soft white material known as hexagonal boron nitride, in which the atoms are arranged in sheets of hexagons, just like the carbon atoms in graphite. This sheetlike structure gives it lubricating properties, like graphite. But unlike graphite it is not electrically conducting, because the band gap between the (filled) valence and (empty) conduction bands is large.

Hexagonal BN has been known since the early part of this century, and many researchers had speculated that it might be possible, by analogy with graphite and diamond, to convert it into a denser form in which the atoms are connected in a three-dimensional crystal lattice rather than in two-dimensional sheets. This form of BN, they reasoned, might have diamondlike hardness. In 1956, Robert Wentorf from the GE diamond-making team showed that hexagonal BN could be converted to a crystalline form with cubic symmetry, called cubic boron nitride or cBN, by heating it with alkali metals (such as lithium, calcium, and magnesium) and their nitrides. Cubic boron nitride is the second hardest material known (although even then the formal measure of hardness is only half that of diamond—by so much does the latter exceed all rivals). Variations of Wentorf's method are now used to manufacture cBN for industrial use as an alternative to diamond; under trade names such as Borazon and Amborite, it is used in cutting tools and as an abrasive. Cubic BN is particularly useful for cutting materials that react with diamond. As steel is one such (the iron in steel will dissolve carbon readily), cubic boron nitride has been able to occupy a niche that synthetic diamond has conveniently left behind.

The story of cubic boron nitride has evolved along very similar lines to that of synthetic diamond, albeit without the controversy that dogged the early years of the latter. Having first been synthesized at almost the same time as diamond itself, by one of the same teams and under similar conditions of high temperature and pressure, it is now being sought after in thin-film form from low-pressure CVD methods. So far, this has proved to be a refractory problem—there are still no well-supported and reproducible claims for the synthesis of cBN films by CVD. Several groups have reported partial success, synthesizing very small cBN crystallites (perhaps 10 to 20 nanometers across) embedded in a matrix of hexagonal or amorphous (disordered) BN. But no one has synthesized large-area films of crystalline cBN comparable to those that can now be made of diamond by CVD. Because cBN can be made semiconducting by doping—and because the semiconducting form has many of the advantages that diamond exhibits, such as high mobility of charge carriers and resilience to high temperatures—researchers hope that when this thorny problem is finally cracked, cBN films might be used in microelectronic devices.

Silicon nitride (Si_3N_4) is made simply by heating the two elements together: silicon powder and nitrogen gas. It is hard and crystalline (with several different crystal structures), and remains stable in air up to around 1,400 degrees Celsius. These properties led to hopes in the 1950s that silicon nitride would be a good material for fabricating moving parts for engines and turbines. But subsequent research has shown that getting reliable high-temperature performance from the material is not easy, and so far it has been mainly used in wear-resistant mechanical parts that operate at rather lower temperatures.

Silicon nitride reacts with alumina to give strong, corrosion- and oxidation-resistant ceramics called *sialons*, mixtures of silicon, aluminum, oxygen, and nitrogen in various proportions. These materials have been finding increasing use in high-temperature engineering applications since the 1990s; they are relatively cheap and retain their hardness and toughness up to temperatures of at least 1,000 degrees Celsius and sometimes significantly higher still.

If you look at the periodic table you'll see that the elements that comprise just about all of these materials (corundum and diamond included) cluster together in the top of the right-hand columns. It's an odd thing among the elements here:

their compounds tend to be one extreme or the other, gases or ultrahard solids. The oxides of carbon are gaseous—carbon monoxide and dioxide—while that of silicon, below carbon, is (in one form) quartz. Nitrogen oxides are gases too; but with silicon, aluminum, and boron, nitrogen forms hard solids. What, however, of the compound formed between carbon and nitrogen? These two elements are so essential to life, so central to organic chemistry, that one might anticipate something soft instead—a plastic polymer perhaps?

As far as we know, carbon nitride is not found in nature. But in 1985 Marvin Cohen from the University of California at Berkeley gave some thought to what such a material might be like. The carbon–nitrogen bond is a strong one, and the atoms are small and can pack closely. Research on hard materials has made it clear that hardness tends to require short, strong bonds. So Cohen reasoned that if one could form an extended solid in which carbon and nitrogen atoms are linked by single covalent bonds, it might have a robustness akin to diamond's. In 1989 Cohen and his colleague A.-Y. Liu constructed a hypothetical carbon nitride on a computer, and performed calculations, based on what is known about the way that these two elements form bonds, to predict what its properties might be like. They found that its hardness should be comparable to, and perhaps even greater than, that of diamond.

Cohen and Liu assumed that the carbon nitride material would have the same arrangement of atoms as silicon nitride (Si_3N_4), with each silicon replaced by carbon—this was a reasonable assumption, because silicon and carbon are in the same column of the periodic table, and both like to form four bonds to other atoms, disposed in a tetrahedral arrangement. There are in fact two different crystalline forms of silicon nitride, designated α and β. Cohen and Liu's hypothetical carbon nitride took the latter form, and so was designated β-C_3N_4 (fig. 8.12).

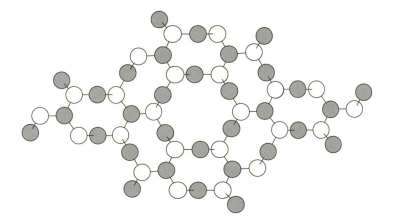

FIGURE 8.12 A new kind of superhard material was postulated by Marvin Cohen in 1985. He pointed out that carbon and nitrogen should be able to form a covalently bound solid which, because of its extremely high packing density of atoms, might be as hard as, or perhaps even harder than, diamond. This hypothetical material is designated β-C_3N_4. But can it be made?

The prospect of a material harder than diamond posed a very tempting challenge for experimentalists—but how should one make it? The obvious approach was to try something akin to the CVD technique used to make diamond at low pressures: to vaporize carbon and nitrogen compounds into their constituent atoms, and deposit them on a substrate. He-Xiang Han and Bernard Feldman of the University of Missouri in St. Louis tried such an approach in 1988—not in an attempt to test Cohen's earlier idea but to try to make *amorphous* carbon nitride, a noncrystalline material in which the atoms were arranged in a disorderly fashion. This material had been predicted four years earlier to be a potential component of the atmosphere of Saturn's moon Titan. Han and Feldman used a method that was essentially plasma-assisted CVD, splitting apart methane (CH_4) and nitrogen (N_2) molecules in a plasma and letting the fragments fall onto an aluminum or quartz substrate. And indeed they found a polymer-like, disordered mixture of carbon, nitrogen, and hydrogen—but there was no sign of a crystalline carbon nitride like that hypothesized by Cohen. Two later experiments involving pyrolysis and shock-wave compression of organic compounds containing carbon, nitrogen, and hydrogen similarly failed to generate the hard crystalline material that had been predicted.

In 1993 Charles Lieber and colleagues at Harvard University decided that the problem might lie with the high energy barrier to the formation of the carbon nitride material. Although it should be relatively stable once formed, β-C_3N_4 would require a substantial input of energy to break apart all of the existing bonds that the carbon and nitrogen atoms had formed in the precursor compounds. Lieber's group decided that it would be best to eliminate hydrogen from the system altogether, since it bonds readily to both carbon and nitrogen. They opted to create a vapor of pure carbon by blasting a graphite target with a laser. The Harvard group mixed this carbon vapor with a beam of nitrogen atoms, formed by shaking nitrogen molecules apart using intense discharges of radio waves (fig. 8.13).

The Harvard team found that they could grow thin films of a material containing just carbon and nitrogen. But although β-C_3N_4 itself would contain about 57 percent nitrogen, Lieber and colleagues were able to incorporate no more than 45 percent in their films. Most researchers in this game soon concluded that there was no strong evidence for β-C_3N_4 in these films. Despite several other claims of hard, crystalline carbon nitride materials by groups in China, Mexico, the United States and elsewhere in subsequent years, the materials scientist Robert Cahn of Cambridge University in England was moved to comment in 1996 that "up to now, Liu and Cohen's proposed superhard phase has proved a will-o'-the-wisp."

But my guess is that before very much longer, someone will succeed in making either β-C_3N_4 or one of the other dense, hard carbon nitrides predicted subsequently. Whether or not such materials will then start to replace diamond at the literal cutting edge of industry is another matter entirely, however—not only does diamond have a substantial head start, but it comes loaded with over a millennium of associations as the most desirable of prizes that the Earth has offered us.

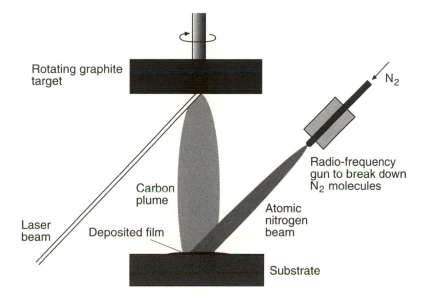

FIGURE 8.13 An attempt to make β-C$_3$N$_4$ by Charles Lieber and colleagues at Harvard University in 1993 involved laser ablation of a carbon source. The researchers fired laser pulses onto a graphite target, heating the material sufficiently to eject a plume of gaseous carbon from its surface. This plume was then intersected by a beam of nitrogen atoms, made by passing radio-frequency waves through molecular nitrogen gas (N$_2$) as it passed under pressure through a nozzle. The carbon and nitrogen atomic vapors intersected at the surface of a cool substrate, onto which carbon- and nitrogen-containing films of material condensed. But Lieber's group was unable to find convincing evidence for the superhard crystalline material.

Chain Reactions

THE NEW POLYMERS

> His classic study of large molecules spanned the decade of the
> twenties and brought us directly to nylon, which not only is a
> delight to the fetishist and a convenience to the armed insurgent,
> but was also, at the time and well within the System, an
> announcement of Plasticity's central canon: that chemists were
> no longer to be at the mercy of Nature. They could decide now
> what properties they wanted a molecule to have, and then
> go ahead and build it.
>
> —Thomas Pynchon, *Gravity's Rainbow*

Since the beginning of the century, polymers—long chain-like organic molecules—have become an increasingly prevalent part of our lives. As cheap, tough plastics, they are a symbol of the modern age. But unlike these "commodity" materials, the new polymers are finely engineered substances with precisely controlled molecular architectures. The control that we have over the structures of polymers enables their design for new uses, ranging from strong fibers to light-emitting displays.

IT WAS ONCE one of the attractive aspects of materials science that you could make a mint out of it without having much idea at all of what it was you were doing. Some might say that this is still possible; but at the start of the century it was pretty much a general rule. In 1905 a Belgian chemist named Leo Baekeland found his way to fame, fortune, and the front page of *Time* magazine for his invention of Bakelite—a material that neither he nor anyone else recognized as the first synthetic example of what we now know as a polymer.

I will risk castigation if I do not make clear that Baekeland was no fool. He had already achieved a degree of celebrity (and wealth) in 1900 with his invention of a photographic paper called Velox; and to turn the resinous substance obtained by heating a mixture of phenol and formaldehyde into the reproducible commercial product that was Bakelite involved no small amount of chemical ingenuity. But what Bakelite consisted of was anyone's guess.

That's not to say that the concept of a polymer did not exist when Baekeland was making his fortune. As we saw in chapter 4, commercial polymers based on the natural products cellulose and rubber became available halfway through the nineteenth century, and the word "polymer" itself dates from 1833, when it was coined by the Swedish chemist Jons Jacob Berzelius. But no one knew what a polymer was. The prevailing idea at the start of this century was that polymeric materials were aggregates of small molecules held together by an unknown force. It was only through the diligent work of the German chemist Hermann Staudinger and the American Wallace Hume Carothers in the 1920s and 1930s that we came to understand that polymers such as Bakelite, rayon, and polystyrene consist of huge chainlike molecules—macromolecules in Staudinger's lexicon—made up of small molecules (monomers) linked by chemical bonds. Armed with this understanding, the polymer industry has never looked back.

The ability of this industry to develop new materials has gone hand in hand with a growing understanding of what it takes to build a macromolecule. Just as the effect of a poem depends on the way that the words are chosen and arranged, so the properties of a polymer are an immediate consequence of the way in which the monomers are put together. The exquisite functions of biological macromolecules are the result of an equally exquisite arrangement of their constituent parts. The poetry of synthetic polymers is far less sophisticated, because we are, if you will, far less chemically literate than nature: we are simply not able to realize molecular arrangements of such complexity. Our human-made polymers are largely doggerel.

Nonetheless, they are the stuff of a multibillion-dollar industry. Today commercial polymers are most commonly associated with the materials known colloquially as plastics, although in fact these materials represent just one of the three major classes of industrial polymers. Plastic is one of those adjectives turned nouns: the word properly describes a property rather than a type of material. If you bend a plastic material, it stays that way. In this sense, most metals are plastic, and indeed materials scientists talk quite happily about plastic deformation of metals, glasses, and ceramics. But for most bulk applications where plasticity is required, organic polymers are the material of choice: they are relatively cheap, lightweight, and durable. The most common polymeric plastics in use today are polyethylene (used in a flexible low-density form in packaging film, insulation, domestic items such as toys and kitchenware, and in a more rigid high-density form in pipes, sheets, and containers), polypropylene, polyvinyl chloride (PVC), and polystyrene. While world consumption of metals grows at a steady, slow rate, the consumption of plastics is booming exponentially, and now far exceeds that of metals (fig. 9.1).

Elastic is another adjective appropriated (inaccurately) as the colloquial noun for a material. The "elastics" of hosiery and haberdashery are more properly known as elastomeric polymers, which have the property that they regain their

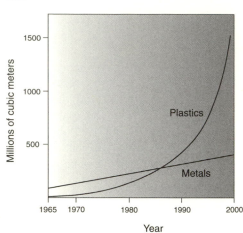

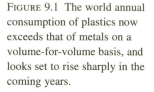

FIGURE 9.1 The world annual consumption of plastics now exceeds that of metals on a volume-for-volume basis, and looks set to rise sharply in the coming years.

shape when severely deformed and then released. (Truly elastic, rather than elastomeric, materials can recover only for a very small degree of deformation.) Elastomeric materials are also called rubbers, although this too is unsatisfactory since rubber is itself a specific kind of *natural* elastomer described in chapter 4. Synthetic elastomers include silicone rubber, polyurethanes, and polyisoprene (an artificial approximation to natural rubber), as well as "mixed polymers" (copolymers) whose chains contain more than one kind of monomer unit, such as poly(styrene–butadiene).

The third major class of industrial polymers is fibers, of which rayon and nylon are familiar examples. Rayon is derived from a natural, fibrous polymer (cellulose), while nylon is a purely synthetic polymer called a polyamide, a crude attempt to copy natural silk. Nylon and polyesters (polymers made by linking together carboxylic acid monomers) together make up 70 percent of the synthetic fiber market.

These polymers, which generally perform structural functions (packaging, holding things together, providing protective coatings, and so on), account for around 90 percent of the output of the polymer industry. Yet for the most part I am now going to ignore them, because they are old-style materials—low-value-added, usually with fairly crudely defined compositions and little subtlety in performance, mostly made by what might be rather impolitely called bucket chemistry. They are regarded as "commodity materials," which says it all. The new advanced polymers are not like this. Their structures are often complex but carefully controlled, and their properties and functions are fantastically diverse.

Traditionally, polymer science is a science of chainlike molecules. When polymer scientists reach for analogies to describe their materials, worms, snakes, and spaghetti make a frequent appearance. These analogies have even become embedded in the formal lore of the field: the motion of individual polymer molecules in a molten polymer is termed "reptation," a reference to the snakelike creeping of the chain ends through the tangled mass of neighboring chains. But in recent years researchers have started to find ways of linking monomer units into new architectures that belie the "straight" image of polymer science. Ball bearings, trees, and sheets are just some of the similes that are beginning to enter the vocabulary of polymer scientists. The ability to link monomers together into structures like this has been acquired by dint of much chemical ingenuity.

A few of these advanced polymers are beginning to find their way into the marketplace, but most are still the focus of primarily academic rather than industrial attention. They are mainly high-value-added materials, expensive to produce and generally in search of small-scale niche applications. They will not supplant commodity polymers, but they will go where no polymer has gone before: into applications such as microelectronics and biomedicine, which have traditionally relied on other high-value-added materials. In small but crucial ways, these new polymers will be ever more a part of our lives.

Strong Threads

One of the most bizarre aspects of warfare in the late Middle Ages must surely be that many of the combatants could barely see and barely move. They needed small cranes to lift them onto horseback (although Hollywood has not yet shown me how they got back down again), and if they fell over in battle then they could only flounder in the mud like a bug on its back. These medieval knights would surely have appreciated the plastic age.

Today body armor is made primarily of lightweight plastics: of polymer fibers embedded in a resin matrix. These fibers have a tensile strength comparable to or better than steel, at a fraction of the cost in weight. They are usually composed of Kevlar, a synthetic polymer introduced by the Du Pont chemicals company in the 1970s. Composites containing Kevlar fibers function as strong materials in all manner of applications where weight is at a premium, particularly in the aeronautical and aerospace industries, while the pure fibers are woven into ropes and cables capable of anchoring oil-drilling platforms.

A clue to attaining high strength in polymer fibers is provided by silk, one of the strongest natural fibers known. (It is no coincidence that silk was used as padding for metal armor by the Chinese in the sixth century A.D., or that it was reintroduced in Russia as a protective material against firearms shortly before the October Revolution.) Silk has a high degree of crystallinity: its protein strands are highly aligned along the axis of the threads, and this alignment means that the chains put up considerable collective resistance to tensile stresses. We saw in chapter 4 that spiders create this oriented polymer by spinning it from a protein solution in which the chains are oriented by the flow of the liquid through a narrow channel.

But some polymers will become aligned in the liquid state even when it is motionless. Such an aligned state is called a *liquid crystal*. To be able to form a liquid crystal in the melt or in solution, a polymer must have fairly stiff chains. If they flop about like cooked spaghetti, they are more likely to form a disordered, tangled mass, which has a low stiffness. The high-strength polymer fibers developed in the 1970s—Du Pont's Kevlar and the X-500 fibers produced by Monsanto—contain chains stiffened by the presence of so-called aromatic groups, whose central component is the benzene ring. Kevlar belongs to a class of compounds called aramids (a condensation of "aromatic amides"), the chains of which contain benzene rings linked through amide bonds like those that join amino acids in proteins (fig. 9.2). The benzene rings make the chains rather in-

FIGURE 9.2 Kevlar is an aramid polymer, whose chains are relatively stiff. This stiffness causes a high degree of alignment in Kevlar fibers, giving them great strength.

flexible, and these stiff rods form liquid crystals in a melt or a concentrated solution of the polymer. Fibers spun from these highly viscous fluids retain the high degree of molecular alignment, giving Kevlar its tremendous strength. The fibers are limited, however, to applications where only strength under tension is required—that is, to situations where they will be pulled rather than squeezed. Under compression their strength is very low: Kevlar fibers can anchor an oil rig down, but you won't find Kevlar girders holding it up. Like most strong polymer fibers, Kevlar also suffers from fibrillation: that is, from split ends, which arise because the polymer chains are only weakly bound to one another.

During the 1970s and most of the 1980s, Kevlar and related aramid fibers such as Twaron (produced by the ENKA company in the Netherlands) had no competitors, but from 1986 onward a number of strong, light fibers made from polyethylene entered the market. Although they suffer from the limitation of a low melting point, these materials have a tensile strength even higher than Kevlar, and indeed higher than any other organic material known. The key to this strength is again the achievement of a very high degree of alignment between the polymer chains. Polyethylene chains are not nearly rigid enough to support a liquid crystalline phase—the alignment results instead from a sophisticated processing method in which a gel-like precursor is first extruded and then drawn out into fibrillar threads. These fibers are highly resistant to degradation, which makes them well suited to applications such as permanent suture threads in surgery.

ORGANIC ARCHITECTURE

Gaining Control

A large part of the diffculty in synthesizing polymer molecules to precise specifications is that conventional methods make a whole bunch of them simultaneously while each competes for the same components parts. The result is a very heterogeneous crowd: some chains are relatively short, some disproportionately long. In the crush, some chains fuse together, while others sport dangling branches. Highly illustrative of this problem is the case of free-radical vinyl polymerization, one of the main routes by which the commercially important polyvinyl polymers (such as PVC or PVA) are made. The monomers are relatives of ethylene (C_2H_4) in which one or more of the hydrogen atoms is replaced by another chemical group. Polymerization here is a matter of bursting open the double bond between the two carbon atoms and using this freed-up bonding power to link the vinyl units together (fig. 9.3a). Since the monomers are stable molecules, we have to introduce some unstable species to get the ball rolling: these so-called initiators are free radicals, molecules with a single "lone" electron—one that is not paired up with another. They will form a bond with just about anything they encounter.

The thing about a lone electron is that you can't just get rid of it—it is like a piece of sticking tape, which you can transfer from one finger to another but can't shake off. When a free radical attaches itself to a vinyl monomer, the product itself becomes a free radical. So it too latches onto another monomer, forming a longer-chain free radical (fig. 9.3b). And so the process continues: the free-radical chain acquires ever more monomers, but always ends up with a free-radical group at the chain end. This is called, sensibly enough, chain propagation.

The only way a free radical can eliminate its lone electron for good is to find another free radical, so that the two can pair up their spare electrons in a chemical bond. This is called a termination step and may take place either between a growing chain and an initiator or between two growing chains. Since there is no way of controlling when termination steps will occur in this process—they depend on chance encounters between radicals—the chains in the final polymer product will acquire a variety of lengths. Moreover, a free-radical chain end can also terminate by pulling a hydrogen atom off the middle of another chain, thereby creating a lone electron in the *middle* of the other chain. This chain will then begin to acquire monomers in its middle, giving rise to a branched chain. This is called a chain transfer or branching step (fig. 9.3b).

Free-radical addition is not the only way to link monomers together; in particular, they can become joined by a so-called condensation reaction, in which a small molecule (commonly water) is expelled as two reactive groups in the monomers join up. This is what happens when a peptide bond is formed in protein synthesis, and it is also the mechanism by which synthetic polyesters are made. But here too there is little possibility for control of the macromolecule's length or structure in the conventional syntheses used industrially.

One of the most important advances in the control of polymer synthesis came in the 1950s, when the German chemist Karl Ziegler discovered a class of catalysts that facilitates the joining up of ethylene into polyethylene at low temperatures and pressures and without extensive branching reactions. Ziegler's catalysts, which were compounds containing transition metals coordinated to organic groups, provided polyethylene macromolecules that were all straight chains, allowing them to pack together efficiently into a material called high-density polyethylene that was denser, tougher, and with a higher melting point than conventional free-radical-polymerized polyethylene. Shortly afterward, the Italian

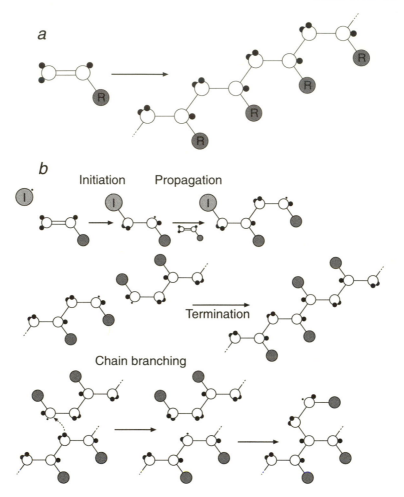

FIGURE 9.3 Many polymers are made by free-radical polymerization of the monomer units. A common class of polymers made this way are the vinyl polymers (*a*), made by linking together units in which two carbon atoms are joined by a double bond. Here R is a generalized side group—for instance, in poly(vinyl chloride) it is a chlorine atom, in polystyrene it is a benzene ring. The monomers link together in a reaction (*b*) that involves free radicals, highly reactive molecules with an unpaired electron (here represented by a small dot). A free-radical initiator (I) gets the process going by adding to one end of the double bond; the other end becomes a free radical and latches onto a second monomer, which in turn acquires a free-radical end. This process of addition of monomers (propagation) continues, with the chain getting ever longer, unless one of two other processes intervenes. If two free-radical chain ends encounter each other, their lone electrons pair up and the chains link up end to end (termination). In effect, termination involves the mutual annihilation of a free-radical pair. Alternatively, a free-radical chain end can pull a hydrogen atom from some part of another chain. Then the former chain stops growing, but the latter chain acquires a free-radical group in the middle of the chain. Addition of monomers at this point gives the chain a new branch.

chemist Guilio Natta showed that Ziegler's catalysts could permit even finer control of the structure of a vinyl polymer chain. For the class of polymers shown in figure 9.3 there are several possible ways in which the substituent groups R on the polyethylene-like chain can be arranged. This is not obvious from the figure, because it shows a two-dimensional, flat representation of what is in fact a three-dimensional structure: the four bonds around each carbon atom point to the corners of a tetrahedron. When represented this way, we can see that there is more than one possible relationship between the R substituents on alternate carbon atoms: each can be on one side or the other of the zigzag chain. In free-radical vinyl polymerization, one has no control over the arrangement of R groups—they are arranged on one side of the chain or the other purely at random. This is called an atactic structure (fig. 9.4a).

Natta's discovery was that catalysts like those developed by Ziegler could produce linear vinyl chains in which all of the substituents lie on the same side (fig. 9.4b). This arrangement is called *isotactic*. The atomic composition of the atactic and isotactic chains is identical, but no amount of twisting of bonds can convert one into the other. Until this discovery, it had proved impossible to polymerize propylene (CH_2CHCH_3) successfully, because the free-radical methods that were available would produce an atactic arrangement of the methyl (CH_3) groups along the backbone. This destabilized the product because the random arrangement prevented efficient packing of the chains. In 1954 Natta showed that

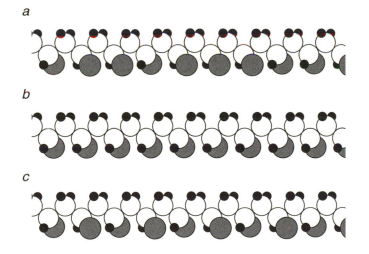

FIGURE 9.4 In free-radical polymerization, the R substituents of vinyl polymers (gray spheres) are generally arrayed at random on either side of the carbon backbone. This arrangement is called atactic (*a*). Notice that it is the three-dimensional, tetrahedral arrangement of the chemical groups around each backbone carbon atom that allows for the distinction to be made between one side of the backbone and the other. A catalyzed polymerization process discovered by Guilio Natta allows the R groups to be placed all on the same side, called an isotactic arrangement (*b*). A third possibility is a syndiotactic arrangement (*c*), in which the R groups alternate from one side to the other.

Ziegler's catalysts could be used to polymerize propylene into isotactic polypropylene. This is now one of the major commercial plastics.

The disorder in an atactic vinyl polymer chain may prevent the polymer from crystallizing: thus, while atactic polystyrene is a glassy material, isotactic polystyrene will form crystals in which the chains are packed closely together. Thus the ability to control a polymer's molecular structure affords some control over its material properties.

Today over eighty billion pounds of polymers are made each year using the catalysts that Ziegler and Natta developed. But in the mid-1980s, the German chemists Walter Kaminsky and Hans Brintzinger developed a new class of polymerization catalysts that allow even greater control over the structure of polyethylene and polyvinyl chains. Their catalysts were based on so-called metallocene compounds, in which a metal ion (such as titanium, zirconium, or hafnium) is sandwiched between two five-carbon rings (known as cyclopentadienyl rings). Whereas Ziegler–Natta catalysts produce mostly isotactic polypropylene, metallocene catalysts allow researchers to vary the tacticity of polypropylene at will, simply by changing the arrangement of substituents around the metal ion. In particular, they make possible the formation of a third form of polypropylene: *syndiotactic*, in which the methyl groups alternate from one side to the other along the chain (fig. 9.4c). Previously, this form of the polymer was only a very minor by-product of the Ziegler–Natta method. It is softer than isotactic polypropylene, but also tougher and clearer, and so is a potentially valuable material. The metallocene catalysts can also allow some control over the length of the polymer chains. They are much more expensive than traditional Ziegler–Natta catalysts, and so have not yet found industrial-scale application; but many polymer companies, including Exxon, Dow, and Hoechst, are working on pilot projects aimed at reducing the overall cost of the products, and they expect that this kind of catalysis will soon be competitive with the older methods.

Staying Alive

The biggest obstacles to a uniform, well-controlled product in traditional free-radical polymer synthesis are the processes of branching and chain termination. At any stage, once the reaction is underway, a growing chain can suddenly find itself next to a free radical that will cap off the reactive chain end and stop further growth. Or the reactive part of a chain might latch onto a neighboring chain and create a branched molecule.

There is a trick for getting around these problems, and it goes by the curious name of *living polymerization*. We shall see shortly why it is so called, but I should say here that "living" is purely a figure of speech: the polymers are certainly not alive in the way that a bacterium is alive. It took the polymer community some time to wake up to the benefits of living polymerization, because although examples of it were known in the 1930s (and were even then used to make a commercial elastomeric material), and despite a description of some of its key characteristics by Ziegler in 1936, it was not until the 1950s that the approach

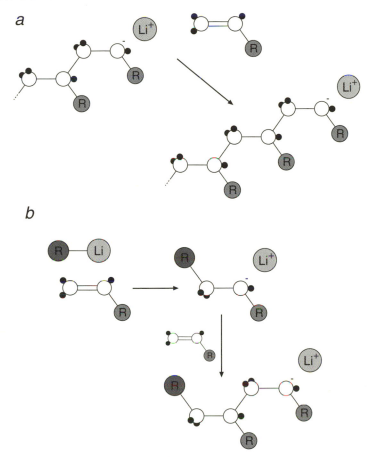

FIGURE 9.5 In living polymerization, one end of a growing polymer chain is kept "alive" (reactive) while being "chaperoned" by a complementary chemical species that ensures that the chain grows at a steady rate, without branching or termination processes. The result is a polymer whose linear chains are all of more or less the same length. In anionic living polymerization, the reactive chain end is a negatively charged (anionic) carbon atom, chaperoned by a positive metal ion such as lithium (a). Polymerization is initiated by an alkyl lithium compound (R–Li): the R group attaches to one end of the double bond of a monomer, while the other end becomes anionic and activated for chain growth (b).

became widely recognized and used. The basic idea is to protect the reactive end of a growing polymer chain by pairing it up with a kind of chaperone that allows the attachment of new monomers but nothing else.

The early examples of living polymerization generally involved the use of a reactive metal such as sodium or lithium to initiate the linking up of monomers containing double bonds between carbon atoms. The metal initiator passes an electron to the double bond in the monomer, creating an ion pair: a positively charged metal ion and a negatively charged monomer (fig. 9.5a). Ionized, carbon-based organic molecules are highly reactive and are important intermediates in many organic reactions; when they are negatively charged, they are called carboanions, and when positively charged, carbocations.

An alternative kind of initiator commonly used in living polymerizations is a so-called alkylmetal compound, in which a hydrocarbon group is attached to the metal. In these compounds the metal is in the uneasy position of being bound to the alkyl group by an essentially covalent bond, whereas its preference is to form an ionic compound instead. Faced with a double bond between carbon atoms, the metal dumps its alkyl group on one end of the double bond and forms a carboanion at the other (fig. 9.5*b*).

Once initiated in this way, the polymerization reaction proceeds by the sequential addition of monomers onto the carboanionic chain end. The metal ion, meanwhile, stays in the vicinity of this chain end and stabilizes it somewhat by virtue of its positive charge. In this way, the negatively charged end of the growing chain remains reactive enough to allow for the attachment of new monomers, but not so much as to indulge in side reactions such as chain branching. In fact, the carboanion/metal ion pair is stable enough that, if the reaction runs out of monomers, the chains will simply sit there like a patient creature awaiting more food. It is in this sense that the polymer chains remain "living"—addition of more monomer will restart the polymerization process immediately. In conventional free-radical polymerization, in contrast, the reactive chain ends simply don't have this patience—they will seize almost anything to cap off the end and stop the reaction for good. Michael Szwarc first elucidated the details of the living polymerization process in 1956, and it was he who introduced the term.

Most living polymerization reactions involve carboanions, but in recent years there has been some success in creating other kinds of reactive chain ends: in particular, positively charged ends (carbocations), which are chaperoned by negatively charged ions. The polymerization reactions of carbocations are harder to control, however, because they are more reactive than carboanions and are thus more easily "killed." For this reason, the chain ends typically are not carbon atoms with a full-blown positive charge, but instead carbon atoms bound to the chaperone by a partly covalent bond, which, however, has a strong degree of polarization with more positive charge at the carbon end and more negative charge at the other. The first successful carbocationic polymerization, conducted in 1984 by Japanese chemists T. Higashimura and colleagues, yielded a polyvinyl ether and used a tri-iodide (I_3^-) initiator to create this kind of part-ionic, part-covalent reactive chain end (fig. 9.6). Living polymerization can also operate with a chaperone that is fully covalently bound to the chain end. This might seem strange, as you might anticipate that such a union simply forms a stable molecule; but the trick is to find an end group that is compliant enough to fall off, enabling the attachment of a new monomer, and then to stick back on again.

Living polymerization processes permit a great deal of control over the products. All of the chains grow at more or less the same rate, since all involve the sequential addition of monomers one by one. If, then, all of the chains are initiated at the same time (which can be ensured by making sure that the initiator acts fast

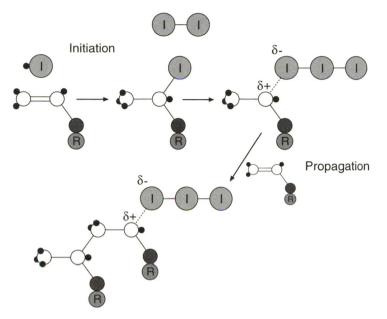

FIGURE 9.6 In cationic living polymerization, the activated end of the growing chain is a carbon atom with a positive charge. Generally this atom and the negatively charged chaperone are not fully ionic—there is a chemical bond between them that is partly covalent (resulting from sharing of electrons, shown as a dashed line) and partly ionic (giving each a partial charge, denoted $\delta+$ and $\delta-$). Here I show the cationic polymerization of a vinyl ether, in which the R substituent is attached via an oxygen atom; these monomers are particularly well suited to this kind of polymerization. The reaction is initiated by a mixture of hydrogen iodide (HI) and iodine (I_2), which forms a negatively charged tri-iodide (I_3^-) chaperone at the cationic chain end.

relative to the time it takes for the chains to grow appreciably), the lengths of all of the living chains will be more or less the same at any stage during the reaction. The final length of all the chains will be determined simply by how much monomer we add. So living polymerization offers a route to polymers with a very narrow range of chain lengths rather than the traditional mélange of molecules of very different sizes. This is particularly valuable when creating polymer architectures other than the usual straight chain; that is, for introducing branching *controllably*. Consider, for example, adding monomers to a molecule that has several reactive groups. If chains grow from each of these groups at the same rate, the result is a many-armed molecule with arms of equal length, called a star polymer (fig. 9.7a). Because they have a compact, globular shape (the arms tend to crumple up), star polymers have very different flow properties than linear polymers: they can roll over each other rather than getting entangled. Star polymers can be useful for altering the viscosity and flow characteristics of straight-chain polymers of similar chemical composition.

Another possibility is to start with a linear polymer that has several reactive groups inserted along the chain, and to begin to grow side arms from each of these

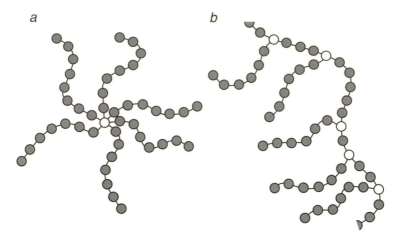

FIGURE 9.7 Living polymerization and other synthetic strategies offer new, nonlinear polymer architectures, such as star (*a*) and comb (*b*) polymers. Here the white monomers provide the focus of branching points from the linear chains of gray monomers.

groups. Living polymerization will ensure that all of these arms have the same length, as well as making sure that they do not join up with the ends of adjacent arms. The result is a polymer comb (fig. 9.7*b*), in which the side arms may have a different composition and different properties from the main backbone. These materials find uses as elastomers and adhesives.

Two in One

Because the ends of living polymers remain active when polymerization stops (that is, when the monomer supply is exhausted), it is straightforward to resume polymerization by adding a *new* type of monomer. This gives rise to chains containing two blocks of different composition—called diblock copolymers. Of course, the end of the second block will also remain active, so we can go on adding further blocks, which can be the same as or different from the first two blocks. These materials, known generally as *block copolymers*, have an extremely useful and interesting range of properties. Many of these can be considered to derive from the fact that the materials are hybrids containing two or more polymers in each single macromolecule. The material must then find a way to accommodate the fact that the two blocks will behave differently.

It is generally the case that one polymer is not miscible with another: because each monomer unit tends to prefer the company of others like it, a mixture of two polymers separates out into layers like unshaken salad dressing. This is not possible, however, when the chains are linked by strong covalent bonds, as they are in block copolymers. The blocks nevertheless try to separate as far as they may, by clustering into microscopic regions whose dimensions are limited by the length of the blocks, typically to between 5 and 100 nanometers. This produces a material that is highly nonuniform on the microscopic scale.

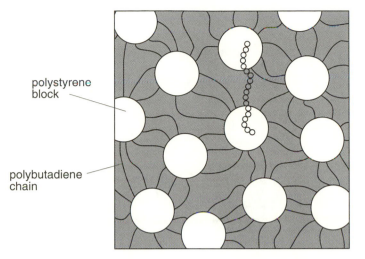

polystyrene
block

polybutadiene
chain

FIGURE 9.8 A triblock copolymer of butadiene and polystyrene, in which a block of the former monomers sits between two blocks of the latter, forms a thermoplastic rubber. The polystyrene blocks (white) form dense, rigid aggregates that are linked together by the more flexible polybutadiene blocks (gray) to give a rubbery material. (Here I show just one of the copolymer chains in full.) Because there are no strong chemical bonds holding this structure together, the material will melt to a plastic, moldable form when heated, but will set to the rubbery state when cooled again.

One of the most commercially important block copolymers consists of a block of polybutadiene sandwiched between two blocks of polystyrene. The polystyrene units aggregate into clusters linked together via polybutadiene chains (fig. 9.8). The clusters are stiff, like bulk polystyrene itself; but the polybutadiene chains are flexible, and the material behaves like a rubbery elastomer. But unlike rubber itself, the copolymer will melt and become plastic and moldable when heated, because the chains are not cross-linked by covalent bonds. This material is an example of a thermoplastic rubber, and is used to make the soles of sneakers.

One of the most remarkable properties of block copolymers is their ability to form *ordered* heterogeneous structures, in which the regions of clustering of the blocks are arranged in a manner that repeats regularly throughout the material. These ordered phases can be considered to be a strange kind of crystal, disordered on the atomic scale (since the individual polymer chains remain a tangled mass) but ordered on a much larger scale. The nature of the ordered structure depends primarily on the length of each block. For example, consider a diblock copolymer consisting of a block *A* made up of a certain number of monomers A and a block *B* made up of B monomers. If *A* is much longer than *B*, most of the material consists of chains of A's, and the short *B* segments will cluster together in this sea of *A*, forming little spherical aggregates. These *B* clusters could be dispersed randomly throughout the sea of *A*, but just as the energy of atoms in crystals is optimized by packing them together regularly, so the *B* clusters might prefer to

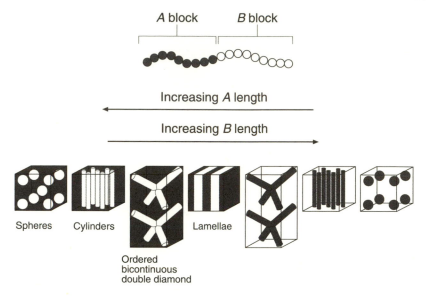

FIGURE 9.9 Diblock copolymers can form a variety of ordered structures, in which the individual chains are tangled and disordered but the aggregates that the two blocks form are ordered over long ranges. The structure of the segregated domains depends on the relative lengths of the two blocks. Some of the common ordered phases of diblock copolymers are shown here.

adopt an ordered arrangement (fig. 9.9). For longer B blocks these clusters are larger, but there comes a point at which the freedom of the A chains in between the clusters starts to be compromised by their size. Then the material takes on another structure, typically one in which the B aggregates form cylinders rather than spheres. As the length of the B blocks approaches that of the A blocks, however, there is no longer any reason why the system should support isolated regions of B surrounded by a sea of A; instead, it will undergo another transition to a state in which the A and B regions are equivalent—a lamellar structure in which layers of A and B alternate. This is a kind of polymer heterostructure, analogous to the semiconductor heterostructures encountered in chapter 1, and with layers of comparable thickness. But the difference is that, while the latter must be built by painstaking deposition of each layer using sophisticated atomic-vapor technology, the polymer structure builds itself spontaneously! Who knows what uses might be found for such structures if they are made from blocks of the kind of electrically or optically active polymers described later in this chapter?

   As the length of the B blocks is increased to exceed that of the A blocks, we find the same sequence of transitions in reverse: first the formation of cylinders of A, and then spheres of A. This sequence of events is by no means exhaustive, since there are many other ways of separating the A and B blocks into ordered domains. For polystyrene–polyisoprene diblock copolymers, for instance, another ordered phase can be identified between the cylindrical and lamellar stages, called the gyroid phase. This is an example of an ordered bicontinuous phase, consisting of two independent, interlocking periodic networks of one block type that extend continuously through a "matrix" of the other block type. The ordered bicontinuous phase shown in figure 9.9 is known as the double-diamond phase, for which the gyroid phase of polystyrene-polyisoprene was initially mistaken when first identified in 1988.[1] Polyisoprene, incidentally, is rubbery, whereas polystyrene is a hard glassy substance; so in all of these ordered phases we find regular arrays of materials with different mechanical properties, providing composite materials with highly controlled microstructures. At present, the uses of such ordered composites is largely a matter for speculation, but it would be surprising if applications do not soon materialize.

While the blocks of copolymers are not miscible with each other, they *will* be soluble in a melt of the respective pure polymers: that is, the *A* block of an *AB* copolymer will be compatible with a polymer of pure *A* chains, and likewise for *B*. This makes diblock copolymers entirely analogous to the surfactants discussed in chapter 4, in that they have two ends that are soluble in mutually incompatible liquids. So these polymers can act as surfactants for polymer mixtures, helping to stabilize a dispersion of one kind of polymer in another by coating the interface between them, just as soaps stabilize emulsions. This allows for the formation of stable polymer blends, "alloys" of polymers in which the two components are segregated into microscopic domains. The ability to blend polymers is important for plastics recycling, since a wide variety of different plastics is recovered from discarded items. (There are some limitations on the possible combinations of blocks in copolymers, however, so this is not a panacea for recycling—some sorting of plastic wastes into different types is still needed.)

Starburst Molecules

The flagships of the new paradigm in nonlinear polymer architecture are the macromolecules called *dendrimers*. They are related to star polymers, with several arms that branch from a single core; but as the name implies (*dendros* is Greek for "tree"), each arm itself branches repeatedly. The branching structure can be controlled precisely, such that every macromolecule of a given dendrimer is more or less identical.

Imagine a tree on which each branch divides into two new branches after a certain length. A single branch forks into two, and then those two each fork into two more, and so on. At each branching stage the number of branches doubles. Before long, a single branch will have diverged into a mass of limbs. This is the nature of a dendrimer molecule, in which two or more trunks are themselves united at a central vertex (fig. 9.10). In real trees branching is invariably accompanied by a reduction in scale, so that each new branch is thinner and shorter than the one from which it diverges. But generally all the branches are identical in

---

[1] A spectacular example of another ordered, bicontinuous network formed by block copolymers is shown on the cover of my book *Designing the Molecular World* (Princeton University Press, 1994).

FIGURE 9.10 Dendrimers are highly branched polymers in which the branches radiate out in a regular manner from a central core. Note that this illustration of the structure is highly schematic—it shows a typical branching sequence, but the real three-dimensional shape of most dendrimers is compact and near-spherical.

length and thickness in a dendrimer—each consists of an identical, short molecular chain. This is hard to represent in a two-dimensional illustration, because the branches soon start to get crowded and overlap. But the molecule itself is three-dimensional, with a densely packed, roughly spherical shape, akin to that of globular proteins (chapter 4). Within this globular mass, the successive "generations" of branches form concentric, more or less spherical shells.

The key to developing this complex but well-defined shape is the ability to control the process of adding branches. Each generation is created by adding a branched unit to the ends of the existing branches. If this addition process is not controlled, different branches would grow at different rates and the molecule would end up as a highly branched but random polymer. So the strategies for making dendrimers generally consist of a sequence of steps in which each new generation of branches is added completely but in a form that cannot react further, so prohibiting the spontaneous addition of still further generations. Once a new generation has been added, the branch ends are "activated" in some way to allow the addition of a new layer of monomers.

There are two ways of achieving this degree of control, which are called divergent and convergent syntheses. In the divergent approach, the central vertex of tree trunks is formed first, and the trees are then grown outward by the successive addition of new generations of branches (fig. 9.11). The first dendrimers were made this way by Donald Tomalia and colleagues of the Michigan Molecular Institute, who initially named them starburst molecules. Typically one starts with a vertex molecule, called the initiator core, which has several (perhaps three or four) arms at the ends of which are reactive chemical groups. To this core unit branched monomers are added, with the crucial feature that only one of the monomer arms is able to react—this sticks to an arm of the initiator core, while the other arms of the monomer are inert. Once a monomer has become attached to each of the arms of the core unit, the molecule can grow no further.

The next step is to render the outer branches of the molecules reactive, so that further generations of branches can be added. This involves some chemical reaction that activates the outer branch tips so that they can react with a fresh batch of monomers. Again one monomer unit is appended to each branch end, exposing new branch tips that remain inert until activated. Repeating the cycle of branch-

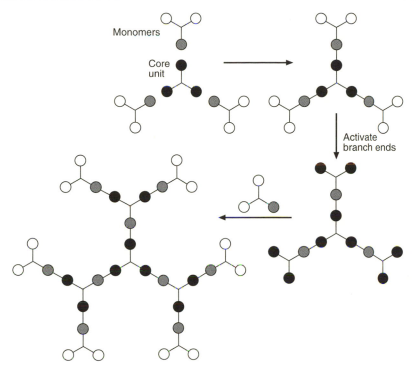

FIGURE 9.11 The divergent approach to dendrimer synthesis begins with a multiarmed core unit, to which branched monomers are attached. Here gray branch ends will react with black ends, but white ends are unreactive. They can, however, be "activated" by a chemical reaction that converts them to black ends. Successive rounds of monomer addition and branch-end activation build up layers of monomers one by one around the central core.

end activation and monomer addition creates successive generations of branches in a controlled, layer-by-layer manner.

In theory this sequence of steps could be continued indefinitely, until the ball-like polymer molecule gets so big that we could see it with the naked eye. But in practice, each new generation of branches is more crowded than the last, even though the dendrimer branches grow outward in three dimensions. Each new generation of monomers increases the radius of the macromolecule by roughly the same amount, but it doubles the number of branches at the surface of the globular molecule; so eventually the surface area of the molecule does not increase quickly enough to accommodate all the new monomers. Therefore the addition of new monomers at the branch ends starts to become slow and irregular after many (typically ten or so) generations, because the monomers simply can't get to the activated branch ends.

The convergent approach to dendrimer synthesis was developed by Jean Fréchet and colleagues at Cornell University, and in parallel by a group at AT&T Bell Laboratories in Murray Hill, New Jersey. The idea here is to make each of the trees "backwards," starting at the outermost branches and working down. One

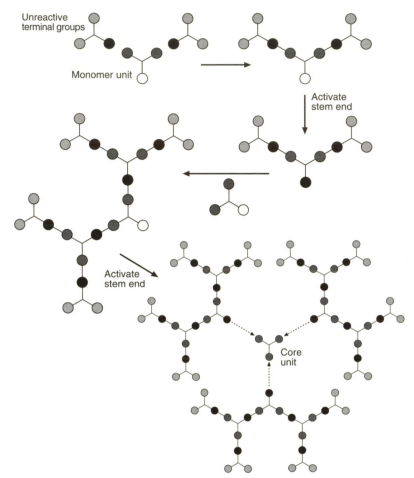

FIGURE 9.12 In the convergent synthesis of dendrimers, the branching "trees" (dendrons) are made from their tips down. Again, black units will react with gray, but white ends are unreactive. The light gray units are different again—they remain inert throughout, ensuring that no monomers are added to these dendron tips. In the final step, fully completed dendrons are attached at their stems to a central core.

begins with monomers that have one arm activated for reaction and the other arms inactive. If one mixes these molecules with branched monomers in which all arms are activated except one, the monomers will acquire "capped" branches on all arms save one (fig. 9.12). Then, by activating this remaining arm, one can stick this entire branch to the arm of another monomer. Each cycle in this reaction sequence produces a more highly branched tree (a dendron), which converges to a single limb that presents the only reactive site on the molecule. And importantly, each cycle induces the same transformation in all the molecules, so all the dendrons end up the same size. These are finally linked together at their roots into a dendrimer by adding a core unit in which all of the arms are activated (fig. 9.12).

These two approaches can be used to create dendrimers in which different regions have a different chemical composition. In the divergent approach, for example, one might use different monomers for different generations of branches. This gives rise to a layered dendrimer, like a multiflavored gobstopper (fig. 9.13a). A two-flavored dendrimer with an inside that is different from the outside is analogous to a micelle (chapter 4), which has a hydrophobic interior and a hydrophilic coat. George Newkome of the University of South Florida has made dendrimers that mimic this structure, having hydrocarbon branches on the inside but an outer coat of carboxylic acid groups. These molecules are a kind of permanent single-molecule micelle, which Newkome calls micellanes. They are soluble in water and can "dissolve" small hydrophobic molecules in their interiors.

The convergent approach, meanwhile, can be used to assemble different dendrons into composite dendrimers with different segments (fig. 9.13b). The properties of such macromolecules then vary as one travels around a circumference, rather than, in the case of layered dendrimers, as one travels from inside to out. Some particularly intriguing examples of such molecules have been made by Jean Fréchet's group. They consist of two hemispherical dendrons, one hydrophobic and one hydrophilic, linked at the core (fig. 9.13c). These molecules are spherical amphiphiles, and might be expected to assemble into giant versions of the

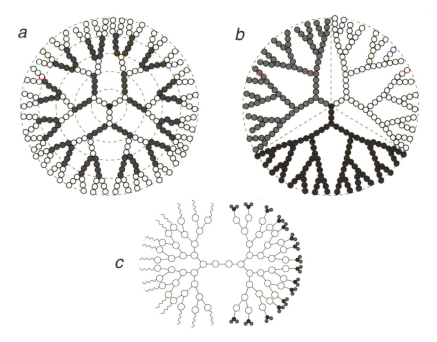

FIGURE 9.13 The composition of dendrimers can be varied throughout the molecule in a systematic way. For example, using different monomers for the addition of each layer in the divergent synthesis gives a dendrimer with distinct, concentric shells (a). And linking different dendrons to the core unit within the convergent approach gives a segmented dendrimer (b). Amphiphilic dendrimers, with two hemispheres that have different water solubility, can be made by the latter scheme, by linking a hydrophobic and a hydrophilic dendron to the core (c).

structures that surfactants form, such as micelles. They can act as surfactants to stabilize emulsions of water and oily organic liquids.

The insides of dendrimers, like those of micelles or vesicles, can provide a microenvironment that is distinct from the environment outside. In particular, the interior can act as a solvent that dissolves small molecules that wriggle their way in between the branches. Because the properties of this solvating interior can be varied continuously by altering the dendrimer's size, there is the possibility of fine-tuning the course of reactions between small molecules trapped at the dendrimer's center, since it is well known that the products of a chemical reaction may be different in different solvents. Moreover, reactants encapsulated in a small space react faster because they encounter each other again and again rather than drifting away after the first encounter. Speeding up reactions by conducting them within micelle-type containers is a well-established practice; dendrimers may offer a new way to do this.

E. W. Meijer and colleagues of the Eindhoven University of Technology in the Netherlands have shown that some small molecules can be trapped inside dendrimers by forming the latter in the presence of the former, and then giving the dendrimers a densely packed outer shell of bulky chemical groups. They refer to these rigid-shelled dendrimers as molecular boxes. Meijer's team found that some trapped molecules could eventually find their way out of the box whereas others could not. The ability to escape depended on the trapped molecule's shape, which determined its ability to wriggle between the branches and squeeze its way through the dense outer layer. Thus dendrimers might be able to release trapped molecules in a selective and gradual way, making them potentially useful carriers for drug delivery (see chapter 5). The macromolecules would be loaded with entrapped drug molecules and injected into the body; the rate at which the trapped molecules escape could then be controlled by varying both the structure and the number of generations of the branches.

Incorporating metal ions into dendrimers further extends their range of possible applications. Researchers in Russia have actually used metal ions *as* the core, while Ed Constable at the University of Basel in Switzerland has shown that metals can make good linking units between branches. Metal atoms are often good catalysts, and are found in the active sites of many enzymes. The ability to control the permeability of dendrimers to small molecules means that they might be designed to mimic the catalytic functions of enzymes, with a metal ion at their center surrounded by a polymeric sheath that, like the protein-based cavities of enzymes, will admit only molecules of the right shape.

Dendrimers occupy an intermediate position between discrete molecules and macroscopic materials: they *are* single molecules, and can be dissolved without sedimenting out of solution, but they are also nanometer-scale particles, big enough to be seen individually in the electron microscope. Dendrimers can therefore be made large enough to be extracted from solution by very fine filters. This might allow them to be used as extraction agents: if their surfaces could be coated with groups that recognize and bind specific molecules, these could be extracted from a mixture of others simply by filtering out the bound complexes. This would be particularly attractive if the receptors on the dendrimer surface can recognize and bind just one of the two forms (left- and right-handed) of chiral molecules; separation of these two forms of chiral molecules is a major challenge in drug production.

Catalytic dendrimers, meanwhile, would combine the assets of both of the two traditional classes of catalysts—homogeneous and heterogeneous. Homogeneous catalysts stay in the same state as the reactant molecules—usually this means that both the catalyst and reactants are in solution. Such catalysts can be highly selective in the products that they generate, but have the drawback that it is not easy to separate the products from the catalyst. Heterogeneous catalysts, which are used in a different (usually solid) state to the (usually gaseous or liquid) reactants, are generally less selective and versatile but are easy to keep separate from the products. Catalytic dendrimers could function as homogeneous catalysts insofar as they are soluble, but could be separated from a solution of product molecules by filtration. This possibility was demonstrated in 1994 by a Dutch team led by Gerard van Koten of Utrecht University, who created dendrimers with complexes of nickel ions at the ends of the branches. As discrete small molecules, these nickel groups were known previously to be effective homogeneous (soluble) catalysts for the so-called Kharasch reaction of carbon-based molecules (fig. 9.14). They performed the same function when bound to the branch tips of van Koten's dendrimers, but in this case the products could then be separated from the catalysts (and the latter recycled) by filtration.

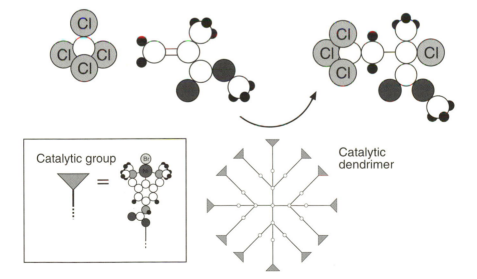

FIGURE 9.14 Dendrimers with catalytic metal-containing groups at their tips represent a new kind of catalyst—soluble, so that they can mix intimately with the molecules whose conversion they are catalyzing, but large enough to be filtered so as to separate them from the products. Nickel-containing dendrimers prepared by a group at the University of Utrecht were shown to be efficient catalysts for the reaction depicted here. The catalytic groups (gray triangles) are positioned at the branch tips; I show the dendrimer framework in schematic form, with white circles representing silicon atoms.

Branching Out Like Crazy

Dendrimers are commonly paraded as an example of the beauty and orderliness of modern polymer chemistry, but they have a group of rather more disheveled cousins called *hyperbranched polymers*. These too have highly branched structures, giving the macromolecules an essentially globular rather than linear shape; but in contrast to dendrimers, the branching of hyperbranched polymers is random. This is the result one would get from the polymerization of branched monomers without the careful activation steps and the capped ends that lead to dendrimers.

Young Kim and Owen Webster at the Du Pont laboratories in Wilmington, Delaware, reported the first hyperbranched polymers in 1988. The monomers that made up their macromolecules had three reactive arms—two of the same type (call them A) and one that could react and link up with the A's (call them B). In such a case, each time a new monomer becomes attached to a growing macromolecule, the polymer acquires one more potential branching point: it loses one through the A–B reaction, but gains two because of the other two reactive groups in the newly added monomer (fig. 9.15). Hyperbranched polymers are formed by inducing a simultaneous frenzy of these linking reactions, leading to an irregular,

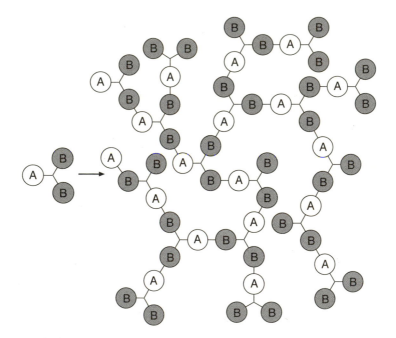

FIGURE 9.15 Hyperbranched polymers are made by the random polymerization of branched monomers. Here A and B groups will readily react to form a chemical bond between them. The branching is uncontrolled, and the highly branched polymer molecules adopt a compact, globular three-dimensional shape like that of dendrimers. Some reactive branch ends become shielded from further reaction by the chains packed all around them.

complex architecture. Like dendrimers, globular hyperbranched polymers might be useful as viscosity modifiers for conventional straight-chain polymers.

There is a good chance that, during the branching frenzy, some of the reactive B groups may become surrounded by branches before they have a chance to react with another monomer, so they lie dormant within the polymer's globular mass. This buried potential for reactivity could be put to good effect. If the "spare" reactive groups can be exposed, for example by warming the polymer so that the branched chains become more flexible and more loosely packed, the material's dormant potential for bonding can be activated. Jean Fréchet's group at Cornell has created a hyperbranched polyurethane that can be thermally activated to expose reactive isocyanate groups. These materials might find application as thermally activated adhesives.

GENETIC PLASTICS

The great variety of structures and functions exhibited by proteins, which are by and large linear chains of amino acids, demonstrates that even traditional straight-chain polymers can potentially be highly versatile. A protein's function is dictated by the folded structure of its chain, which in turn is determined by the protein's composition—its amino acid sequence. (One might add that many proteins are extensively modified, by addition of auxiliary functional groups and sugars, to convert them into their active form; but that is another matter.) The sequence is uniquely specified right down to the last amino acid. This means that not only does each one of these sit at the right position along the chain, but all of the chains of a particular protein are (in principle) identical. Even with living polymerization methods, we still can't put together synthetic polymers this accurately. If we could define the precise monomer sequence of a polymer, we might then hold out the hope of preprogramming individual macromolecules to fold up, to crystallize, or to function in specific ways.

The price that biological systems pay for this control of polymer structure is that they need large data banks to hold the information. To make a linear chain from identical monomer units does not require much information: the instruction is simply "join the ends together—again and again." We can say that the structure of such a molecule is highly redundant, in that it encodes the same operation over and over again. But to make a protein chain requires a separate instruction for joining each amino acid monomer, since in principle nature has several (twenty) choices of amino acid each time a new peptide bond is forged, and it matters which choice she makes. So whereas making a linear chain of the same monomers (a homopolymer) requires just one instruction (along with perhaps two more to specify when to start and when to stop), making a linear chain of a nonrandom sequence of different monomers requires as many instructions as there are monomers. This is why every cell has to have its own genetic data bank holding the blueprints for every protein it requires. Yet that is a small price to pay, because the data banks can be crammed into a bag just a tenth of a micrometer across.

Thanks to advances in biotechnology over the past few decades, the blueprint for proteins—the genetic database of the DNA molecule—can now be redrawn. That is to say, researchers can change the nucleotide sequence of a gene and thus alter the amino acid sequence of the protein that it encodes, creating a new engineered protein with a modified function. And more: they can build their own DNA sequences and insert them into genes, or even contemplate building entire artificial genes. On the whole, biotechnologists want to do this so that they can develop new drugs, peptide molecules based on natural proteins but with subtly different structures and thus behaviors that might correct a physiological imbalance or activate or inhibit other genes.

But some polymer chemists have now started to regard these tools in a new light. For here, they say, are polymer factories of infinite potential. The biomolecular machinery responsible for translating genes into proteins—the transcription and translation apparatus—is merely a set of devices for carrying out polymer synthesis monomer by monomer, for polymers of any specified sequence. The only limitation is that the monomers have to be naturally occurring amino acids (and even that limitation can be tested, as we shall see). The biotechnologists, who are concerned on the whole with making protein molecules with enzymelike functions, would not yet dare to start from scratch, by making a purely synthetic gene to encode them—we simply don't know enough about how the performance of these complex molecules is related to their molecular structure, and besides, they typically contain hundreds or thousands of monomers with no long-ranged regularity of sequence. But *structural* proteins, whose properties leave materials scientists so awed—they are another matter. Not only do we have a better grasp of how their structures (exhibiting motifs such as α-helices and β-sheets) are determined by their amino acid sequences, but we know that these sequences are largely repetitive, containing short fragments repeated again and again. So can we use biotechnology to make new materials?

That was the question pondered by polymer chemists David Tirrell and Joseph Cappello, who in the late 1980s and early 1990s developed a strategy for making synthetic polypeptides, analogs of structural proteins, using the protein-synthesizing machinery of bacteria. The central idea of this approach is to design an artificial protein consisting of a repeating short sequence of amino acids, which one hopes will enforce a specific secondary structure or material property, and then encode the plan for the designed protein in an artificial gene that is spliced into the genetic material of a bacterium. If all goes well (and this involves a number of important "ifs"), the cellular machinery of the organism will translate the gene into copies of the polymer with high fidelity.

Recall that the structure of a protein is encoded in a DNA gene such that every three nucleotides in DNA represent one amino acid. The gene is copied onto an RNA molecule, and the protein is made on this RNA template through the cooperation of transfer RNAs and a protein/nucleic acid complex called the ribosome.

To make an artificial DNA sequence that would encode the repeat sequence of the designed protein, Cappello and colleagues at Protein Polymer Technologies Inc. in San Diego used the methods of DNA synthesis developed for biotechnology in the 1980s. The synthetic DNA consists of the two complementary strands of a double helix, but with overhangs at each end where one strand extends a little beyond the other. These sticky ends allow the strands to be linked together by ligation enzymes (ligases). In this way, short fragments of synthetic DNA can be linked together into long repetitive chains—the artificial genes that the bacteria will ultimately express as proteins. Thanks to the polymerase chain reaction (chapter 4), one can make as much of an arbitrary stretch of DNA as one needs.

Unlike our own DNA, that of bacteria is not densely packaged into separate chromosomes but comes in the form of double-stranded rings. In addition to a main DNA ring, some bacteria possess much smaller rings called plasmids, and it is into these that Cappello's team inserted their artificial genes. They used restriction enzymes to snip open the plasmid rings at a specified location—these are able to recognize certain sequences in a DNA chain and to cut the chain in those places. Then ligation enzymes were employed to splice the population of synthetic DNA strands in place. The PPTI team put the plasmids back into the cells of their original host bacteria, *Escherichia coli*, and the bacteria proceeded to translate the synthetic genes into polypeptides (fig. 9.16).

The PPTI researchers first used this scheme to make an artificial silklike protein, using a constructed gene that encoded a short (six-amino-acid) sequence known to be responsible for β-sheet formation in natural silk. Stimulated by the success of this enterprise, they then aimed for a more ambitious target: a designed protein that would combine the good structural properties of their artificial silk with the useful function of a biological protein called fibronectin. This biomolecule helps cells to stick together in the body's tissues. Such cell adhesives are useful for the kind of biomedical tissue engineering described in chapter 5, for which one might need to attach human cells to synthetic substrates such as cultured tissue cells. Existing methods for doing this make use of either natural cell-adhesion proteins extracted from blood or from animals, or purely synthetic peptides that contain the amino acid sequences most directly responsible for adhesion. The latter are more stable than their natural counterparts, but also less proficient at promoting adhesion.

The PPTI team hoped to combine the advantages of both, by making a protein that was part natural (based on fibronectin) and part synthetic. They devised an artificial protein in which silklike structural blocks of six amino acids repeated along the chain, interrupted every ninth block by a single fibronectin-like block of sixteen amino acids. This 9+1 sequence was itself repeated thirteen times in the full protein (fig. 9.17).

They made this designed polypeptide by splicing the corresponding synthetic gene into plasmids of *E. coli*. They found that a coating of the protein would indeed ensure good cell attachment to a variety of substrates, including polystyrene; in some cases the artificial protein performed better than fibronectin itself (fig. 9.18). The structural blocks, meanwhile, ensured that the protein stayed stable in temperatures at which the natural proteins would lose their activity. The protein is now marketed commercially by PPTI under the name Pro-Nectin F.

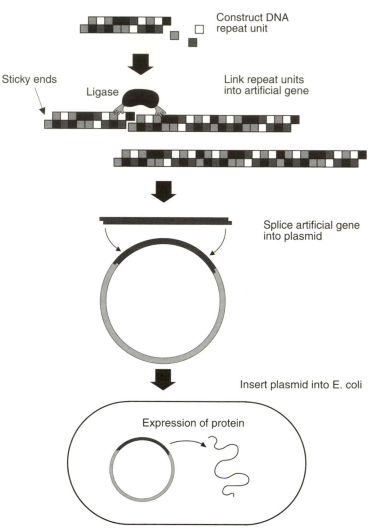

FIGURE 9.16 Polypeptides can be made to order with great accuracy by inserting into the DNA of bacteria "synthetic genes" that encode the polypeptide's structure in the same way as regular genes encode a protein's structure. Biotechnological techniques are used to synthesize and insert the artificial genes. Like structural proteins, polypeptides with useful material properties generally have repetitive amino acid sequences, and so the respective genes are made by linking together short genetic sequences that encode the repeat sequence of the polypeptide. These DNA sequences can be synthesized chemically, with each triplet of nucleotides encoding a single amino acid according to the rules of the genetic code. The DNA sequences are given sticky ends so that they can be linked together by enzymes called ligases. The strands can be separated according to their number of repeat sequences, and those with the same number are then spliced into bacterial plasmids, circular strands of DNA that all bacteria carry. The bacteria will then "express" these synthetic genes in fermentation vats, translating the DNA sequence to the corresponding repetitive polypeptide.

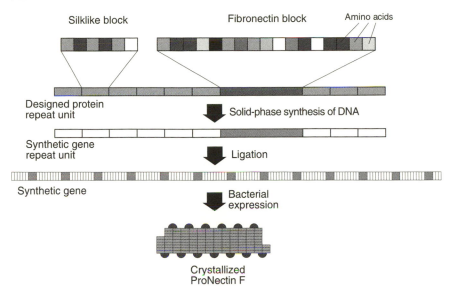

Silklike block   Fibronectin block   Amino acids

Designed protein repeat unit

Solid-phase synthesis of DNA

Synthetic gene repeat unit

Ligation

Synthetic gene

Bacterial expression

Crystallized ProNectin F

FIGURE 9.17 ProNectin F, a polypeptide prepared by researchers at Protein Polymer Technologies Inc. in San Diego, helps cultured tissue cells to adhere to supports. It contains a sixteen-amino-acid sequence that promotes cell adhesion, copied from the natural cell-adhesion protein fibronectin, interspersed with repetitive sequences of six-amino-acid blocks that form a folded silklike structure. Each chain of the polypeptide contains thirteen of these bifunctional units.

FIGURE 9.18 Cell adhesion to a polystyrene support in the presence of ProNectin F (*right*) is better than that in the presence of the natural adhesion protein fibronectin (*left*). (Photographs courtesy of Joseph Cappello, Protein Polymer Technologies Inc., San Diego).

Other groups have now used this approach to make a variety of synthetic proteins with structures similar to those of natural fibrous proteins. While at the Army Natick Research Center in Massachusetts, David Kaplan has used bacteria to express artificial genes based on those that the orb-weaving spider uses to make its silk. In principle this approach might make silklike materials much more plentiful, as it could be grown in fermentation vats rather than being harvested from

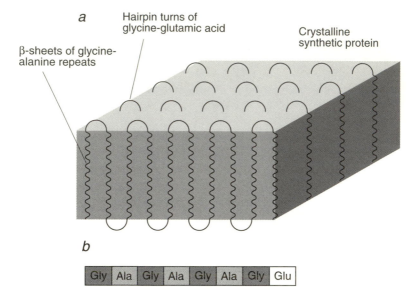

a Hairpin turns of glycine-glutamic acid

Crystalline synthetic protein

β-sheets of glycine-alanine repeats

b

| Gly | Ala | Gly | Ala | Gly | Ala | Gly | Glu |

FIGURE 9.19 Synthetic polypeptides that fold up into β-sheetlike films have been made by David Tirrell and colleagues (*a*). The polypeptides contain glycine-alanine repeat units that are known to form β-sheets in structural proteins, punctuated every three pairs by a glycine–glutamic acid pair that introduces a hairpin turn into the chain. The β-sheetlike crystalline layers therefore have a thickness determined by the distance between these hairpin turns. The full repeat unit of the polypeptide is an eight-amino-acid sequence (*b*).

silkworms. A team at Allied Signal in New Jersey has made synthetic collagen-like proteins, while Dan Urry at the University of Alabama has made artificial proteins similar to elastin, the elastomeric material found in the arteries, ligaments, lungs, and skin. Urry believes that these materials might prove useful as elastic, biocompatible materials for wound repair and as superabsorbents for diapers.

Whereas these studies have focused on making artificial proteins that are closely related to natural structural proteins, David Tirrell and colleagues at the University of Massachusetts at Amherst have been exploring the possibilities for designing new kinds of secondary structure. They have created highly ordered, crystalline sheets in which the polypeptide chains form β-sheets that bend back on themselves like hairpins (fig. 9.19*a*). Tirrell reasoned that the straight portions of such a folded polymer could be made of alternating glycine and alanine residues, as these are known to stack into β-sheets. After every third glycine–alanine repeat sequence he introduced a glycine–glutamic acid pair, since these are known to introduce hairpin turns into polypeptide chains (fig. 9.19*b*). Analysis by X-ray diffraction showed that these synthetic polypeptides, created in *E. coli*, do indeed pack into corrugated sheets.

Tirrell has explored the use of nonnatural amino acids as building blocks for designed bacterial polypeptides. This could potentially expand the range of materials that the bacterial method can produce. In particular, Tirrell is interested in using fluorinated amino acids, chains of which might resemble polytetrafluoroethylene (PTFE)—better known as the nonstick coating Teflon (chapter 10). Because fluorinated amino acids are hydrophobic (in other words, they tend to avoid water), Tirrell reasoned that it might be possible to control the way in which these residues are partitioned in the folded form of the polypeptide: if the chain is rich in water-soluble (hydrophilic) residues, it should fold up so that the fluorinated groups are segregated at the polymer surface to escape this hydrophilic interior, producing a kind of non-stick synthetic protein.

But cells cannot normally use amino acids different from the twenty found in natural proteins. So realizing this plan required a touch of bacterial husbandry as well as genetic engineering. A fluorinated amino acid was encoded in the synthetic genes by the same triplet of DNA bases as that which encodes the natural variant; but Tirrell and colleagues had to breed a new strain of bacteria, using a kind of forced Darwinian selection, that could use the fluorinated amino acids for protein synthesis and that would not synthesize their own unfluorinated amino acids instead. The fluorine-containing group was a modified version of the amino acid phenylalanine, which has a benzene-ring side group to which fluorine atoms can be attached. In this way, Tirrell's team has been able to persuade bacteria to express a fluorinated polypeptide. The fact that this polymer is rather insoluble in water suggests that the fluorinated groups may indeed migrate to the surface of the folded polymer, but this has yet to be confirmed.

David Tirrell sees almost limitless opportunity in this newfound ability to direct the biosynthetic pathways of bacteria. In particular, he envisages making modified proteins tagged with a synthetic polypeptide chain that makes the biomolecule a more tractable material. This could be achieved by inserting the DNA sequence for an artificial polypeptide right next to the sequence for the natural protein in the bacterial plasmid. Once the protein-building machinery has made the protein, it will continue reading along the DNA sequence and so attach the artificial polypeptide substituent (in general, this reading process stops only when the machinery—the ribosome—encounters a sequence that encodes the "stop" instruction).

For example, the protein might be given an artificial side chain that will fold up into a crystalline film (fig. 9.20a), just like the hairpin polypeptides that Tirrell's group has already built. Then the proteins should spontaneously form thin films on a substrate, which would immobilize them for use in devices such as bioreactors and biosensors (see chapter 5). Or a protein might be equipped with an optically or electrically active substituent, made up of suitably modified, nonnatural amino acids, that could allow us to control the protein's function with light or electricity. Tirrell and colleagues have developed a strain of bacteria that will use the compound 3-thienylalanine in place of the natural amino acid alanine itself. The thienyl side group of this nonnatural amino acid is the monomer unit for

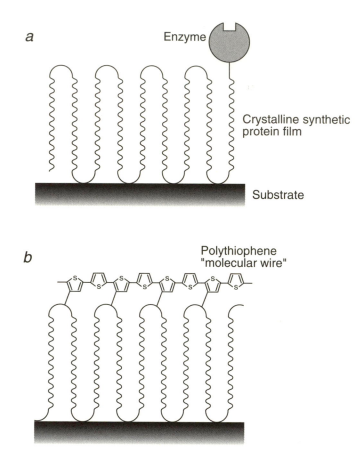

FIGURE 9.20 New peptide-based materials might be created by linking a synthetic, structural peptide sequence with functional substituents. A synthetic gene spliced next to the gene for an enzyme might generate a "film-forming" enzyme, in which the designed polypeptide sequence folds into a crystalline film with the enzyme attached at its surface (*a*). Such a synthetic/natural composite polypeptide might be useful for biotechnological applications. By incorporating into polypeptide chains nonnatural amino acids, their range of functions might be enhanced. David Tirrell and colleagues have made bacterial polypeptides containing thienyl side groups, which might ultimately be linked together into a polythiophene molecular wire at the surface of the synthetic polypeptide sheet (*b*).

polythiophene, a polymer that conducts electricity (discussed later in this chapter). Their hope is ultimately to make a polypeptide film in which many of these thienyl groups bristle from the surface, where they can be cross-linked into an electrically conducting layer (fig. 9.20*b*). Such a material might provide a biocompatible conductor, which could be valuable in biomedical devices. New polypeptides like this will lead to an ever diminishing distinction between natural and synthetic materials.

BREAKING THE MOLD

Most of the plastic objects that we encounter on a daily basis have been formed by injection molding: a molten polymer is squeezed under pressure into a mold and cooled until it solidifies. Now this idea is being reduced to the microscopic scale: "micromolding" technologies are being developed for fashioning polymeric structures too small for the eye to see (fig. 9.21). The approach relies on our ability to create finely patterned microscopic molds, and that is something that has blossomed only recently.

But some researchers are now boldly asking, "Why use a mold at all?" This proposal is stimulated by the realization over the past few decades that molecules can organize *themselves* into a great variety of structures—we don't have to impose the organization on them. In particular, amphiphilic molecules can, under the right conditions, self-assemble into sheets, vesicles (hollow cell-like struc-

FIGURE 9.21 Delicately patterned polymer nets have been synthesized by George Whitesides and coworkers using a micromolding approach. Using photolithographic techniques developed for making microelectronic structures, he patterned a silicon wafer into shallow channels about a micrometer deep and a few micrometers in width, and used this as the template to make a mold from the elastomeric polymer poly(dimethylsiloxane). The polymer net was then created by placing the mold face down on a flat substrate and allowing liquid methyl methacrylate to be sucked into the spaces between the mold and the substrate by capillary action. The liquid was then photochemically cross-linked into poly(methyl methacrylate), and the elastomeric mold was removed. The resulting patterned polymer networks might be useful for filtration or for polymer-based optoelectronics. (Photograph courtesy of George Whitesides, Harvard University.)

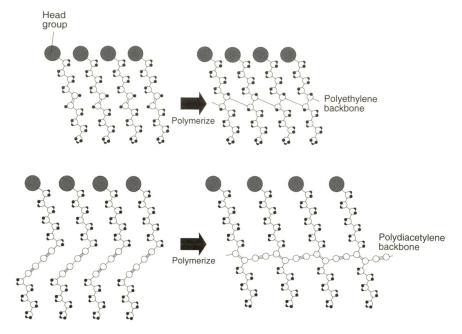

FIGURE 9.22 Amphiphilic molecules containing unsaturated (double or triple) bonds in their hydrocarbon tails can be cross-linked into polymerized films. Double bonds give rise to polyethylene backbones, and double groups of triple bonds to polydiacetylenes, which are strongly colored and electrically conducting. This cross-linking allows self-assembling amphiphilic aggregates such as vesicles to be converted into robust polymeric structures.

tures), cylinders and tubes, and more complex three-dimensional networks. Several of these structures were described in chapter 4: for example, amphiphiles will form films one molecule thick on the surface of water, or vesicles whose walls are bilayers. In all of these assemblies there is a high degree of molecular order: the amphiphiles arrange themselves side by side, head to head and tail to tail.

These loosely bound aggregates are, as we have seen, useful in their own right—for example, artificial phospholipid vesicles (liposomes) are used for drug delivery. But by giving the molecules the ability to cross-link one can "freeze" these assemblies into permanent polymeric structures. Random cross-linking between the adjacent amphiphiles turns the aggregate into a single polymer molecule with a preformed shape. To allow for cross-linking, the most common approach is to incorporate double or triple bonds between carbon atoms into the hydrocarbon tail of the amphiphiles, which can be burst open and cross-linked in just the same way that ethylene and acetylene molecules are linked up into polyethylene and polyacetylene (fig. 9.22).

This approach was pioneered in the early 1980s by Helmut Ringsdorf's group at the Johannes Gutenberg University in Mainz, Germany, and by Steven Regen's group at Lehigh University in Pennsylvania. The German group cross-linked films of phospholipid molecules containing vinyl or diacetylene groups in their tails to make well-ordered two-dimensional polymeric sheets, while Regen's team showed that polymerizable phospholipids could be linked up in the bilayer walls of vesicles, creating hollow polymer sacs that were less easily broken down than the unlinked versions. The cross-linking in these systems is generally induced photochemically with ultraviolet light. Perhaps the most complicated polymer structures made so far by cross-linking self-organized lipids are those of David O'Brien and colleagues at the University of Arizona, who have succeeded in capturing the labyrinthine "cubic" phases formed from interpenetrating channels of bilayers (fig. 9.23). These structures, whose channels are about 6 nanometers across, might ultimately find uses as molecular-scale filters and selective catalysts.

If the cross-linking group in these lipid assemblies is a diacetylene unit, the result of polymerization is a material with a particularly valuable property. Polydiacetylene itself is richly colored, a consequence of the double and triple bonds that alternate along the backbone: these make the polymer strongly light-absorbing. The precise color, however, depends on the ambient conditions. A

FIGURE 9.23 David O'Brien and colleagues at the University of Arizona have cross-linked polymerizable amphiphiles in the ordered cubic phase, a structure in which curved bilayers form a network of channels. The result is a polymer material with an ordered network of regular pores, revealed here by electron microscopy. Scale bar: 100 nanometers. (Photograph courtesy of David O'Brien, University of Arizona.)

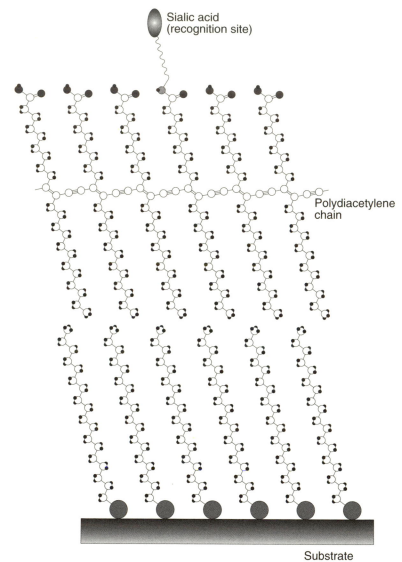

FIGURE 9.24 Bilayer films with a polydiacetylene backbone linking the amphiphilic constituents can serve as color-change sensors for the binding of molecules to recognition sites at the film surface. In this example, devised by researchers at the Lawrence Berkeley National Laboratory, a bilayer film with a fatty acid top layer was prepared on a solid substrate. The fatty acids have diacetylene groups in their tails that can be cross-linked by ultraviolet light. To some of the fatty acid head groups are linked sialic acid molecules, which serve as recognition sites to which a surface protein of the influenza A virus will bind. When binding occurs, the conformation of the polydiacetylene chain alters and the film changes color.

change in acidity or in temperature can trigger a change in the conformation of the backbone, which has a corresponding effect on the wavelength at which the material absorbs light—it switches between blue and red. The same effect can be seen in lipid assemblies cross-linked via diacetylene groups, which essentially become polydiacetylene films with side chains dangling from the backbone. In 1993 Deborah Charych and colleagues at the Lawrence Berkeley National Laboratory in California exploited this color change to develop a chemical sensor for the influenza virus. Using the Langmuir-Blodgett technique (page 415), they prepared a bilayer film in which the top layer consisted of diacetylenic fatty acids. When cross-linked (fig. 9.24), this film turned blue. To some of the fatty acid head groups the researchers attached sialic acid groups, to which the influenza A virus will bind. When the film was exposed to the virus, the docking of the bulky virus at these binding sites caused a change in the conformation of the polydiacetylene backbone running through the film, altering the absorption properties and turning the film red (plate 11). This provides a rapid visual means of detecting the virus. The same system works when the cross-linked bilayer films instead constitute the walls of liposomes. These systems are now being explored for medical diagnostic applications.

Polymers Get Active

Imagine a television that you could roll up and put in a briefcase. Something from the hallucinatory worlds of William Burroughs or Salvador Dali? No, this surreal image may soon be realized, thanks to a new breed of light-emitting plastics.

The first flat televisions will probably be made from liquid crystals, not polymers. These materials—part liquid, part crystal—have already given us the credit-card pocket calculator, the fold-away screen for laptop computers, and the wrist-watch television. They are remarkable substances in their own right, being able to retain a degree of ordering between their constituent molecules even though they flow like a liquid. As I mentioned earlier in this chapter, most liquid crystals are rodlike molecules that can line up in the liquid state much as logs might do while floating on a river. This preferential alignment enables liquid crystal films to block out polarized light if the plane of polarization is not lined up with the molecular rods. If the molecules have an imbalance of charge along their length, their orientation can be switched by an electric field, and so an electric signal can be used to switch on and off the passage of polarized light through the film. In this way, arrays of liquid crystal cells can provide pixelated display screens. Simple, monochrome displays of this sort have been appearing in digital watches and clocks for years now. But the flat-screen TV is taking much longer because of a number of technical difficulties: achieving fast switching speeds and multicolor screens, and figuring out how on earth to wire up all those pixels in a full-sized TV screen.

Light-emitting plastics have a number of potential advantages for flat-screen display technology. First, there is literally their flexibility: plastic light-emitting devices have now been made that are barely thicker than a sheet of paper, and which can be bent just as easily. Second, full-color displays might be easier to make, because the color of the emitted light depends on the chemical nature of the plastics themselves, and can be tuned right across the visible spectrum from red to blue. Furthermore, it is now possible to make plastic devices that emit light of different colors simply by changing the voltage driving them. Third, the fabrication methods could potentially be much simpler, involving nothing more than a printing process to imprint the various polymer layers of each device.

How do you make a polymer shine? In much the same way as the semiconductors such as gallium arsenide used in today's light-emitting diodes: by the creation of energetic states that lose their energy by emitting photons. We saw in chapter 1 that two approaches are commonly used to make semiconductors glow. One is to shine light on them, whereupon photon absorption pumps an electron from the occupied valence band into the (largely) empty conduction band, creating an electron–hole pair (recall that the hole is the gap left in the valence band, which acts rather like a positively charged particle). Emission occurs when electrons and holes recombine, or in other words when an excited electron falls back into a hole in the valence band. The light emission induced in this way is called photoluminescence. The other way to stimulate light emission is to inject electrons and holes into, respectively, the conduction and valence bands by applying an electric field across them. The process by which these injected charges recombine and emit a photon is called electroluminescence.

Light-emitting polymers work in the same way, because they too are semiconductors. Molecular materials are commonly nonconducting, and indeed polymers such as polyethylene are such good insulators that they are used to coat electric cables. For this reason there is a common perception that plastics don't conduct electricity. To rationalize this very crudely, we can say that electrons have no easy way to travel through most molecular materials: they rattle around the molecules themselves in molecular orbitals, but they can't get from molecule to molecule. So a current cannot pass.

In semiconducting polymers, however, the electron clouds (orbitals) on each polymer chain overlap, providing a continuous pathway for electrons throughout the material. These polymers usually contain so-called delocalized chemical bonds, which are made up of molecular orbitals that stretch along the entire length of the polymer chain. Polyacetylene is one such material, in which the delocalized orbitals are the result of double bonds between carbon atoms, which alternate with single bonds along the backbone. When the chains lie tangled together in the solid polymer, these smeared-out orbitals can overlap to provide a potential conduction pathway.

Another way of saying the same thing is to consider that the overlap of the smeared-out orbitals forms extended electronic bands. In polyacetylene the delocalized orbitals contain their full capacity of electrons, so the bands formed from their overlap are also fully filled. But the next highest, empty electronic band (which is also the result of this orbital overlap) is not too far above—the band gap is small—so the material is a semiconductor. Actually, polyacetylene has a rather large band gap as semiconductors go, so its conductivity is not very great. It can be vastly increased, however, by adding dopant molecules to the polymer: either donors, like metal atoms, which can donate electrons into the conduction band, or acceptors, like iodine, which can accept electrons from (and leave holes in) the valence band. In this way, polyacetylene can be made highly conducting—a plastic metal.

The first visible-light-emitting polymer device was made from a semiconducting plastic called poly(phenylene vinylene) or PPV (fig. 9.25). In 1990 Rich-

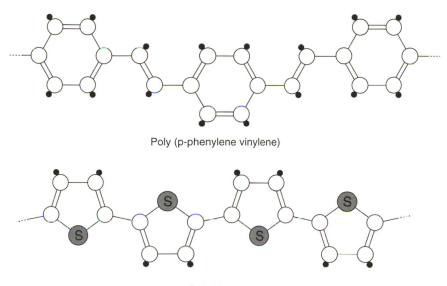

Poly (p-phenylene vinylene)

Polythiophene

FIGURE 9.25 Electroluminescent polymers emit light when charge carriers (electrons and holes) are injected into them. The light emission stems from recombination of the charge carriers. Polymers of this sort have a degree of electrical conductivity. The first polymer light-emitting diodes (LED's) were made from poly(phenylene vinylene) (*top*), which emits yellow-green light. Polythiophenes (*bottom*) can provide blue-light LEDs.

ard Friend and colleagues at Cambridge University found that PPV will exhibit photoluminescence in the yellow part of the spectrum when illuminated with green light. Having spent several years building diodes from conducting polymers such as polyacetylene, the Cambridge group reasoned that they might be able to make light-emitting diodes (LED's) by stimulating electroluminescence from PPV.

Why were Friend and colleagues making plastic diodes in the first place? One of the great challenges in exploiting conducting plastics is to use them in real microelectronic devices. Electronic circuitry made from plastics would be lightweight, tough, and potentially cheap and easy to make, and Friend's group had already made a step in this direction in 1987 when they fabricated a solid-state diode that had a polymer as the semiconducting material.

It did not take them long to adapt their fabrication technology to make a PPV light-emitting diode. Semiconductor diodes have layers of semiconducting material into which charge is injected at a positive and negative electrode. In the PPV light-emitting diode a film of the polymer less than a micrometer thick was sandwiched between a lower contact of indium tin oxide (ITO, a transparent inorganic conducting material) and an upper contact of cadmium metal. These rather exotic electrode materials were needed rather than conventional silver or copper contacts because of the need to match their electronic bands with those of the polymer into which charges would be tipped by an applied voltage. The researchers were not enthusiastic about the need for a cadmium electrode, as this metal is reactive and toxic and must be itself coated with an inert polymer to protect it from air.

The PPV diode worked as hoped: when a voltage of 12 volts was applied across the contacts, electrons and holes entered the polymer and recombined to generate photons of yellow light. The driving voltage was just about small enough to be practical (although typically liquid-crystal display panels run on 6 volts), but there were several other factors that made this very much a laboratory prototype rather than a practical device. First, the LED was not very bright, because most of the holes and electrons injected decayed by routes other than radiative recombination. For every photon of yellow light emitted, about 5,000 electron–hole pairs had to be injected (in contrast, a light bulb generates a photon for every ten electrons that pass). Second, the device did not run for long before its performance degraded. The polymer was simply not chemically stable enough to last for long in an electronic device.

Work on polymer LED's is now being carried out in many laboratories worldwide, and the potentially lucrative goal of flat-screen displays makes for a competitive race. One of the first objectives was to extend the range of colors that the polymers would emit. The color is determined by the band gap, which in turn depends on the polymer's chemical structure; so making new colors means tampering with the chemical makeup of the material. This has required close collaboration between physicists and polymer chemists. In 1991 a group led by Alan Heeger at the University of Santa Barbara was able to make red-light LED's from polymers related to PPV but with a smaller band gap. Blue light is harder—that requires a large band gap, which is not easy to obtain in delocalized polymers. But several large-band-gap polymers have now been used successfully in blue-light LED's, including poly(*p*-phenylene) and compounds based on polythiophene (fig. 9.25). Olle Inganäs and colleagues at the University of Linkoping in Sweden have made multicolored LED's by using a blend of polythiophene-based polymers, intimately mixed like a marble cake, with different band gaps and thus different emission colors. Because the voltage at which each polymer "turns on" increases as the band gap gets larger, they could fabricate an LED made from a blend of two such polymers that emitted yellowish light when driven at 22 volts and blue light when driven at 28 volts. Ultimately, the Linkoping group hope to be able to make polymer LED's that can be switched between red, blue, and green, the three colors of television pixels. They could then make full-color plastic screens from a single kind of pixel, greatly simplifying the problem of wiring them all up.

Some of the most compelling demonstrations of the practical potential of polymer LED's have come from advances in their method of fabrication. Alan Heeger and coworkers at the UNIAX company in Santa Barbara have built LED's from all-plastic materials, rather than using brittle ITO as part of the device. They deposited a layer of the electrically conducting plastic polyaniline onto an inert, flexible polymer substrate, and then laid down a chemically modified version of PPV on top of this. The polyaniline acted as the positive electrode of the device, which injected holes into the PPV. On top of the PPV they deposited thin strips of calcium metal as the negative electrode, which completed the diode structure by providing electrons to recombine with the holes inside the PPV layer. When switched on, the device could be bent every which way while continuing to glow (plate 12). Thus the UNIAX researchers were able to realize one of the potential advantages of plastic LED's—unlike conventional LED's in use today, which are made from hard, brittle materials, these devices were soft, pliable, and resilient. In 1994 Francis Garnier and colleagues from the Laboratory of Molecular Materials in Thiais, France, built an all-polymer transistor—not a light-emitting device, but another type of electronic component constructed from conducting-polymer layers—using methods developed for printing fabrics. The fabrication of previous polymer devices had involved costly high-vacuum and high-temperature processes to lay down the metal electrodes, whereas Garnier's devices could be turned out by simple and cheap processing techniques. It is now conceivable that entire full-color plastic display screens might eventually be fabricated in this way, bringing a whole new meaning to the concept of wallpaper TV.

Electrically conducting polymers are finding applications ranging from anti-static coatings for hi-fi equipment to "plastic" batteries, and researchers are also working to develop magnetic polymers. Plastic LED's are perhaps the most dramatic example of the trend in plastics toward *active* materials. No longer are polymers content in the role of carrying shopping goods, packaging lettuces or insulating attics and electric cables. The advanced polymers of today *do* things: they respond to stimuli in a smart manner (chapter 3), they hold optical information (chapter 2), they switch photonic circuits (chapter 1), they catalyze chemical reactions. They are materials that make things happen.

Face Value

SURFACES AND INTERFACES

> In 1757, being at sea in a fleet of 96 sail bound against Louisburg,
> I observed the wakes of two of the ships to be remarkably smooth,
> while all the others were ruffled by the wind, which blew fresh.
> Being puzzled with the differing appearance, I at last pointed it out
> to our captain, and asked him the meaning of it? "The cooks,"
> says he, "have, I suppose, been just emptying greasy water through
> the scuppers, which has greased the sides of those ships a little,"
> and this answer he gave me with an air of some little contempt as
> to a person ignorant of what everybody else knew.
>
> —Benjamin Franklin, *Philosophical Transactions of the*
> *Royal Society of London*, 1776

Many properties of materials are determined by the nature of their surfaces—whether they are rough or smooth, inert or reactive, sticky or repel-
lent. Engineering surface properties is an old art, relevant to disciplines as
diverse as microelectronics, mechanical engineering, and medicine; but
today we can do it with molecular-scale precision.

SOMEWHERE in our past we acquired a disdain for surfaces. To be unable to see
beyond the surface of a matter is regarded as a failing, and no one wants to be
considered superficial. The natural state of most surfaces is dirty, yet we expend
a great effort (some of us) in trying to make them otherwise. But when I started
doing scientific research on surfaces I realized that, on the contrary, surfaces are
everything. It takes highly sophisticated apparatus ever to see beyond the surface
of things, so most of us never do. And why should we? For it is at surfaces and
interfaces that we conduct almost all of our interactions with the world. When I
play the piano, the surfaces of my fingers meet the surfaces of the keys; they need

get no more intimate than that, for this action is enough to send several other sets of surfaces impacting on and sliding over one another, until a felt hammer strikes the surface of a metal wire, and acoustic waves tickle the surfaces of my ear drums. The response of each of these components—the weighted "feel" of the key, the vibration frequency of the wire—is a consequence of the properties of the bulk materials, such as density and hardness; but all of the action happens at surfaces and interfaces.

We would be unwise to rely on our experiences of a material's surface as an indicator of what lies within the bulk material. Most often, the structure, composition, and properties of a surface will differ from those that we would see if we could probe a cross-sectional layer buried deep within a substance. I will look in more detail later at the reasons why this is so, but for now I might illustrate the point by analogy. Agriculturalists know all about the uniqueness of boundaries, of interfaces where different ecosystems meet. At the edge of a field, where it borders onto a forest, you will find a diversity of life that far exceeds that either in the field itself (all too commonly now a crop monoculture) or in the depths of the forest. There are species at the border that cannot survive to either side. Partly this will be the result of human intervention—wildflowers and small shrubs absent from the field find refuge at the edges because the farmer cannot be bothered to chop them down where his tractor cannot plough, or because herbicides are less intensively applied there. But even in wild meadows the species diversity of both flora and fauna is likely to be higher at the borders. This, of course, makes these interfaces all the more interesting.

The same thing happens in demography—border communities have special characteristics. In the past (and even today in some regions), border folk had to be hardier—because the boundary might be among a mountain range, or because the people had to live off the sea rather than the land, or because raids from the adjacent country were common. The surfaces of materials must often be adapted to hardship too: they might be exposed to air (a highly corrosive substance, let's not forget), to impacts, to frictional wear and tear. In many materials applications, the ideal material for the job does *not* have a sufficiently hardy surface, and it must be protected with a thin film of some other, more resilient material.

So scientists of all persuasions have found the study of surfaces to be a particularly rich playground, one that brings to light new fundamental phenomena that often have extremely important practical ramifications. Whenever two materials are brought together, whether that be in fabricating a transistor, painting a wall, or applying a Band-Aid, we cannot take it for granted that the interface will behave in a predictable manner; rather, we must expect the unexpected. I would hazard to say that the behavior of surfaces and interfaces is a central consideration in the practical application of *any* material. We must ask: Will it tarnish? Will it stick? Will it explode? In this chapter I want to explore some of the ways in which we can look at surfaces, what we have learned about their nature, and how we can now modify their structure and properties. In doing the latter, we can bring about new kinds of marriages between materials—some of which are effected, like those in a romantic thriller, through the use of superficial disguise.

THROUGH THE LOOKING GLASS

As the importance of processes at surfaces and interfaces has become more and more apparent in the past few decades, researchers have wanted to explore their structure and composition in ever increasing detail. The techniques that they have developed are so central to the materials science of surfaces that it is worth looking at some of the more prominent of these in a little detail before we see what they have told us about the places where materials meet and end.

We don't have to look very hard to discover that surfaces may not be all they seem. Primo Levi tells us how gazing through a microscope that could magnify a mere two hundred times revealed a world that was at once more beautiful and more ugly, and certainly more astonishing, than his eyes alone suggested: "The skin of the fingertip was difficult to observe, because it was almost impossible to keep the finger still in relation to the lens; but when one managed to do so for a few instants, one saw a bizarre landscape, which recalled the terraces of the Ligurian hills and ploughed fields: large, translucent, pinkish furrows, parallel but with sudden bends and bifurcations." If this is what happens once we blow things up just two hundred times, it is no wonder that surfaces revealed under modern microscopes, which can magnify at anything up to a billion times, soon become landscapes of pure fantasy. The smooth surface of a glass plate becomes a range of streaklike hills and gullies; the luster of a metal alloy dissolves into a patchwork of ragged domains. And if we can achieve magnification high enough, we may see the elegant, Platonic order of rows of atoms, like oranges on a greengrocer's stall. These details of surface structure too small for the eye to see are not mere flourishes or embroidery but may play a central role in determining the surface's properties, at the microscopic scale and sometimes at the everyday scale, too.

But a light microscope like Levi's will never show us atoms. Because the wavelength of light (hundreds of nanometers) is so many times greater than the size of an atom (about a hundredth of a nanometer), trying to "see" a visible-light image of atoms is a little like trying to paint a miniature using a wallpaper brush. The lower limit on the size of an object that a microscope can resolve is more or less equal to the wavelength of the light used to illuminate the sample.

Yet there is now an impressive array of techniques that can show us the positions of individual atoms on a surface. These techniques do away with light as the imaging medium and instead paint with a finer brush. Although the atomic portraits that they generate are nothing more than vague blobs, nonetheless they enable us to see at last the fundamental building blocks of matter first postulated by the Greek Democritus in the fifth century B.C.

Arguably the first successful attempt to see atoms was that of Max von Laue, who found in 1912 that atoms packed together in regular arrays in crystals scatter beams of X rays in a manner that leaves a distinctive, symmetrical pattern on photographic plates exposed to the scattered rays. Thus van Laue invented *X-ray crystallography*, still the most important technique for investigating the atomic

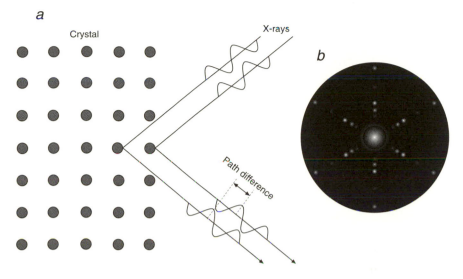

FIGURE 10.1 (*a*), In an X-ray diffraction experiment, a beam of X rays is bounced off a crystal. The waves reflected from different atomic layers interfere with one another: at some angles the peaks and troughs reinforce each other (constructive interference), creating a bright spot, while at other angles they cancel each other out (destructive interference). (*b*), When the diffracted beam strikes a photographic plate, it therefore produces a pattern of bright spots, whose symmetry properties reflect that of the crystal structure. By analyzing mathematically the positions of the spots, the relative positions of the atoms in the crystal can be deduced. Shown here is the diffraction pattern produced by a silicon crystal, the atomic structure of which has hexagonal symmetry along some directions.

structure of crystalline solids. Crystallography exploits the phenomenon of diffraction, which occurs when waves are scattered by a periodic array of "scattering centers" whose separation is comparable to the wavelength of the scattered waves. A beam of X rays incident on a crystal will be scattered by the successive equally spaced layers formed by planes of atoms in a crystal, and a ray scattered from one layer will interact with those from the other layers (fig. 10.1*a*). For certain scattering angles, the peaks of the rays scattered from one layer will coincide with the troughs of the rays from the next layer, and the rays will therefore cancel each other out: they are then said to interfere destructively, and will produce an X-ray intensity of zero at that angle. At other angles, the peaks of the rays from each layer will coincide, reinforcing each other: this will produce a bright scattered beam at these angles. So the result of the scattering process is an X-ray beam that is bright at some angles and dark at others. When this scattered beam falls on a photographic plate, the bright parts cause the plate to darken, whereas in the dark regions the photographic emulsion remains unchanged. The photographic plate thereby acquires a pattern of dark spots corresponding to angles at which constructive interference occurs: these are called diffraction peaks, and are usually depicted in photographic negative as bright spots on a dark background (fig. 10.1*b*).

The angles at which the diffraction peaks occur are determined by the wavelength of the X rays (which is known in advance) and the distance between the atomic planes in the crystal (since this determines the relationship between the peaks of one scattered ray and those of the ray scattered from the next layer down). By determining these angles, William Bragg and his son W. Lawrence Bragg showed, a few months after von Laue's discovery, that one can calculate the distance between the atomic planes. Since there are many different planes that pass through the crystal lattice at different angles, the diffraction pattern is often a rather complex pattern of spots, and decoding this pattern to identify the various crystal planes then enables one to figure out where the atoms sit in the crystal structure.

But for surface scientists, X-ray crystallography is not very useful. Because X rays penetrate rather deeply into materials before they are scattered, most of the scattered beam bounces off atoms far from the surface. To look at surfaces, the beam of X rays must be replaced with a beam of electrons.

This may seem strange, since diffraction is something that happens to waves, yet a beam of electrons is a beam of particles. But it has been known for over seventy years now that, for tiny objects like electrons, the distinction between wave and particle is at best a fuzzy one. An electron beam is very different from a blast of grapeshot: the electrons can under certain circumstances exhibit wave-like behavior, and can be considered to have an associated wavelength that depends on their energy. This dual behavior is one of the curious manifestations of the world of quantum mechanics, the theoretical framework invented at the start of the century to describe the subatomic world. Wave-particle duality was formalized by Count Louis de Broglie in 1924, who proposed a relationship between the velocity, mass, and wavelength of quantum particles.

So, electrons can be diffracted when scattered from a set of atomic planes, if their wavelength is comparable to the distance between planes. Diffraction of electrons was observed three years after de Broglie put forward his wavelength equation, by the American scientists Clinton Davisson and Lester Germer.

The scattering of electrons by an atom is a consequence of the interaction of the negatively charged electrons with the atom's own electron clouds and with its positively charged nucleus. This strong electrostatic interaction means that only a few layers of atoms are needed to scatter an electron beam strongly, so it does not penetrate far into a material. The diffraction pattern in the scattered electron beam therefore encodes the structure of the surface layers. The use of electron diffraction for surface analysis was developed during the 1970s, using electrons accelerated in a field of around 150 volts—this gives them a wavelength of around a tenth of a nanometer, comparable to the interatomic distance in crystals. Compared with other electron-beam analytical techniques such as electron microscopy, these electrons have relatively low energies, and so the technique is known as *low-energy electron diffraction*, or LEED. Today LEED is a standard tool of surface scientists, providing a quick and efficient means of identifying the structure of atomically ordered surfaces.

Just as electrons can be used like X rays for crystallography, so they can be used like light for microscopy. I mentioned above that the resolution of a microscope is comparable to the wavelength of the illuminating beam. As electrons have wavelengths comparable to the spacing between atoms, the electron microscope can see features on this scale. In the most common manifestation of the electron microscope, an electron beam is sent hurtling through a thin sample, and the way that the transmitted beam is scattered reveals the structure of the sample on a scale of perhaps 10 angstroms or so. This technique, called *transmission electron microscopy*, has been refined to such a degree that it is often possible to see the atomic-scale structure of a material. The electron-scattering power of an atom increases as its mass increases, since larger atoms have more electrons and protons and so their interaction with the electron beam is stronger. This means that heavy atoms show up clearly, whereas lighter ones are harder to see; in practice, oxygen atoms are about the lightest atoms that can be seen routinely by today's transmission electron microscopes.

Because the scattered beam passes right through a thin sample, the features in the microscopic image represent a cross section—a kind of projected "shadow" of the atomic layers—and says little or nothing specific about the sample's surface. The technique is valuable, however, for showing interfaces between materials in cross section if these are placed parallel to the electron beam. It is much used to study the interfaces in layered heterostructures used in microelectronics and photonics (fig. 10.2). Here one can investigate the uniformity and thickness of the layers, and also the details of the atomic structure where one layer meets another. These interfaces may have a crucial effect on the electronic properties of the material, because if the atomic periodicities of two different layers do not match up (that is, if they are not epitaxial), defects may arise that degrade the electrical conductivity.

FIGURE 10.2 Transmission electron microscopy can be used to investigate the structures of interfaces between materials. The interfaces here, between indium gallium arsenide quantum wells (dark layers) and indium phosphide (light layers) in a photonic heterostructure, have been placed parallel to the electron beam, so that its cross section shows up in the microscopic image. Each of the dark spots corresponds to just one atom. (Photograph courtesy of Tony Cullis, University of Sheffield, U.K.)

Electrons *will* show specifically the surfaces of materials in a different form of electron microscopy known as *scanning electron microscopy*. Here a focused electron beam is scanned across a surface, and these electrons knock secondary electrons out of the surface atoms in a kind of quantum-mechanical billiards match. The secondary electrons are detected and converted into an image in which the intensity of the gray scale reflects the intensity of electron emission. This latter varies depending on the accessibility of the surface to the scanning electron beam—microscopic potholes will be at least partly "in shadow," for instance, and so they will appear as dark areas in the image. The images from a scanning electron microscope therefore have the appearance of a surface under

FIGURE 10.3 The scanning electron microscope forms an image from secondary electrons knocked out of the surface of the sample by an electron beam. As more secondary electrons are knocked out of exposed regions of the surface, these appear brighter—so the image shows a landscape of bright surfaces and shadowy recesses. Here I show a micrograph of a micromotor fabricated by researchers at the University of Wisconsin. Techniques such as lithography are shrinking the scale of mechanical engineering beyond the range that the eye can resolve: the central rotor is 150 micrometers in diameter. (Photograph courtesy of H. Gückel, University of Wisconsin.)

illumination—they have bright, exposed areas and darker, shadowed regions, and the contrast and detail can be very sharp (fig. 10.3). But the resolution is much lower than for transmission electron microscopy—one can distinguish only features more than about 50 nanometers across, far too large to see atoms.

The Atomic Vista

But let's now put aside such sober realities and take a ride across the atomic landscape of a surface. We are going to skim the most prosaic of surfaces, that of a coin—a dime perhaps—as it might appear freshly plucked from your pocket. As we swoop down toward the gleaming surface, new sights greet us at progressively smaller scales. Scratches that are invisible to the naked eye become visible as deep scars, and then we can make out the mosaic patchwork of nickel and copper domains in the alloy. Some of this we can see through an optical microscope; a scanning electron microscope will show us finer details. But if we imagine going beyond the resolution limits of these instruments, descending toward the surface like a swooping hawk, we start to pick out the most fundamental features of the landscape. Microscopic scratches are now towering canyons, stretching up out of sight, and we are down amidst the pebbles of the valley floor. Here the irregularity of the topography suddenly gives way to order, for these pebbles are arrayed in perfect rows, packed together like eggs in an egg box, for as far as the eye can see. This is the atomic lattice of the metal surface.

We can distinguish two types of pebble—for the surface in this nickel domain is in fact nickel oxide, the metal having reacted with oxygen in the air. Occasionally the regularity is broken by a further kind of pebble poking above the surface: a contaminant atom of some other element. In some places the landscape descends in broad, terraced steps; in others it plunges over abrupt cliffs. And there is life here: we might encounter the polyhedral shell of a virus, stalking the land like a Platonic predator, and organic debris—large, irregular disfigurations of carbon-based molecules—lies scattered and flattened against the inorganic ground. And if we look very closely at the pebbly surface, we might just be able to discern the subtle ripples that spread across it from irregularities and step edges: a sign of the delocalized electron sea that washes ethereally through the land (plate 13).

Fifteen years ago this journey would have been conducted only in the imagination of a surface scientist, who would know more or less what was down there but wouldn't expect to be able to see it in all its panoramic glory. Now, we can take the trip for real. In the early 1980s, surface science was transformed by a new microscopic technique that literally brought the atomic structure of surfaces into focus.

The scanning tunneling microscope (STM), developed by Gerd Binnig and Heinrich Röhrer of IBM's research laboratories in Zurich, Switzerland, is not like other microscopes. It does not use wave optics of any sort to form an image. The image that it produces is, in fact, wholly the creation of a computer, which builds the picture by decoding a signal from a probe that scans the surface. The micro-

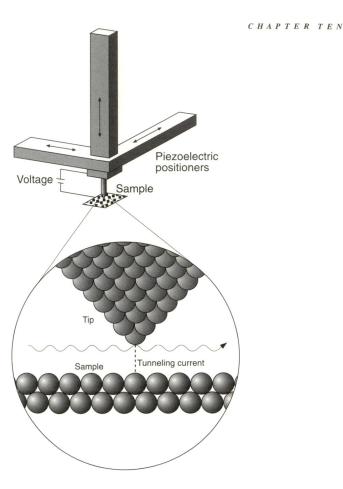

FIGURE 10.4 The scanning tunneling microscope takes atomic snapshots. A metal needle with an extremely fine tip is scanned across the surface of a sample with great precision, made possible by using the kind of piezoelectric positioners discussed in chapter 3. By applying a voltage between the tip and the sample, electrons can be pulled one way or another across the gap when it is no wider than a few atomic diameters. This "tunneling current" varies very sharply with the distance between the tip and the surface. So the topography of the surface— bumps and dips caused by the stacking arrangement of the surface atoms or by molecules stuck to the surface—can be mapped out by monitoring how the tunneling current varies from place to place. Commonly, a feedback device couples the tunneling current to the vertical positioner, and the tip–sample separation is altered to keep the tunneling current constant: we can picture this as an up-and-down movement of the scanning tip such that it mimics the ups and downs of the surface.

scope's probe is not a beam of light or electrons but a needle-like tip of metal only a few atoms across at its apex (fig. 10.4). The tip of the STM is moved back and forth across the surface, typically just a few atoms' widths from contact, under the control of very delicate positioning machinery. In a sense the tip "feels" the topography of the surface as a series of bumps. The tip is given an electrical charge, either positive or negative relative to the surface below. At such a small distance from the surface, the charged tip has the ability to pull electrons out of the surface atoms or molecules (if it is positively charged) or to inject electrons into the surface (if it is negative). The tip is a little like a thundercloud, traveling over the ground and discharging electrical current as it goes—with the important distinction that this discharge is continual, not in sudden bursts.

But this "classical" description ignores the fact that the process of electron transfer between tip and surface actually stems from a peculiar property of tiny particles with wavelike properties: they can travel *through* barriers rather than over them. Whereas a football will not pass from one side of a wall to the other unless it is given enough energy to take it over the top, an electron, which possesses the wave nature postulated by de Broglie, can traverse such a barrier even if it does not, classically speaking, possess enough energy to do so. It does this by "tunneling" through the barrier, an effect that is predicted by quantum mechanics.

For the purposes of the STM, the crucial characteristic of this tunneling process is that it depends very sensitively on the separation between the tip and the surface. If there is some protrusion on the surface, such as a foreign atom or molecule that has become stuck there, the separation decreases and tunneling takes place more readily. A measurement of the variation in tunneling current from place to place over the surface therefore generates a map of the topography of the surface, showing its atom structure as an array of bumps.

The interpretation of an STM image is not, however, always quite as straightforward as this, since the tunneling current is not just a function of the tip–sample separation. It also varies with the chemical nature of the surface: in general, the very existence of a tunneling current requires that the region beneath the tip be electrically conducting—otherwise, there is no electrical circuit and so no flow of current. The STM is therefore able to generate an image of the atomic structure of the surface of metals and semiconductors (fig. 10.5a), but not of insulators such as most plastics.

So immediate and profound was the effect of the STM on surface studies that it won Binnig and Röhrer the Nobel Prize for physics in 1986, just five years after they first announced its invention. The invention was all the more startling for its simplicity, for no one would have expected that an instrument of this sort would be sensitive enough to image individual atoms. For the first time, it was possible to see directly that molecules really do look like the schematic pictures that chemists draw: benzene rings, for example, are truly like doughnuts (fig. 10.5b). Of course, there had been no doubt previously that this was so—diffraction experiments had already revealed these structures—but never before had there been such a direct window into the atomic world.

The one major drawback to the STM was its restriction to surfaces and surface-bound molecules from which one could extract a tunneling current. But this limitation was overcome in 1986 when Gerd Binnig and Christoph Gerber at IBM, working with Calvin Quate of Stanford University, developed a related scanning probe microscope that does not rely on electron tunneling to produce an image. Whereas the STM feels the shape of a surface by the rather indirect means of

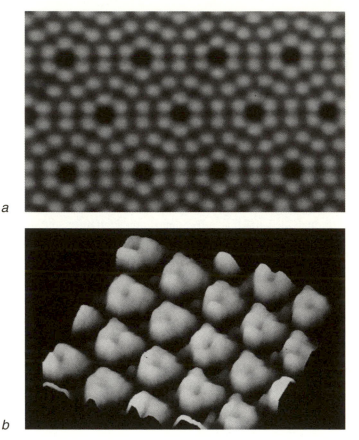

a

b

FIGURE 10.5 The STM shows the atomic-scale structure of surfaces and of molecules adsorbed on them. (*a*), The surface of silicon. Each of the bright spots corresponds to the position of an atom (although this 1:1 correspondence is not always the case). This is one of several "reconstructed" surfaces that silicon exhibits, called the 7×7 reconstruction. (Photograph courtesy of Roland Wiesendanger, University of Hamburg, Germany.) (*b*), Benzene molecules lying flat on a graphite surface are revealed by the STM as ringlike molecules with a hole in the center. (Photograph courtesy of IBM Research.)

pumping electrons into it (or sucking them out), the device invented by Binnig, Gerber, and Quate is a more truly tactile instrument. Called the atomic force microscope, it literally feels the force exerted by the surface on the probe tip of the microscope. Most materials exert a mutual attraction at small enough distances owing to so-called van der Waals forces. (At still smaller distances the interaction becomes strongly repulsive as the two surfaces come into contact.) The strength of the attraction depends on the distance between the surfaces, as well as their chemical nature. So a probe scanned at a constant height across a surface will feel an attractive force that rises and falls in concert with the rise and fall of the surface topography, since hills will be relatively nearer the probe and valleys relatively further away.

A measurement of the force felt by a scanning probe tip, reasoned Binnig and colleagues, will therefore reveal the corrugations of the packed rows of surface atoms and the abrupt plateaus of adsorbed molecules. The idea is almost trivially simple in principle. But to put it into practice required immense ingenuity, because the forces involved are distressingly small: generally the force felt by an AFM tip is equivalent to the weight of a few dozen bacteria. To measure these minuscule forces of attraction, the tip is attached to the end of a long, thin strip of a hard, springy material such as tungsten or silicon nitride. This strip is called a cantilever, and it is flexible enough that even a tiny force on the tip will cause the strip to bend slightly (fig. 10.6). The bending can be detected in several ways—typically by bouncing laser light off the back of the cantilever's end and measuring changes in intensity owing to interference between the incoming and reflected

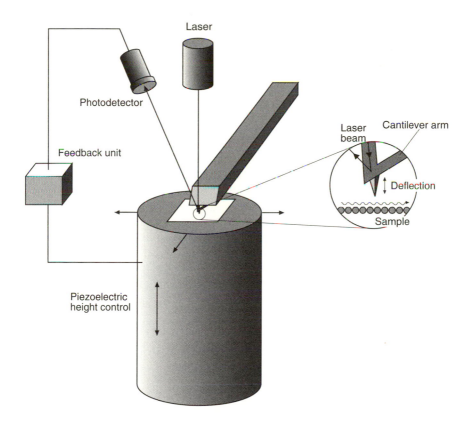

FIGURE 10.6 The atomic force microscope senses the forces between a surface and a fine tip attached to a flexible cantilever. The magnitude of the force determines how much the cantilever bends. This displacement can be measured very accurately in several ways—for example by interferometry, where the change in path length of a laser beam reflected from the cantilever end causes a change in light intensity measured by a photodetector. Commonly the sample is maintained at a constant distance from the tip by a feedback unit connected to a piezoelectric height controller, so that it is the degree of compensatory displacement required to maintain this constant separation that is the ultimate measure of the tip–sample forces.

beam (a method called interferometry). One of the biggest attractions of the AFM, relative to the STM, is that it can be used to image biological and organic specimens, which are usually electrically insulating and therefore harder to probe with the STM.

All microscopic techniques show a picture of sorts; but the atoms, molecules, or other topographic features or contrast variations in these images are not conveniently labeled "copper," "benzene," or "DNA." In general, identifying the chemical nature of surfaces requires the use of *spectroscopic* techniques, which tell us about the atomic identity of the constituents by probing their energies. I won't go into these, except to say that they are manifold and are as valuable to the surface scientist as microscopic methods. It is not without good reason that most scientists are reluctant to accept any attempt to characterize a surface until at least two independent methods have been used, of which one will often be microscopic and one spectroscopic. As I've said already, at surfaces you can never be quite sure what to expect, and it pays to take a look from several perspectives.

WHY THINGS GET WET

A silver pepperpot loses its luster and turns black with age. My windows acquire a patina of London grime (which is removed with regrettable infrequency). The head of my tape recorder gets a rusty red film that reduces the sound of my cassettes to a blur.

These sad inevitabilities, and a great deal else in surface science, can be understood on the basis of a very simple principle: surfaces do not, in general, particularly want to be surfaces. They are generally less stable and more reactive than the material buried deep below, and a consequence of this is that substances impinging on a surface will commonly stick there, sometimes by undergoing chemical reactions with the surface material.

It isn't hard to see why this should be so. That materials stay together at all, instead of flying apart into a gas of individual atoms or molecules, is the consequence of forces of attraction between these fundamental constituents. The attractive forces take various forms. In ionic crystals such as salt, for example, they are the electrical forces of attraction between atoms bearing opposite charges; in covalent solids like diamond, they are chemical bonds between atoms, resulting from the sharing of electrons. Liquid water is held together by hydrogen bonds, and even an inert, electrically neutral substance like the rare gas argon can be liquefied at low temperatures as a result of the weak attraction—the van der Waals force—that every atom or molecule exerts on any other.

Atoms, ions, or molecules deep within the bulk of a material are surrounded by others, and so will feel these attractive forces from all sides, as well as exerting such forces on their neighbors. But atoms at the surface of the material will experience relatively fewer attractive interactions, because they have fewer neighbors. As a consequence, the interior of a material is more stabilized by interatomic attractions than is the surface. The surface can be considered to have an *excess*

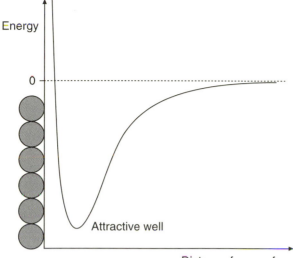

FIGURE 10.7 Surfaces (represented here by a row of atoms) exert an attractive force extending a few nanometers into space. In other words, there is an energy well at the surface into which molecules may fall. Molecules held at the surface by this physical force of attraction are said to be physisorbed.

energy, which can be expended by binding foreign atoms or molecules that impinge on it—by burying the surface, in other words. It is an expression of the same thing to say that the surface exerts an attractive force extending a short distance out into the space beyond—a kind of well, into which other atoms or molecules can fall and become trapped (fig. 10.7).

If the surface is deformed—if, for example, we were to create protrusions and valleys by plucking clusters of atoms from the flat surface layers—this involves exposing still more surface, and so increases the surface excess energy. A "surface tension" therefore exists that has a tendency to pull the surface back to smoothness. In other words, because creating surfaces costs energy, materials will prefer to arrange themselves so as to minimize their surface area.

The idea of a surface tension is a familiar one for liquids: we all know that it is what pulls the surface of water in a glass or a pond into mirrorlike flatness, and raindrops into spheres. But the idea that solids too possess a surface tension may seem a little odd—after all, if we scratch the surface of a metal, the grooves do not instantly vanish to restore the surface's mirror sheen. But this is simply because the forces that hold a metal together are strong enough to permit the atoms scarcely any mobility—a rough surface has a higher energy than a smooth one, but the surface tension is not strong enough to wipe away the furrows. If we heat up a rough metal surface so that the atoms become more mobile, the roughness will then start to disappear even before the bulk metal melts.[1]

---

[1] There is a further wrinkle, so to speak. Some surfaces roughen spontaneously when heated—they acquire a disorderly, ramified surface at the microscopic scale, scattered with pits and peaks. This takes place abruptly at the so-called roughening temperature, and is a consequence of the disruptive effects of thermal vibrations, which oppose the smoothing influence of surface tension. The surfaces of most liquids, including water, are "rough" on this microscopic scale, exhibiting tiny undulations called capillary waves.

When the cohesive forces in a material are covalent chemical bonds, the surface atoms are left with spare "dangling" bonds. These atoms are therefore eager to form chemical bonds with just about anything that encounters the surface; diamond surfaces, for instance, will pick up a layer of hydrogen atoms to cap the dangling bonds (see chapter 8). Similarly, a silicon surface will quickly be oxidized in air, reacting with the oxygen to form a layer of silicon dioxide. Because silicon dioxide is insulating, surface oxidation is an important process in semiconductor technology: it "passivates" silicon, leaving poor electrical contact with a material deposited on top.

Rearranging the Surface . . .

Undergoing chemical reactions is one way in which surfaces cope with their excess energy. Another is to alter the structure of the bare surface. In the interior of a crystalline material, atoms arrange themselves so as to minimize their energy. The optimal arrangement depends on a number of factors, such as the sizes of the atoms, their preferences for forming bonds in particular directions or geometries, and the temperature. But there is no reason at all why the packing arrangement of atoms in the bulk is the one that will best suit the surface atoms too. It is therefore common for the surfaces of materials to undergo a *reconstruction* to a structure that is different from merely an abrupt termination of the bulk structure. These changes in surface structure often involve the shifting of atoms from layers below the outermost one—this will increase the energy of the depleted lower layers, but this may be more than compensated for by a stabilization of the surface layer.

Probably the best-studied examples of surface reconstruction are those that take place at the surface of silicon, because of the central importance of these surfaces in microelectronic technology. The nature of the reconstruction depends on the nature of the surface on which it occurs: cutting a crystal like silicon will expose a different initial surface structure depending on the direction in which the cut is made (recall fig. 4.35). These different surfaces are denoted in general by a sequence of numbers that refer to the different symmetry axes of the crystal. In figure 10.8a I show the silicon (100) surface, which alongside the (111) surface is one of the two most important in silicon technology.

Both of these two surfaces undergo reconstructions to a new arrangement of surface atoms. The surface energies of the reconstructed layers vary with temperature, so a different reconstruction may take place at one temperature than at another. Figure 10.8b shows one of the surface reconstructions that takes place on the silicon (100) surface; in figure 10.5a I showed an STM image of an important reconstruction of the silicon (111) surface, called the 7×7 reconstruction. Understanding the factors that control these reconstructions is of critical importance for silicon technology because the deposition of layers of other materials on silicon is highly sensitive to the arrangement of surface atoms and the extent to which this matches the atomic structure in the deposited layer.

a   (100)   b   (100) 2x1 reconstruction

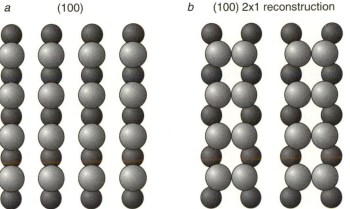

FIGURE 10.8 (*a*), The structure of the so-called (100) crystal face of silicon (here I show just the top two layers). (*b*), The atoms at the (100) surface may rearrange themselves to lower the surface excess energy. One of the most common surface reconstructions of this sort is the 2 × 1 reconstruction, in which atoms of silicon pair up in rows.

. . . and Getting Stuck There

In materials science, as in the material world, we find endless marriages of surfaces—endless *interfaces* between substances, be they solids, liquids, or gases. On the whole these marriages do not require much coercion—it's often hard to prevent them!—because most solid surfaces are sticky. It may not seem that way—my finger encounters no tackiness when I press it against the window pane—but that is because the stickiness is, on our human scale of measuring things, very weak and short ranged. Sadly, to a particle of dust it is all too apparent, so that these gradually coat the windows and just about every other surface in the house unless I do something about it. Outside, my car is pronouncing just as clearly that, to the dust and grime of the city air, it is like flypaper.

But this is no more than I should expect if I remember that attractive well lying just above the surface (fig. 10.7). The surface is saying: here I am—come and stick to me. Reduce my excess energy with a soothing patina. And that is what happens: gas molecules or small solid particles will fall into the well, a process called *adsorption*. This process has two complexions: in *physisorption*, foreign molecules adhere to the surface via the relatively weak van der Waals forces between them and the surface atoms, whereas in *chemisorption* chemical bonds are formed between the adsorbed molecules and the surface atoms, holding the former more firmly in place. The oxidation of a silicon surface is an example of the chemisorption of oxygen, which is called the adsorbate.

Adsorption processes are of tremendous importance in industrial chemistry, where they constitute the first stage in the operation of many of the catalysts used to accelerate chemical reactions. In materials science, adsorption is often a nuisance, since it results in surface contamination. This is particularly to be avoided in semiconductor technology, where the smooth deposition of a thin film of material atop another requires very clean surfaces. Such deposition processes, and many studies of surface structure, have to be conducted under high vacuum to exclude gas molecules that would otherwise stick to and obscure the surface. It is not uncommon for adsorption to actually alter the nature of the surface, by inducing reconstructions to structures that, in the presence of an adsorbed layer of gas, have lower surface energy.

On the other hand, sometimes we *want* a material to become adsorbed at a surface. This is, after all, precisely what one is trying to achieve when preparing a sandwich-style semiconductor device by chemical vapor deposition (chapter 1). The aim here is to create a film of constant thickness, which is free from structural defects and impurities. Avoiding defects is one of the biggest challenges that the microelectronics industry faces.

Materials scientists therefore spend a lot of time trying to develop ways to control absorption—to avoid that of unwanted impurities and contaminants, while ensuring that adsorption of deposited layers takes place readily and evenly. The most critical factor in determining the latter is whether or not the adsorbed film *wets* the substrate.

Wetting sounds as if it should be too old-fashioned and qualitative a phenomenon to command attention at the forefront of materials physics. But like most everyday words commandeered for scientific use, it means something subtly different to scientists than to others. One substance is said to wet another if it spreads across it spontaneously in a uniform film. That sounds obvious enough, and you might imagine that this definition implies that all liquids would wet all solids (provided that they don't dissolve them). Not so, however. Think of water in a nonstick frying pan: it beads up, forms little rivulets that skitter across the surface, and seems to do everything it can to avoid spreading evenly. The frying pan is most probably coated with Teflon, and water clearly does not wet Teflon. Indeed, Teflon is used in nonstick kitchenware for this very reason: other substances too will refuse to spread over and stick to it.

At this macroscopic scale, wetting behavior is important in all sorts of applications. It determines how the rain runs off your windshield. It provides the science behind water-repellent, wipe-clean wallpaper. It makes the difference between a wound dressing that becomes ugly with body fluids and one that stays clean. At the microscopic scale, the same phenomenon will determine whether or not we can effect thin-film deposition of one substance onto another. If an adsorbed material does not wet the substrate, it will form clumps separated by bare surface, rather than a smooth film. These islands of material are the counterparts of the beads of water on a plastic surface.

Making Contact

So what determines whether one material will wet another? Clearly, this has no direct relation to whether the deposited material forms a solid or a liquid film—

atomic vapor deposition can produce either a smooth or a clumpy solid film, and a liquid can either spread in a uniform film or bead up. The wetting behavior is controlled by the balance between each of the surface energies in the system.

If we consider the case of water on a flat solid surface, there are three different kinds of interface: between the surface and the air, the water and the air, and the surface and the water. Each of these interfaces has a different energetic price. The wetting behavior is then determined by the question: Do we stand to pay a lower overall cost in surface energy by letting the water spread to eliminate the solid/air interface and replace it with solid/water, or by keeping the area of the solid/water interface to a minimum by having the water bead up on the surface? This depends on the relative magnitudes of the various surface energies, which in turn will depend on the chemical composition of the materials concerned. If we stand to gain (that is, to pay a lower energetic cost) by covering up the solid surface with the water—that is, if the sum total of the solid/water and water/air surface energies is less than that of the solid/air interface—then the water will wet the surface.

If the reverse is true, the droplet will not spread completely on the surface. But it is all a matter of degree—the less favorable it is to put water between the substrate and the air, the less the droplet will spread. This idea was put forward in qualitative terms by Thomas Young in 1805, who introduced the idea of a contact angle. Simply speaking, this is the angle between the substrate and the edge of a droplet (fig. 10.9*a*). When the droplet wets the surface completely, the contact angle is zero: the droplet forms a flat film. At the opposite extreme, the contact angle is 180 degrees, which corresponds to the case of a spherical droplet perched on the surface (fig. 10.9*b*). In this case, the droplet does not wet the surface at all, and could in principle be plucked off to leave a totally dry surface.

a

b

Figure 10.9 (*a*), A liquid droplet on a solid surface spreads to a degree quantified by the contact angle, θ. If θ is zero, the droplet is flat—it spreads completely and the liquid is said to wet the surface. (*b*), In the case of complete non-wetting, the contact angle is 180 degrees, and the liquid forms a spherical droplet perched on the surface. Here a water droplet has a contact angle of close to 180 degrees on the surface of an organic material, whose non-wettability is due to a highly ramified "rough" texture at a scale that is not visible in this photograph. (Photograph courtesy of T. Onda, Keio University.)

Contact angles are a very convenient measure of this balance of surface energies, because they can be sized up at a glance. The larger the air/substrate surface energy, the more likely it will be that this can be lowered by contaminants attaching to and spreading over the surface. So to design a nonstick or dirt-resistant surface, one needs to give it a low surface energy.

But how can we achieve this, given that the surface atoms are inevitably going to be less happy than those in the bulk? The usual strategy is to arrange for the surface to be covered with chemical species that do not interact very strongly with *anything*, thereby ensuring that we gain very little by covering up this surface with foreign substances. The polymer molecules that make up Teflon are long chains of carbon atoms studded with fluorine (fig. 10.10); the fluorine atoms are

FIGURE 10.10 Teflon is the Du Pont trade name for polytetrafluoroethylene (PTFE), a polymer in which a carbon backbone is studded with fluorine atoms (gray). The fluorine atoms give Teflon a very low surface energy, so that it is not readily wet by liquids.

small and compact, shunning the company of other atoms. Therefore the surface of Teflon exerts scarcely any attractive force on adsorbates, and so they cannot adhere or spread easily. This is the same as saying that Teflon has a very low surface energy. (This is perhaps too facile an explanation, however, because polyethylene, which has still smaller hydrogen atoms in place of the fluorines, nevertheless is less nonstick—that is, it has a higher surface energy. It is closer to the truth to say that fluorine atoms hold onto their electron clouds more tightly than any other atom, whereas the van der Waals forces through which such molecules interact with others rely on a certain floppiness of the molecular electron clouds.)

Teflon is the trade name used by the Du Pont chemicals company for the polymer polytetrafluoroethylene, or PTFE. This material, which was synthesized by R. J. Plunkett at Du Pont in 1938, has been the mainstay material for nonstick coatings for decades. But it has several drawbacks, not the least of which is that applying a PTFE coat to a surface involves the use of some unpleasant solvents, as well as a high-temperature curing process that can prove tricky when it is the hull of a ship, not the inside of a pan, that one wants to render water-repellent. In addition, these films generally contain microscopic pores, allowing material to lodge in the cracks.

In recent years, new nonstick coatings have been developed that promise to address these shortcomings. Some of these have been based on silicone polymers—chains of silicon and oxygen atoms with carbon-based side chains attached. The silicon–oxygen backbone provides an anchoring unit, while the side chains provide a low surface energy: in one product, they consist of short PTFE-like strands. Polyurethane elastomers also provide effective nonstick coatings.

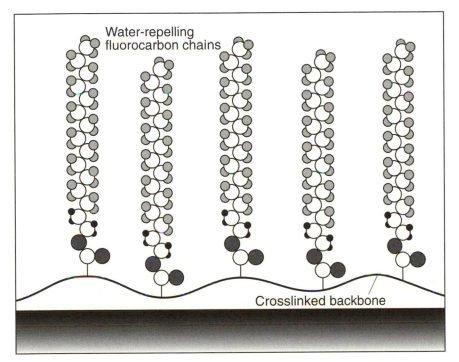

Water-repelling fluorocarbon chains

Crosslinked backbone

FIGURE 10.11 The Dow coating contains cross-linked polymer chains to which PTFE-like side chains are attached. These side chains line up more or less perpendicular to the substrate on which the film is deposited, giving a very-low-surface-energy coating that repels adhesives and contaminants.

The current state of the art, however, is defined by a material concocted by scientists at the Dow chemical company in Michigan. This substance can be applied in a hard, clear coat from which dirt simply falls off. A marker pen leaves no mark—the ink just beads up and can be wiped away (plate 14). When part of a car was treated with the Dow coating, it remained grime-free after driving around the American Midwest for several weeks. And there seems to be no known glue that will stick to this slippery customer.

Like some silicone coatings, the Dow compound consists of chainlike polymers to which short strands of PTFE-like polymers are attached like the legs of a centipede. When laid down on a surface, the fluorine-coated chains all line up like bristles (fig. 10.11). The orderliness in this structure gives it a lower surface energy than PTFE itself. Applying a coat of the new material is relatively simple: the backbone chain contains chemical groups that make it soluble in water, so a solution of the polymer can be sprayed onto a surface. The sprayed film is converted into a robust coating by adding a cross-linking agent that forges bonds between the backbone chains when a moderate degree of heat is applied.

The Dow scientists believe that it should be possible to develop a form of the coating that could be applied with a paintbrush. Provided that the need for heat treatment does not throw up too many problems, one can imagine giving buildings a lick of this transparent paint to make them graffiti-proof.

Holding Fast

Preventing two materials from sticking together is just one side of the coin, relevant when we wish to keep a surface free from contamination. There are countless situations is which we would like very much the converse: that is, for two materials to bind very firmly together. Central to the science of adhesion is again the concept of surface excess energy: two surfaces will adhere if the interface between them has a lower surface energy than those of the two bare surfaces interfaced with air. As this is true for most solid surfaces—in crude terms, marrying like with like by bringing two surfaces together tends to be more energetically favorable than leaving the surfaces exposed—we might expect most solids to stick to most others. This manifestly does not happen, however, unless we apply some kind of glue to bind the two—and that is a different matter entirely. It doesn't surprise us that two sheets of glass will not spontaneously stick together when brought into contact, because experience tells us that this does not happen. But once we take into account the idea of surface excess energy, perhaps we *should* be surprised: after all, could not both of the sheets lose all of the surface energy of one face if they were to adhere together to eliminate the surface entirely, one sheet of glass merging seamlessly with the other?

The reason why we do not, on the whole, observe this kind of spontaneous adhesion is that most surfaces are very far from perfect. Under the microscope, most of them—even that of an apparently smooth glass plate—turn into a landscape of hills and valleys, scratches, bumps and other irregularities. These microscopic mountains are called *asperities*, and because they do not match up when two surfaces are brought together, one will rest on the other like two sheets of rough sandpaper: touching at only a few points, with gaping spaces between them (fig. 10.12). So bringing the two surfaces together does not do very much at all to eliminate the two solid/air interfaces.

FIGURE 10.12 Most solid surfaces are rough at the microscopic scale, so that two surfaces in apparent contact will in fact touch only at a few points, called asperities.

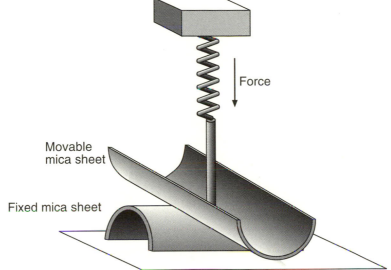

Force

Movable
mica sheet

Fixed mica sheet

FIGURE 10.13 The surface force apparatus measures attractive forces between surfaces as they come into contact. The surfaces are sheets of cleaved mica, which are extremely smooth and flexible enough to be bent into half-cylinders. Two half-cylinders at right angles to one another touch at only a single point. The force between the surfaces is registered by a sensitive spring to which one of the mica sheets is attached.

The smoother we can make a surface, therefore, the more adhesive it should be. It is hard to make large surfaces free from irregularities, but surfaces of smaller area will stick together more readily because the number of asperities propping them apart is smaller. Thus powders can be compacted into dense, hard aggregates (like soluble aspirin tablets) even though each particle retains its integrity. David Tabor and coworkers at Cambridge University showed in the early 1970s that even two plates of a hard mineral such as mica will stick together when they are sufficiently smooth. Mica is a claylike substance containing layered sheets of aluminum, silicon, and oxygen atoms; it can be cleaved between two sheets to present a very smooth surface. Very thin slices of mica can be bent without breaking, and Tabor and colleagues constructed a device, which they called the surface force apparatus, to investigate the nature of the forces between two such bent sheets. To minimize the contact area, and thus the chance of this being interrupted by asperities or other defects in the mica sheets, they brought them together in a crossed geometry, with one at right angles to the other. If both surfaces are perfectly smooth, they will touch only at a single point (fig. 10.13).

The surface force apparatus measures the force of attraction between the two surfaces with highly sensitive springs. The separation of the surfaces can be adjusted to an accuracy of about an angstrom. Tabor and Jacob Israelachvili found in 1973 that the force between the curved mica sheets became very strongly attractive as the distance between them was decreased to about 5 nanometers,

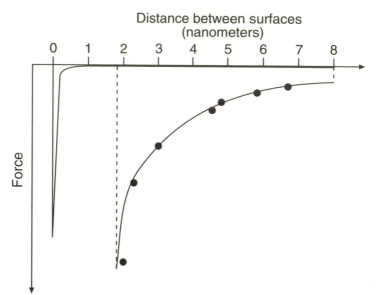

FIGURE 10.14 The force of attraction between two crossed mica cylinders in air becomes very strong when they are only a few nanometers apart. When they are closer than about two nanometers, it is hard to prevent them from jumping into adhesive contact. This compelling demonstration of the stickiness of bare, smooth surfaces was performed in 1973 by David Tabor and Jacob Israelachvili, who made the measurements using the surface force apparatus. Their experimental results are shown here as black circles; the solid curves show the predictions calculated from the standard theory of surface forces. The forces were too great to measure for separations less than around two nanometers. I have shown the force for greater separations than this on an expanded scale (lower curve), because the changes are otherwise too small to be seen relative to the huge attractive forces inside one nanometer.

and that the two surfaces jumped into adhesive contact at shorter distances (fig. 10.14). As Kevin Kendall of Keele University in England says, "Anyone who has observed these smooth solid surfaces leaping into contact must rearrange their ideas about the behavior of objects. One's previous experience is that bodies do not stick together without assistance . . . these ideas must be discarded when we approach small sizes." If we want to make mechanical machines of nanometer dimensions, he points out, we must face the problem that their moving parts will spontaneously stick together.

Aside from making them perfectly smooth, another way to improve the adhesion between two surfaces is to make at least one of them soft and deformable: in this way, it can mold itself around any asperities on the other surface and form an intimate interface without air gaps. This is most apparent when one of the surfaces starts off as a liquid, which is poured over a solid surface and then set by freezing or by evaporation of a solvent. This is how a paint acquires sufficiently close contact with a rough wall to stay put once it has dried. The adhesive power of soft polymer films is also very familiar to us, in the form of the transparent plastic wrapping known in the U.K. as "clingfilm." This material is capable of forming such an intimate interface that it feels almost tacky to the touch. Some highly plastic polymers can act as weak adhesives when pressed to enhance their penetration into the nooks and crannies of a rough surface: the sticky substance known as "Blutack"—the bane of wallpaper in every student's room—is an example of this. A liquid film between two solid surfaces may also act as an adhesive, binding wet hair into clumps and wet microscope slides into stacks.

Films That Assemble Themselves

Against the Flow

Until 1992, the only person to my knowledge who could make water run uphill was the Dutch graphic artist M. C. Escher (fig. 10.15). I don't think even Escher could have actually built his water mill, not least because it violates the Second Law of Thermodynamics. So what were George Whitesides of Harvard University and Manoj Chaudhury of the Dow Corning Corporation up to when in that year they published in the journal *Science* a paper called "How to Make Water Run Uphill"?

Perhaps the most remarkable aspect of this paper, which demonstrated just what the title claimed, was that there was not really any new science in it. We had known all along how to make water run uphill. The trick is a straightforward matter of energetics. Water runs downhill because that way it loses gravitational (so-called potential) energy. Similarly, water spreads on a surface such as glass because that way it reduces the total surface energy (you can see from fig. 10.9*b* that this, rather than gravity, is the driving force for spreading of small droplets). But water will not spread *up* a sloping sheet of a wettable material like glass because the loss in surface energy in doing so does not exceed the gain in gravitational potential energy as it moves higher. (Actually even this much is not quite true, because water *does* rise up the side of a vertical glass to some extent, creating a curved meniscus. The meniscus rises at the edges until the loss in surface energy is balanced by the gain in potential energy.)

To make water run uphill, therefore, all we need do is ensure that the reduction in surface energy as it spreads uphill remains sufficient to more than compensate for the gain in potential energy as the water gets higher. Whitesides and Chaudhury achieved this by applying to a silicon wafer a coating that afforded a graded surface energy, such that the tendency for it to be wet by water increased up the slope. A droplet placed at the foot of the slope therefore moved up it en masse, in order to lower the total surface energy. No matter how high it went, it always found that it could lower the surface energy still further by going higher, and that this more than paid for the gain in potential energy (fig. 10.16). The

FIGURE 10.15 "Waterfall" by M. C. Escher uses an optical illusion to defy gravity and thermodynamics. (Picture reproduced with permission of Cordon Art BV, The Netherlands.)

gradation of surface energy of the silicon wafer was achieved by exposing it to a compound called decyltrichlorosilane (DTS). Molecules of this substance consist of long hydrocarbon chains, like those in paraffin wax, to one end of which are attached silicon atoms capped with three chlorine atoms. These molecules can bind to the surface of silicon via chemical reactions that I shall discuss shortly; it is enough for now to know that these reactions leave the DTS molecules with the

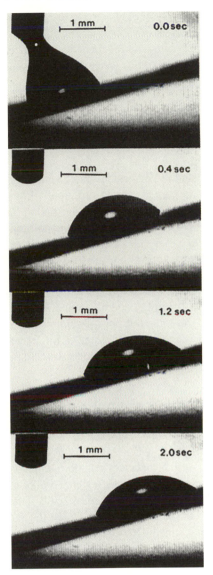

FIGURE 10.16 Water runs uphill. This droplet creeps up a slope of 15 degrees at a speed of a millimeter or so per second, because the slope on which it moves has a graded surface energy that makes it increasingly favorable for the water to wet the surface the higher up it goes. (Photograph courtesy of George Whitesides, Harvard University; reproduced with permission from *Science* **256**, 1539 [1992].)

silicon end down and the hydrocarbon tails poking up. Bare silicon is readily wettable by water, but as it becomes covered with the waxy hydrocarbon chains, it becomes water-repelling (hydrophobic). The greater the number of DTS molecules attached to the surface, the more hydrophobic it is; in other words, the greater is the contact angle between the surface and a water droplet.

Whitesides and Chaudhury achieved a smooth variation in the amount of these molecules across the surface of a silicon wafer by placing the wafer edge-on next to an evaporating droplet of decyltrichlorosilane. The evaporated vapor diffused over the surface of the wafer so that the part of the wafer closest to the droplet got a higher dose than more distant regions. This gradient in vapor concentration was translated into a gradient in concentration of molecules coating the silicon surface as they reacted and became attached by chemical bonds. A water droplet placed on a prepared wafer inclined at an angle of 15 degrees to the horizontal found its way uphill, receding from the less wettable toward the more wettable part of the surface, at a speed of one or two millimeters per second.

Smooth Customers

Close inspection of surface films like that created by Whitesides and Chaudhury reveals that, when the chainlike organic molecules are present in sufficiently high concentrations, they form an orderly layer with the reactive head groups bound to the surface and the long tails aligned like the bristles of a brush. There is nothing particularly remarkable about the fact that these molecules will stick to a glass surface—this is merely a form of chemisorption, adsorption due to a chemical reaction between the adsorbate and the surface. But what is more surprising is the extreme orderliness of the films. The hydrocarbon tails are really rather floppy appendages, and if they are very long they will simply become entangled. But within a certain range of chain lengths—not too long and not too short—these molecules pack together in a highly organized, regular manner, creating films one molecule thick (monolayers) with extremely uniform surface properties. Such films have come to be known as *self-assembled monolayers* (SAMS).

It has been known since the beginning of the century that amphiphilic molecules such as fatty acids (fig. 4.20) will form a monolayer on a hydrophilic surface like glass, if a glass slide is simply dipped into a solution of the amphiphiles. But the recognition that a simple procedure like this can provide ordered, chemically bound self-assembled monolayers dates from 1980, when Jacob Sagiv of the Weizmann Institute of Science in Israel showed that octadecyltrichlorosilane—the same molecules as Whitesides and Chaudhury used to make water run uphill, but with a longer tail—will self-assemble from solution onto a glass slide. The trichlorosilane ($-SiCl_3$) groups are hydrolyzed at the glass surface, which means that a reaction with water replaces the chlorine atoms by hydroxyl ($-OH$) groups. These then react with the silica glass, grafting the amphiphiles chemically to the surface. The remaining hydroxyl groups in neighboring surface-bound molecules react with each other, cross-linking to form a kind of surface-bound silicone polymer (fig. 10.17). The hydrocarbon groups protruding from this surface render it very hydrophobic.

In 1983 Ralph Nuzzo and David Allara of the University of Illinois found that organic alkylthiol molecules, which comprise hydrocarbon chains terminating at one end in a thiol group ($-SH$) will self-assemble in much the same way on the surface of gold. The molecules in these SAMS stand side by side with their chains tilting at a uniform angle (fig. 10.18). This kind of orientational ordering of molecules is characteristic of liquid crystals, but here it occurs within a single layer that is bound to a solid substrate.

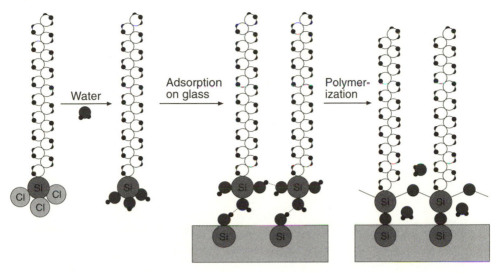

FIGURE 10.17 Alkyltrichlorosilane molecules will assemble on silica glass into well-ordered films in which all of the alkyl groups point upward, making the surface hydrophobic. The chlorine atoms are first replaced by hydroxyl groups in a reaction called hydrolysis; the hydroxyls form hydrogen bonds with similar groups at the glass surface; and cross-linking reactions then take place both between the surface and the bound molecules and between neighboring molecules. The result is a cross-linked film one molecule thick, called a *self-assembled monolayer*.

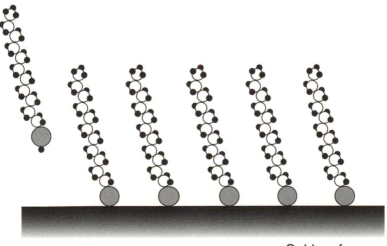

Gold surface

FIGURE 10.18 Self-assembled monolayers of alkylthiols will form on the surface of gold, to which sulfur atoms (gray circles) bind avidly. The alkyl tails of the bound molecules are closely packed in a tilted array.

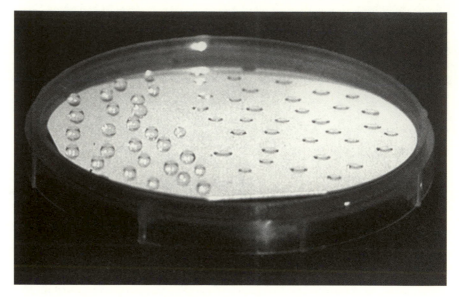

FIGURE 10.19 The surface excess energy (or in other words, the wettability) of a surface can be modified by a self-assembled monolayer. Here the left-hand side of a silicon wafer has been coated with a SAM that presents hydrophobic chain ends at the surface, so that water droplets bead up with a contact angle of close to 180 degrees. The right-hand side of the wafer supports an SAM with more hydrophilic ends, so that droplets flatten out and spread on the surface. (Photograph courtesy of George Whitesides, Harvard University.)

SAMs convey a highly localized (because they are chemically bound) and highly uniform (because they are ordered) modification of the surface properties. The surface energy of the SAM can be varied by altering the chemical composition of the exposed ends of the chains—for example, making them hydrophilic or hydrophobic (fig. 10.19). Thus surfaces can be given exquisitely tailored surface properties (that is, surface excess energies) by applying patterned SAMs. George Whitesides has developed a technique called microcontact printing to imprint parts of a surface with SAMs while leaving the other parts bare. The printing process first involves the formation of a rubber stamp, which, like most rubber stamps, is made from a negative mold. This mold is carved into a solid such as silicon, using the lithographic techniques that microelectronics engineers use to etch circuitry into silicon chips. A liquid polymer is then poured over this mold, and is set into a rubbery (elastomeric) form, which reproduces faithfully all of the nooks and crannies (fig. 10.20).

The rubber master is then peeled from the mold and is coated with an ink of the molecules that will form the SAM (say, alkylthiols). When the stamp is brought into contact with a gold surface, an SAM is formed at only those parts that touch the surface. This film therefore reproduces the shape of the pattern that was etched into the silicon chip.

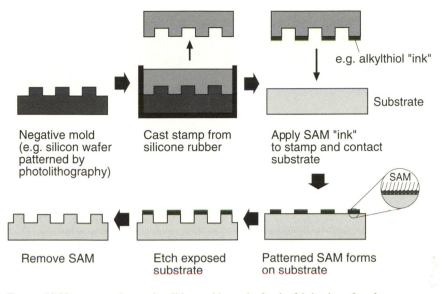

e.g. alkylthiol "ink"

Substrate

SAM

Negative mold
(e.g. silicon wafer
patterned by
photolithography)

Cast stamp from
silicone rubber

Apply SAM "ink"
to stamp and contact
substrate

Remove SAM

Etch exposed
substrate

Patterned SAM forms
on substrate

FIGURE 10.20 SAMS can be used as lithographic masks for the fabrication of surface structures by etching. First a stamp is prepared by casting a rubbery polymer in the channels of a surface imprinted with a pattern by conventional photolithography. The stamp is coated with an "ink" of the amphiphilic molecules from which the SAM will be formed, and is stamped onto the surface to be patterned. Where the stamp touches the surface, a SAM is formed; elsewhere, the surface remains bare. Etching will then eat away the exposed parts of the surface, but will leave the parts covered by the SAM "resist" untouched. In this way, surface patterns can be reproduced again and again from the rubber stamp by a simple printing process, rather than having to use photolithography to fabricate a patterned resist on each surface.

Chapter 1 showed some of the astonishing feats of which microfabrication technology is now capable—microelectronic components can be made small enough for an entire circuit board to fit on a pinhead. Conventionally, these tiny devices are fashioned using photolithography. Although this process is automated and inexpensive enough to support an industry with an annual turnover of hundreds of billions of dollars, Whitesides's microcontact printing process offers a vastly simpler and cheaper alternative. Once a rubber stamp is made, one can use it again and again to imprint a patterned SAM onto a surface. As SAMs of alkylthiols on gold will confer resistance to chemical etching of the metal below, patterned SAMs can be used as a mask for turning a gold film into a patterned circuit (fig. 10.21).

This printing process will reproduce features just 0.2 micrometers across, a resolution that generally requires electron-beam etching in conventional lithography. And Whitesides has demonstrated that the scale of the SAM patterning can be reduced by merely squeezing the rubber stamp: compressing the block to reduce the length of each side by a half will reduce the dimensions of the pattern elements by the same amount, thereby doubling the resolution in a single stroke.

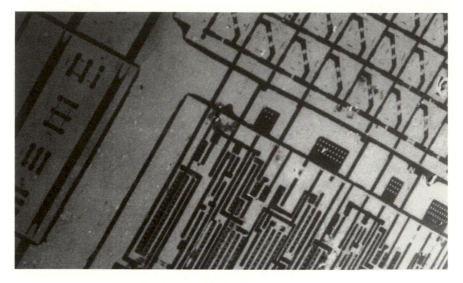

FIGURE 10.21 Microcontact printing can be used to imprint extremely fine circuit patterns in a gold film. (Photograph courtesy of George Whitesides, Harvard University.)

Going for a Dip

The deposition of thin organic films on substrates has a long history: ancient Chinese ship builders smeared oils such as tung oil on wood to preserve it and render it water-resistant (this practice still goes on today). But the systematic study of such films originated in the late nineteenth century with Lord Rayleigh, who explored the possibilities for altering the surface energies of materials by spreading them with fatty acids. Rayleigh was particularly interested in films formed at the surface of water, where they could spread to a layer just one molecule thick. Building on studies of such monolayer films by Agnes Pockels in the 1890s, Rayleigh proposed that they could be used to estimate the size of molecules, and his estimate of the length of the fatty acid molecule stearic acid was remarkably close to that measured by modern techniques.

But the underpinnings of the modern science of thin organic films were laid by the American chemist Irving Langmuir, who is generally regarded as a founding father of the whole of surface science. In 1912, he and his student Katharine Blodgett showed that monolayer films of fatty acids could be transferred from the surface of water to the surface of a glass slide dipped into the solution. Films of amphiphilic molecules lying at the air/water interface with their hydrophobic tails in the air and their hydrophilic heads in the water (see fig. 4.35) are now called *Langmuir monolayers*—arguably an injustice to Agnes Pockels, who explored them first. Langmuir and Blodgett used a movable barrier at the surface of the water to confine these films and thereby alter the packing density of the molecular layer. If a glass slide is dipped into a trough containing a densely packed Lang-

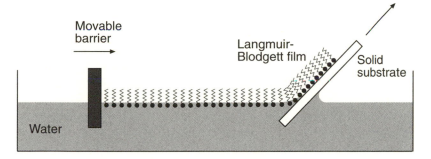

FIGURE 10.22 Single, closely packed layers of amphiphilic molecules at the surface of water can be transferred onto a solid substrate such as a glass plate by a simple dipping process. The solid-supported films are called Langmuir–Blodgett films.

muir monolayer, the carboxylic acid head groups of the amphiphiles will become adsorbed to the hydrophilic surface of the glass. So long as the barrier is moved during the dipping process to keep the film pressed against the glass surface, the fatty acid molecules will coat the surface in a smooth film (fig. 10.22), which remains adhering to the surface as the slide is withdrawn. The surface-bound film, called a *Langmuir–Blodgett (LB) film*, is rather similar to a self-assembled mono-layer in that it consists of amphiphilic molecules with their tails arranged in an oriented, orderly manner; but unlike SAMs, the amphiphiles in LB films are phys-ically, not chemically, bound to the substrate. Moreover, repeated dips of the slide will result in the buildup of additional layers, so that LB films several molecular layers thick can be created.

The bristling hydrophobic tails of a single-layer, "head-down" LB film pro-foundly alter the wetting properties of the substrate, making it less wettable by water. Because of their orderly nature, LB films, like SAMs, create a very homoge-neous modification of the surface energy. Strangely, however, interest in these molecular films languished for decades after Langmuir's work, and it is only recently that researchers are beginning to realize their potential. One of their most attractive features is the ease with which they are formed. A smooth monolayer can be created by the simple dipping procedure in air, in contrast to the high-tech vacuum process needed to make smooth thin layers of semiconducting materials. Some researchers are now exploring the use of LB films in molecular electronic devices, making them from amphiphilic molecules that conduct electricity, or in photonic devices by using molecules with nonlinear optical behavior.

WEAR AND TEAR

There is enough that still remains to be understood about the structure and prop-erties of surfaces and interfaces to keep scientists busy for the indefinite future. But once surfaces start to *move*—then it is another ball game entirely. Moving surfaces are to be found in all reaches of mechanical engineering (to mention but one relevant field), and they give engineers plenty to worry about. When surfaces slide against each other, they wear out. So one of the first concerns is to keep them well oiled, so to speak.

The car driver who has neglected the engine oil or the motorcyclist with a seized piston does not need to be reminded of the importance of lubrication. But it is a critical consideration for *any* mechanism with moving parts, and that applies as equally to a knee joint as to a hydraulic ram, to a recording head skimming over a magnetic disk, or to a drill bit plunging into the Earth's crust from an oil-drilling platform. The flip side of lubrication is wear—the grinding away of material from two poorly lubricated surfaces in dynamic contact. The scientific study of lubrication, sliding, friction, and wear is called *tribology* (from the Greek *tribos*, meaning rubbing). Traditionally it has been an engineering discipline, but in recent years it has become possible to study tribological phenomena at the nanometer scale—the scale of individual molecules—and this has given birth to the field dubbed *nanotribology*. These molecular-scale studies are to traditional tribology what atomic-scale investigations of static surfaces are to the older, more applied disciplines of wetting, adsorption, and adhesion; but the knowledge that they are providing is of more than intellectual importance, since it is becoming increasingly relevant as the scale of mechanical engineering continues to shrink to microscopic proportions (see fig. 10.3).

Lubrication is closely related to adhesion; materials that exhibit low adhesion, such as PTFE, commonly serve as good lubricants too. But while adhesion is a static property, determined by the surface energies of two materials in contact, lubrication is a dynamic effect: it involves interfaces in motion, which is to say, out of equilibrium. Lubrication behavior cannot always be predicted simply by considering the properties of two contacting surfaces at rest.

The kind of surfaces one typically encounters in mechanical devices such as pistons are rather rough, with many pits, grooves, and asperities. As they are dragged over one another, the asperities will tend to interlock and bump into each other, giving rise to a frictional resistance to motion that is localized at several contact points rather than being smoothly distributed over the surface. As a result, the surface will be heated up locally by the rubbing of asperities, and these will tend to gouge into the opposite surface, leaving scars like the tracks of tires in muddy ground. This is the origin of microscopic wear.

These fine details of friction and wear between bare surfaces can now be studied using scanning probe microscopes—the scanning tunneling microscope (STM) and the atomic force microscope (AFM)—whose fine probe tips can be used to mimic a single contact asperity. Here the microscopes become both a tool and a means of visualization: the tip is brought into contact with the surface to investigate the sliding process, and is then scanned in "imaging mode" just out of contact to reveal what havoc it has wreaked. Dragging an AFM tip across a smooth solid surface may leave a tiny scratch, whose depth depends on the force with which the tip is pressed into the surface (fig. 10.23). Repeatedly scanning the tip

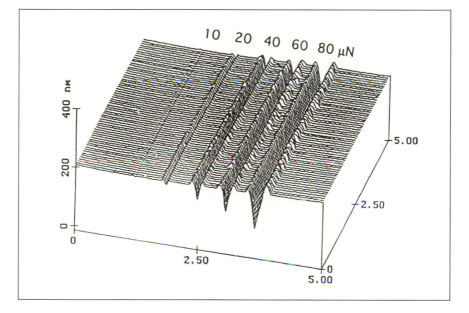

FIGURE 10.23 Scratches in a silicon surface created and then imaged with the tip of an atomic force microscope. The force with which the tip is pressed into the surface increases from left to right. (Figure courtesy of Bharat Bhushan, Ohio State University.)

across the surface while it is in contact will gradually wear away whole regions, leaving a pitted surface.

The frictional force that the tip experiences in these experiments can be measured using a modified version of the AFM developed for this purpose, the friction force microscope (FFM). This instrument detects changes in the frictional force between the tip and the sample as the former moves across the latter. The main distinction between the FFM and the AFM is that, while the latter measures the strength of the force perpendicular to the surface, the FFM measures forces parallel to the surface—it is these lateral forces that are responsible for frictional drag. Applying a lateral force to a cantilevered scanning probe tip will cause the cantilever to twist (fig. 10.24). In the FFM, this twisting is usually measured by optical means: a laser beam shined from above onto the end of the cantilever will be deflected at an angle as the arm twists. By monitoring this angle of deflection, and knowing the springiness of the cantilever, it is possible to calculate the size of the force that produced the twisting.

The friction force microscope provides different kinds of information about a surface from the AFM, and sometimes reveals things that an AFM cannot. For example, it can detect the difference between two regions of the surface that have the same topographic height but, by virtue of a differing chemical composition, different frictional properties (fig. 10.25); and conversely, it ignores changes in surface height that are not accompanied by a change in frictional drag.

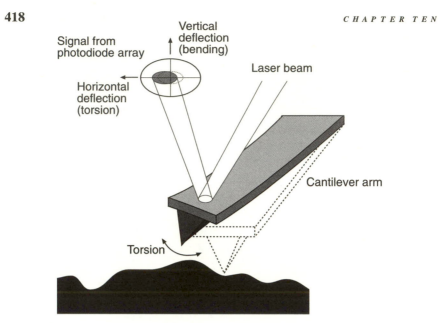

FIGURE 10.24 The friction force microscope is an adaptation of the atomic force microscope, which can measure the horizontal frictional force on the probe tip as it scans across a surface. These frictional forces cause twisting (torsion) of the cantilever arm—the greater the frictional resistance to the tip's horizontal motion, the greater the torsion. There are several ways to measure the twisting of the cantilever. In the first measurement of atomic-scale frictional forces using this device (in 1987), the torsion was detected interferometrically. In the device shown here, twisting of the tip causes deflection of a laser beam reflected from the back of the cantilever, which is detected by a position-sensitive array of photodiodes; in this arrangement, both bending (due to vertical forces) and twisting (due to horizontal frictional forces) can be measured simultaneously. Other detection techniques include the measurement of capacitance changes between the cantilever end and capacitor plates above and to either side of it.

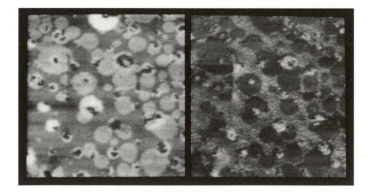

FIGURE 10.25 The friction force microscope sees things that are not necessarily evident to the atomic force microscope. The latter measures surface topography, but does not discriminate between differences in the chemical nature of the surface. The former measures the frictional

Greasing the Surfaces

Atomic-scale studies like these show us exactly what we generally want to avoid: the scratching, deformation, and wearing away of materials when they slide over each other. What happens, at these scales, when we put a lubricant in between the surfaces?

In a car engine, lubrication can be effected by rather crude liquid lubricants such as the inert, low-viscosity hydrocarbons used in engine oil, or by waxy substances such as axle grease. If the surfaces remain far enough apart (or if the forces squeezing them together are small enough) that a thick film of fluid can be interposed between them, the frictional forces can be made very low and depend only on the flow properties of the fluid, not on the specific nature of the surfaces. This is called hydrodynamic lubrication. But as the surfaces get closer together, the amount of contact between them increases, and the frictional forces rise accordingly. In this situation, lubrication depends on the detailed structure and dynamics of the surface regions, in particular on the behavior of a thin film of lubricant molecules—perhaps just one molecule thick—between the solid surfaces. This is called boundary lubrication.

Although Irving Langmuir was primarily interested in the effect of LB films on surface energies and wetting behavior, he noticed that an LB film alters a substrate's friction properties too: it lubricates the surface that it coats. Langmuir wondered whether this might be exploited to good effect for lubricating bearings, such as those in his meters. He had in fact discovered that LB films are boundary lubricants.

More generally, these lubricants tend to be chainlike organic molecules like amphiphiles or polymers attached to the solid surfaces either by physical adsorption or by chemical bonds. The first detailed study of the effects of surface-anchored boundary lubricants was made by William Hardy in the 1920s in work that was extended in subsequent decades by F. P. Bowden and coworkers at Cambridge University. Hardy studied the tribological effects of the types of molecular film that interested Rayleigh and Langmuir: organic compounds such as alcohols and fatty acids smeared on the surfaces of metals. He proposed that in the boundary lubrication regime, when the surfaces are very close together, low friction is maintained by a very thin film of the organic molecules, perhaps just a monolayer, adsorbed at each of the surfaces.

One of the most valuables instruments for studying the molecular details of boundary lubrication is the surface force apparatus. Recall that in this device, two curved cylindrical surfaces are brought close together and the forces between

---

force as the tip moves across the surface, and this may depend on the kinds of molecules deposited on the surface. Here an AFM image of a mixed film of hydrocarbons and fluorinated hydrocarbons (*right*) exhibits slight contrast differences between regions of each, owing to slightly different surface heights; but the FFM sees large contrast variations (*left*), because the frictional forces for the two types of film are very different. (Photograph courtesy of Jane Frommer, IBM Almaden, San Jose.)

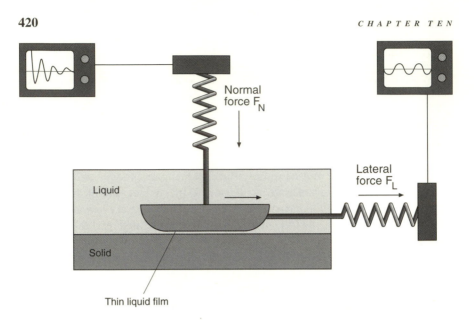

FIGURE 10.26 The surface force apparatus can be modified to measure lateral sliding (frictional) forces as well as normal forces perpendicular to the two surfaces. This modified instrument can be used to study the effects of a liquid lubricant between two solid surfaces in close proximity.

them are monitored. Initially conceived as a device for measuring the static forces of attraction or repulsion between surfaces, the SFA has now been adapted for measuring the frictional forces generated by sliding one of the surfaces over the other. For studying "normal" forces (that is, those perpendicular to the surfaces), one surface is attached to a delicate spring that extends or contracts in this perpendicular direction. To measure lateral forces (those in the plane of the surfaces—the frictional forces generated by sliding), one can simply attach another spring parallel to the surface (fig. 10.26).

Jacob Israelachvili and colleagues at Santa Barbara have studied the effect of surface-attached polymers and amphiphilic molecules using the SFA, and have found a rather complicated range of effects that depend on the nature of the lubricant molecules, the temperature, the rate at which the surfaces slide past each other, and other factors (fig. 10.27). For instance, under some conditions monolayer films of amphiphiles adopt an ordered structure, rather like that of self-assembled monolayers, as the surfaces slide over each other. These films are solidlike, and they induce so-called stick–slip behavior of the sliding surfaces: jerky motion in which the surfaces alternately lock together and slide. Stick-slip motion is common for solid surfaces sliding over one another; it is responsible for the screech of chalk on blackboards, the vibration of a bowed violin string, and the episodic ruptures of earthquakes along geological fault lines. If on the other hand the molecular chains are disorderly, the lubricant films can either act as liquids, giving low friction, or as amorphous hairy coats that have a tendency to become entangled, creating high friction. Israelachvili's group found that at high

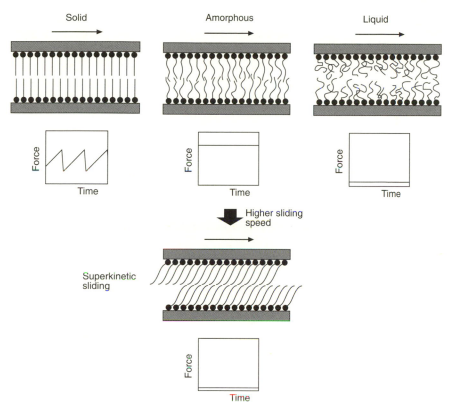

FIGURE 10.27 The frictional force between two surfaces bearing boundary lubricants can be a complex function of temperature, sliding rate, and other factors. Researchers at the University of California at Santa Barbara found that the force can first increase from a low value and then decrease again as the temperature of the system is raised, or as the sliding velocity is lowered. They proposed that these changes in conditions were causing changes in the structure of the chainlike molecules. At low temperatures or fast sliding speeds the chains are ordered into solid films, and the dense packing of the chains in each one prevents the chains of the other from interpenetrating, allowing the films to slide over each other. At high temperatures or slow sliding speeds, the chains are completely disorderly, like a liquid; the boundary lubricant films therefore act much like liquid lubricants, again giving low friction. But in the intermediate regime, the chains have a degree of disorder that is not fully liquidlike, so that rather than flowing past one another the chains in the two films can become entangled, creating viscous drag between the surfaces and increasing the friction force.

sliding speeds some amphiphilic boundary lubricants also show a very-low-friction state. They proposed that in this case the shear forces cause the chains to become aligned, as though one were drawing a comb through the tangle. In this aligned state the two films do not become intertwined, and sliding is easy ("super-kinetic"). But when the shear force is reduced, the chains revert to a disordered, entangled state and the friction increases (fig. 10.27).

With an improved understanding of the effects of boundary lubricants, it may become possible to give moving parts a permanent low-friction coating consisting of chemically grafted molecular films. This would remove the need for using liquid lubricants altogether, so that for example you would never have to give your engine an oil change. But current nanotribological studies make it clear that the behavior of such systems is a rather complex function of many different parameters, so there is a lot of work that remains to be done before we can throw away the oil can forever.

WRITING WITH ATOMS

In 1959, the physicist Richard Feynman gave a startlingly prescient lecture on an imagined technology of the future. "What I want to talk about," he said, "is the problem of manipulating and controlling things on a small scale." By a small scale, Feynman did not mean "electric motors that are the size of the nail on your small finger"—that's nothing, he said, compared with what he had in mind. Feynman was interested in storing information on the smallest scale imaginable, using just a handful of atoms for each bit of information. He suggested that as a conservative estimate we might consider using 125 atoms for each bit. The density of information storage that this allows would make microfilm look as cumbersome as the British Library.

But could this ever be possible? It would require, said Feynman, that we be able to arrange individual atoms whichever way we want them. He clearly considered this a rather controversial, perhaps even heretical thing to suggest: "I am not afraid to consider [it]," he said, but stressed that it was a goal for "the great future." It is a great shame, then, that Feynman—who died from cancer in 1988—did not live to see this challenge met on the cover of *Nature* in April 1990. The cover picture showed a series of bullet-shaped blobs rising in a row from a smooth surface. Each blob was a xenon atom, and they had been arrayed in line by Don Eigler and Erhard Schweizer from IBM's Almaden research center in San Jose. In their *Nature* paper the IBM researchers went further: they wrote their company's three-letter logo on the surface of a nickel crystal using just thirty five atoms (fig. 10.28)—achieving the limit of data storage at a density of one atom per bit, over a hundred times better than Feynman's "conservative" estimate for what might be needed to write in atoms.

Eigler and Schweizer achieved this staggering feat of manipulation using the scanning tunneling microscope. As we saw earlier, this device was not conceived originally as a tool for atomic calligraphy but as an instrument for simply seeing into the atomic world. Toward the end of the 1980s, however, researchers began to suspect that the tip of an STM might be useful not only as a set of eyes for seeing atoms but also as a set of hands for manipulating them: picking them up, putting them down and shunting them around on surfaces.

The first tentative steps in this direction were taken in 1987, when Russell Becker from AT&T Bell Laboratories in New Jersey and colleagues used the STM

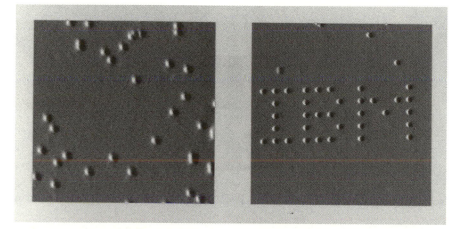

FIGURE 10.28 The scanning tunneling microscope is a tool for atomic calligraphy. Here it has been used to drag xenon atoms over a nickel surface so as to spell out the company logo of the researchers who did the experiment—in letters just 5 nanometers tall. (Photograph courtesy of Don Eigler, IBM Almaden, San Jose.)

to create tiny protrusions on the surface of germanium, a semiconducting material. They held the tip over one point on the surface and briefly increased the voltage across it. This produced a conical feature on the surface, clearly visible when the STM was then used for imaging. Becker and colleagues suggested that the little mound was formed from germanium atoms that had become attached to the STM tip following some earlier contact between it and the surface: the high-voltage pulse blasted them off into a heap.

This approach for evaporating atoms from the tip onto the surface was pursued by John Mamin and colleagues at IBM Almaden, who deposited whole strings of atomic mounds onto a gold surface using a gold STM tip: they used voltage pulses to evaporate part of the tip itself. Each mound was just 10 nanometers across and 2.5 nanometers high, containing several hundred atoms. By 1991 this approach had been refined sufficiently to allow Phaedon Avouris and In-Whan Lyo of IBM's laboratories in Yorktown Heights, New York, to transfer a single atom from a silicon surface to an STM tip, and then to deposit the atom somewhere else on the surface, using voltage pulses to effect both operations (fig. 10.29).

The STM calligraphy of Eigler and Schweizer was an even more delicate piece of work: it involved using the STM tip to drag atoms around on a surface, without removing them. The force required to do this is much less than that needed to pluck an atom from the surface, and the manipulation requires careful adjustment and control of the interaction between the tip and sample. Eigler chose to work with xenon atoms because they are highly unreactive. But in order for them to remain immobile on the surface, the whole system was cooled to the temperature of liquid helium, about 4 degrees Celsius above absolute zero.

Eigler and Schweizer first deposited a small amount of xenon on the nickel surface: enough to give it a random covering of the atoms that they would use as

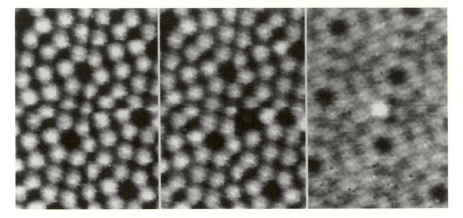

FIGURE 10.29 Atoms can be lucked off a surface and deposited elsewhere, one by one, with the STM. Here an atom of silicon in the 7 × 7 surface reconstruction is transferred from the surface to an STM tip (first and second frames) by applying a voltage pulse to the tip. The atom is then placed elsewhere on the surface (third frame), where, because it stands on top of others, it shows up as a brighter spot. (Photograph courtesy of Phaedon Avouris, IBM Yorktown Heights, New York; reproduced with permission from *Science* **253**, 173 [1991].)

their ink. The underlying crystal lattice of nickel atoms provided a grid for precise positioning of the xenon atoms. The two researchers then lowered the STM tip until the attractive force between it and a xenon atom was just large enough for the atom to be dragged along in the wake of the tip, but not so large that the atom would instead jump onto the tip. Each of the thirty-five atoms was individually dragged into place on the surface grid until they spelled out, in bold capitals, "IBM." It was undoubtedly the world's smallest advertisement, but it was soon staring out from newspapers all around the world.

In 1991 Eigler and colleagues used this same xenon/nickel system to fabricate arguably the smallest ever electronic device—an atomic switch in which the switching operation involved the motion of a single atom. Like Avouris and Lyo (whose paper was published only a few weeks after Eigler's), they used voltage pulses to transfer the atom back and forth between the STM tip and the surface. Because this operation changed the current passing between the tip and the nickel surface (fig. 10.30), Eigler and colleagues could argue that this was truly a kind of electronic switching. This experiment showed that in principle it might be possible to create electronic or memory devices in which the fundamental operations involve individual atoms. (The "device" here has dimensions much larger than the atomic, however, since it incorporates the whole of the STM itself.)

Might one use these principles to make a memory device in which each bit of information is represented by a single atom? It is unlikely that we would ever have much use for a memory in which the information is written out, like the IBM logo in figure 10.28, in alphabetic form; far better to use a binary code written into a checkerboard array of atoms positioned on a flat surface, where the presence of an atom at a given grid point signifies a "1" while a gap—an empty grid point—

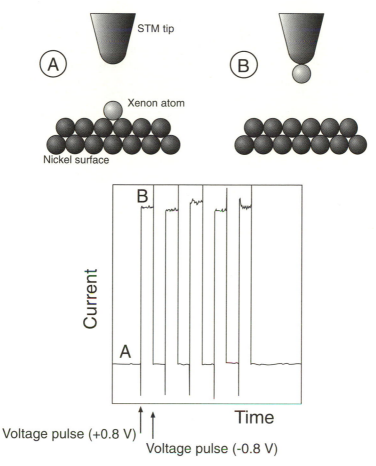

FIGURE 10.30 The transfer of an adsorbed atom back and forth between a surface and an STM tip provides a device whose conductivity can be switched by the movement of a single atom. With the switchable atom on the surface and the tip positioned above it, the tunneling current through the tip is low (A). Applying a voltage pulse transfers the atom to the tip (B), whereupon the tunneling current increases abruptly.

signifies a "0." The ability to pick up and put down atoms at specified locations, and to *see* that this has been done, seems to offer this prospect.

But whether this kind of memory would be sufficiently stable against the jostling of thermal fluctuations to hold information reliably over long times is debatable. Eigler and Schweizer had to use liquid-helium refrigeration to keep their writing stable, and it is doubtful that there is a future in trying to build helium-cooled desktop computers. Greater stability of the stored information might be possible if one uses a small cluster of atoms—perhaps as many as the hundred or so envisioned by Feynman—to store each bit. Since van der Waals forces give all atoms some cohesion—some degree of attraction to each other—thermal effects will be less likely to wipe the information out, provided that the temperature is not too high. At one hundred atoms per bit, such a memory would still provide a storage density hundreds of times greater than that of any magnetic device.

This is all very well, but to position one hundred atoms on a surface would take at least an hour by Eigler's method. A memory in which the data is written in at one bit per hour will not get very far, no matter what density of information it can hold. Many researchers feel that a data storage device based on STM manipulation of surfaces will have to work in a cruder way if it is to be efficient. The deposition of tiny mounds by applying voltage pulses to the tip, as demonstrated by John Mamin and others, is one possibility: in principle it should be feasible to write digital information onto a surface very quickly in this way. A major problem here, however, is that such a memory would not easily be erasable: removing the nano-scale mounds of atoms is possible, by reversing the voltage and using pulses to blast them away, but this is a fairly messy, uncontrolled process that will leave the surface scarred and littered with debris. Problems of this sort continue to pose obstacles to the realization of an STM-based memory device.

Round 'em Up

Where do you go once you have written in atoms? Don Eigler decided to try to use atom dragging to make a useful structure, namely one that would allow him to explore new aspects of surface science. In 1993 he and his colleagues Mike Crommie and Charles Lutz built a circular ring of iron atoms on the surface of copper and were able to see the ripples of electrons bouncing around inside.

Electrons at the surface of a so-called noble metal such as copper, silver, and gold are confined there by electronic effects that prevent them from penetrating into the bulk of the material. So these electrons are free to move only in a two-dimensional sheet at the metal surface. This kind of behavior is similar to that found in quantum-well heterostructures such as those discussed in chapter 1, in which electrons in a thin layer of a doped semiconductor are confined by the presence of a material with a different band gap above and below the layer.

Electrons confined to the surface of a conducting material behave like waves in a shallow dish. When these waves encounter a surface irregularity, such as a single atom adsorbed on the surface or a step in the height of the surface, they are reflected back from the irregularity and interfere to set up standing waves— electron ripples in which the peaks and troughs remain in fixed positions rather than propagating. Eigler and colleagues found that these electron standing waves can be imaged with the STM (see plate 13).

The IBM researchers then set about constructing a specific arrangement of atoms on the surface of copper, to see what effect this had on the surface electron waves. They used the dragging procedure to make a ring of forty-eight iron atoms (plate 15). They found that the electrons bounced back from the iron stockade, setting up a pattern of concentric ripples within a tiny, circular quantum well. Because of this confining effect, Eigler and colleagues called the structure a quantum corral. The ripples provide striking confirmation of the quantum-mechanical wavelike behavior of electrons, and the possibility of making corrals of other shapes raises the prospect of investigating the behavior of confined electrons in unprecedented detail. In corrals of certain shapes, the reflection of electrons is predicted to produce not regular standing waves but instead irregular behavior with many of the characteristics of a "chaotic" system—an electron would bounce from wall to wall along a trajectory that never quite repeats on itself and which is extremely sensitive to the electron's initial motion. A classical object like a billiard ball confined in a corral of this shape would exhibit chaotic motion— impossible to predict at the outset—but quantum-mechanical effects will modify this behavior in the case of an electron, leaving its motion with just the "ghost" of chaos. Exploring such phenomena inside quantum corrals might give physicists new understanding of the similarities and differences between classical and quantum motion.

The emerging new science of nanotechnology is an endeavor to control and manipulate matter at the scale of individual atoms and molecules. Today's nanotechnologists draw much inspiration from Richard Feynman's lecture, and the STM is allowing them to put his speculations into practice. The lecture was entitled "There's Plenty of Room at the Bottom." Now that we have a tool for exploring just how much room there is down there at the atomic scale, the possibilities seem endless.

BIBLIOGRAPHY

General, nonspecialist books about materials almost do not exist, although plenty of text-books will tell you about the fundamentals of materials properties and function. I say "almost" because the honorable and splendid exception is Rodney Cotterill's *The Cambridge Guide to the Material World* (Cambridge University Press, 1985), which traces a clear and accessible path from atomic theory through minerals and metals to the organic world of synthetic polymers and natural materials. And J. E. Gordon's *The New Science of Strong Materials, or Why You Don't Fall through the Floor* (Penguin, 1991; Princeton University Press, 1984) is still a classic, providing a lively and nontechnical account of the structure and properties of strong solid materials. I recommend the companion volume *Structures, or Why Bridges Don't Fall Down* (Penguin, 1991) as an introduction to the use of materials in mechanical and civil engineering.

My book *Designing the Molecular World* (Princeton University Press, 1994) outlines the chemical background for many of the concepts discussed in this book, as well as providing a different perspective on some of the materials that I have considered here (and others that I have not). While this book was being written, Ivan Amato was preparing *Stuff* (Basic Books, 1997), which provides a much-needed description of the development of materials science over the past decades, along with a discussion of some of its contemporary directions.

The magazines *Scientific American* and *New Scientist* continue to provide regular, readable accounts of developments in new materials. At a slightly more technical level, *Nature* and *Science* do the same, as well as publishing much of the groundbreaking original research. More specifically focused on materials are the journals *MRS Bulletin* (published by the Materials Research Society in Pittsburgh, Pennsylvania) and *Advanced Materials* (published by VCH in Weinheim, Germany); the former publishes short review-type articles, the latter a broad range of original research.

My task was aided tremendously by the publication in 1994 of *The Encyclopedia of Advanced Materials*, edited by D. Bloor, R. J. Brook, M. C. Flemings, and S. Mahajan (Pergamon Press, 1994), a four-volume bible of information about new materials. The articles are generally rather technical, but the coverage is comprehensive. Contributions to the encyclopedia are cited extensively below, where I simply denote it as *EAM*.

INTRODUCTION

Materials and the Economy

"Beyond the era of materials," E. D. Larson, M. H. Ross, and R. H. Williams, *Scientific American*, June 1986, p. 24.

"Advanced materials and the economy," J. P. Clark and M. C. Flemings, *Scientific American*, October 1986, p. 43.

Green Manufacturing

"An overview of industrial waste generation and management practices," D. T. Allen; "Introduction to environmentally conscious manufacturing," R. D. Watkins and B. Granoff; "Industrial ecology: the materials scientist in an environmentally constrained world," B. R. Allenby, *MRS Bulletin*, March 1992.

The Greening of Industrial Ecosystems, ed. B. R. Allenby and D. J. Richards (National Academy Press, Washington, D.C., 1994).

Materials Selection

Materials Selection in Mechanical Design, M. F. Ashby (Pergamon Press, 1992).
"Materials in mechanical design," M. F. Ashby, *MRS Bulletin*, July 1993, p. 43.
"Materials selection charts," M. F. Ashby in *EAM*, p. 1448.

CHAPTER 1

Electronic Structure of Materials

"Electronic and magnetic materials," P. Chaudhari, *Scientific American*, October 1986, p. 114.
Electronic Structure of Materials, A. P. Sutton (Oxford University Press, 1993).
The Physics of Semiconductor Devices, 3d ed., D. A. Fraser (Oxford University Press, 1983).
The Solid State, A. Guinier and R. Jullien (Oxford University Press, 1989).

Photonic Materials and Telecommunications

"Photonic materials," J. M. Rowell, *Scientific American*, October 1986, p. 124.
"All-optical networks," V.W.S. Chan, *Scientific American*, September 1995, p. 56.

Optical Fibers

"Infrared optical fibers," M. G. Drexhage and C. T. Moynihan, *Scientific American*, November 1988, p. 110.
"Paving the information superhighway with plastic," R. F. Service, *Science* **267**, 1921 (1995).
"Optical fibers, glass," C. R. Day in *EAM*, p. 1825.
"Optical fibers, polymer," Y. Koike and T. Ishigure in *EAM*, p. 1831.

Semiconductor Microlasers

"Microlasers," J. L. Jewell, J. P. Harbison, and A. Scherer, *Scientific American*, November 1991, p. 86.
"Microstructured semiconductor lasers for high-speed information processing," P. L. Gourley, *Nature* **371**, 571 (1994).
Semiconductor Lasers, ed. G. Agarwal (American Institute of Physics, 1996).
"Quantum cascade laser," J. Faist, F. Capasso, D. L. Sivco, C. Suirtori, A. L. Hutchinson, and A. Y. Cho, *Science* **264**, 553 (1994).
"A news laser promises to put an end to band gap slavery," G. Taubes, *Science* **264**, 508 (1994).
"Cascading electrons," R. Tsu, *Nature* **369**, 442 (1994).
"Wide-bandgap II–VI heterostructures for blue/green optical sources: key materials issues," L. A. Kolodziejski, R. L. Gunshor, and A. V. Nurmikko, *Annual Review of Materials Science* **25**, 711 (1995).

Manufacturing of Semiconductor Devices

"Towards 'point one,'" G. Stix, *Scientific American*, February 1995, p. 90.
"Organometallic chemical beam epitaxy of III–V layers," W. T. Tsang in *EAM*, p. 1855.

Integrated Optoelectronics and Photonics

"Optoelectronic and photonic integrated circuits," M. A. Pollack in *Perspectives in Opto-electronics*, ed. S. S. Jha (World Scientific, Singapore, 1995).
"Quantum well optical devices and materials," P. K. Bhattacharya and N. K. Dutta, *Annual Review of Materials Science* **23**, 79 (1993).
"Key optoelectronic devices," A. W. Nelson in *Electronic Materials*, ed. L. S. Miller and J. B. Mullin (Plenum Press, 1991).
"Optical computing," D. R. Selviah in *EAM*, p. 1820.

Nonlinear Optics

Electronic Materials, ed. L. S. Miller and J. B. Mullin, chapters 20–24 (Plenum Press, 1991).
"Nonlinear optical frequency conversion," M. M. Fejer; "Ultrafast switching with nonlinear optics," M. N. Islam; "Nonlinear optics in semiconductors," E. Garmire; "Nonlinear optics of organic and polymer materials," A. Garito, R. F. Shi, and M. Wu, in *Physics Today*, May 1994.
"Nonlinear optical materials: an overview," D. Bloor in *EAM*, p. 1773.
"Inorganic crystals for nonlinear optical frequency conversion," P. F. Bordui and M. M. Fejer, *Annual Review of Materials Science* **23**, 321 (1993).
"Nonlinear competition heats up," R. F. Service, *Science* **267**, 1918 (1995).
"Devices based on electro-optic polymers begin to enter marketplace," R. Dagani, *Chemical & Engineering News*, 4 March 1996, p. 22.

Luminescent Silicon

"A glowing future for silicon," L. Canham, *New Scientist*, 10 April 1993, p. 23.
"Silicon quantum wire array fabrication by electrochemical and chemical dissolution of wafers," L. T. Canham, *Applied Physics Letters* **57**, 1046 (1990).
"Silicon sees the light," D.A.B. Miller, *Nature* **378**, 238 (1995).
"Progress towards crystalline-silicon-based light-emitting diodes," L. Canham, *MRS Bulletin*, July 1993, p. 22.

CHAPTER 2

Magnetic Materials

"Magnetic materials: An overview," G. Y. Chin in *EAM*, p. 1423.
"Magnetic materials: Basic concepts," J. S. Kouvel in *EAM*, p. 1432.
The Solid State, A. Guinier and R. Jullien, pp. 136–170 (Oxford University Press, 1989).

Magnetic Memory Devices

"Data-storage technologies for advanced computing," M. H. Kryder, *Scientific American*, October 1987, p. 73.
"Magnetoelectronics today and tomorrow," J. L. Simonds, *Physics Today*, April 1995, p. 26.
"Recording materials, magnetic," G. Bate in *EAM*, p. 2236.
"Ultrahigh-density recording technologies," M. H. Kryder, *MRS Bulletin*, September 1996, p. 17.
"Materials for magnetic-tape media," S. Onodera, H. Kondo, and T. Kawana, *MRS Bulletin*, September 1996, p. 35.

Magneto-Optic Memories

"Magneto-optical storage materials," M. H. Kryder, *Annual Review of Materials Science* **23**, 411 (1993).

"Recording materials, magnetooptical," M. H. Kryder in *EAM*, p. 2244.

"Recording materials, optomagnetic," D. D. Stancil in *EAM*, p. 2259.

"Magneto-optic recording materials," T. Suzuki, *MRS Bulletin*, September 1996, p. 42.

Magnetic Multilayers

"Thin films and multilayers, magnetic," S. D. Bader in *EAM*, p. 2852.

"Giant magnetoresistance," S.S.P. Parkin, *Annual Review of Materials Science* **25**, 357 (1995).

"Magnetic recording head materials," J. A. Brug, T. C. Anthony, and J. H. Nickel, *MRS Bulletin*, September 1996, p. 23.

Holographic Memories

"Holographic memories," D. Psaltis and F. Mok, *Scientific American*, November 1995, p. 70.

"Holography in artificial neural networks," D. Psaltis, D. Brady, X.-G. Gu, and S. Lin, *Nature* **343**, 325 (1990).

"Optical neural computers," Y. S. Abu-Mostafa and D. Psaltis, *Scientific American*, March 1987, p. 66.

"Volume holographic storage and retrieval of digital data," J. F. Heanue, M. C. Bashaw, and L. Hesselink, *Science* **265**, 749 (1994).

"Will holograms tame the data glut?" J. Glanz, *Science* **265**, 737 (1994).

"Holographic-data-storage materials," M.-P. Bernal et al., *MRS Bulletin*, September 1996, p. 51.

Photorefractive Materials

"Photorefractive polymers and composites," Y. Zhang, R. Burzynski, S. Ghosal, and M. K. Casstevens, *Advanced Materials* **8(2)**, 111 (1996).

"Polymers scale new heights," W. E. Moerner, *Nature* **371**, 475 (1994).

"Photorefractive materials," G. C. Valley, M. B. Klein, R. A. Mullen, D. Rytz, and B. Wechsler, *Annual Review of Materials Science* **18**, 165 (1988).

"Photorefractive polymers poised to play key role in optical technologies," R. Dagani, *Chemical & Engineering News*, 20 February 1995, p. 28.

Bacteriorhodopsin Memories

"Protein-based computers," R. R. Birge, *Scientific American*, March 1995, p. 90.

"Protein-based optical computing and memories," R. R. Birge, *IEEE Computer*, November 1992, p. 56.

"Bacteriorhodopsin," R. R. Birge in *EAM*, p. 199.

"Mutated bacteriorhodopsins: Competitive materials for optical information processing?" N. Hampp, C. Brauchle, and D. Oesterhelt, *MRS Bulletin*, November 1992, p. 56.

"Mutated bacteriorhodopsins," N. Hampp and D. Zeisel, *IEEE Engineering in Medicine and Biology*, February/March 1994, p. 67.

"All-optical nonlinear holographic correlation using bacteriorhodopsin films," R. Thoma, M. Dratz, and N. Hampp, *Optical Engineering* **34**, 1345 (1995).

CHAPTER 3

Smart Materials and Structures

"Intelligent materials," Craig A. Rogers, *Scientific American*, September 1995, p. 122.
"Intelligent materials," W. C. Michie, B. Culshaw, and P. T. Gardiner in *EAM*, p. 1120.
Smart Structures and Materials. B. Culshaw (Artech House, Norwood, MA, 1996).
U.S.-Japan Workshop on Smart/Intelligent Materials and Systems, ed. I. Ahmad, A. Crowson, C. A. Rogers, and M. Aizawa (Technomic, Pennsylvania, 1990).
"Animating the material world," I. Amato, *Science* **255**, 284 (1992).
"Even aircraft have feelings," C. Friend, *New Scientist*, 3 February 1996, p. 32.
"Sensors, actuators and smart materials," S. Trolier-McKinstry and R. E. Newnham, *MRS Bulletin*, April 1993, p. 27.

Ceramic Sensors and Actuators

"Ceramic actuators: Principles and applications," K. Uchino, *MRS Bulletin*, April 1993, p. 42.
"Photostrictive actuator," Kenji Uchino, *IEEE 1990 Ultrasonics Symposium*, p. 721 (1990).

Shape Memory Alloys

"Shape memory alloys," C. M. Wayman, *MRS Bulletin*, April 1993, p. 49.
"Shape memory alloys," T. Tadaki, K. Otsuka, and K. Shimizu, *Annual Review of Materials Science* **18**, 25 (1988).

Magnetostrictive Materials

"Magnetostrictive materials," K. B. Hathaway and A. E. Clark, *MRS Bulletin*, April 1993, p. 34.
"Magnetostrictive materials," A. E. Clark in *EAM*, p. 1445.

Electrorheological Fluids

"Electrorheological fluids," T. C. Halsey and J. E. Martin, *Scientific American*, October 1993, p. 58.
"Electrorheological fluids," T. C. Halsey, *Science* **258**, 761 (1992).
"Electrorheological fluids," D. A. Brooks in *EAM*, p. 747.

Smart Polymers and Polymer Gels

" 'Intelligent' polymers in medicine and biotechnology," A. S. Hoffman, *Macromolecules Symposia* **98**, 645 (1995).
"Molecular bioengineering of biomaterials in the 1990s and beyond: A growing liason of polymers with molecular biology," A. S. Hoffman, *Artificial Organs* **16**, 43 (1992).
"Gels," T. Tanaka, *Scientific American*, January 1981, p. 124.
"Intelligent gels," Y. Osada and S. B. Ross-Murphy, *Scientific American*, May 1993, p. 82.
"Environmentally sensitive polymers and hydrogels," A. S. Hoffman, *MRS Bulletin*, September 1991, p. 42.
"A novel approach for preparation of pH-sensitive hydrogels for enteric drug delivery," L.-C. Dong and A. S Hoffman, *Journal of Controlled Release* **15**, 141 (1991).
"Shape memory in hydrogels," Y. Osada and A. Matsuda, *Nature* **376**, 219 (1995).

CHAPTER 4

Protein Structure

Biochemistry, J. D. Rawn (Neil Patterson Publishers, Burlington, NC, 1989).
Biochemistry, 2d ed., D. Voet and J. G. Voet (John Wiley, 1995).
Introduction to Protein Structure, C. Branden and J. Tooze (Garland, 1991).

Protein Engineering

"Designer proteins," R. Lipkin, *Science News*, 10 December 1994, p. 396.
"Engineering enzymes and antibodies," D. Hilvert, *MRS Bulletin*, November 1992, p. 48.
"Design of peptides and proteins," W. DeGrado in *Advances in Protein Chemistry*, vol. 39, ed. C. B. Anfinsen, J. D. Edsall, F. M. Richards, and D. S. Eisenberg (Academic Press, 1988).
"Design of a 4-helix bundle protein: Synthesis of peptides which self-associate into a helical protein," S. Ho and W. DeGrado, *Journal of the American Chemical Society* **109**, 6751 (1987).
"Design and characterization of 4-helix bundle proteins," T. Handel in *Nanotechnology: Research and Perspectives*, ed. B. C. Crandall and J. Lewis (MIT Press, 1992).

DNA and Genetics

Biochemistry, J. D. Rawn, chapters 22–30 (Neil Patterson Publishers, Burlington, NC, 1989).
Genetics, 2d ed., P. J. Russell (Scott, Foresman and Co., Glenview, IL, 1990).
"Molecular structure of nucleic acids," J. D. Watson and F.H.C. Crick, *Nature* **171**, 737 (1953).

The Polymerase Chain Reaction

"Primer-directed enzymatic amplification of DNA with a thermostable DNA polymerase," R. K. Saiki et al., *Science* **239**, 487 (1988).
"Recent advances in the polymerase chain reaction," H. A. Erlich, D. Gelfand, and J. J. Sninsky, *Science* **252**, 1643 (1991).
Making PCR: A Story of Biotechnology, P. Rabinow (University of Chicago Press, 1996).

Cells and Biological Membranes

"The biological membrane," M. D. Alper, *MRS Bulletin*, November 1992, p. 53.
Biochemistry, J. D. Rawn, chapters 1 and 9 (Neil Patterson Publishers, Burlington, NC, 1989).
"The molecules of the cell membrane," M. S. Bretscher, *Scientific American*, October 1985, p. 86.
"The molecules of the cell matrix," K. Weber and M. Osborn, *Scientific American*, October 1985, p. 92.
The Machinery of Life, D. S. Goodsell (Springer-Verlag, 1993).

Artificial Membranes and Surfactancy

Physical Chemistry of Surfaces. 5th ed., A. W. Adamson, chapters 13 and 14 (John Wiley, 1990).
Order in Thin Organic Films, R. H. Tredgold, chapter 8 (Cambridge University Press, 1994).

Phospholipid Bilayers: Physical Principles and Models, G. Cevc and D. Marsh (John Wiley, 1985).

Membrane Mimetic Chemistry, J. H. Fendler (John Wiley, 1982).

Liposomes from Physics to Applications, D. D. Lasic (Elsevier, Amsterdam, 1993).

"Higher order self-assembly of vesicles by site-specific binding," S. Chiruvolu et al., *Science* **264**, 1753 (1994).

Structural Proteins

Biochemistry, J. D. Rawn, chapter 4 (Neil Patterson Publishers, Burlington, NC, 1989).

Structural Biomaterials, J.F.V. Vincent (Princeton University Press, 1990).

"Biologically related materials," R. M. Turner, J.F.V. Vincent, and G. Jeronimidis in *EAM*, p. 244.

"Protein composite materials," P. Calvert in *Protein-based Materials*, ed. D. Kaplan and K. McGrath (Birkhäuser Publishers, 1996).

Silk

"Spider webs and silks," F. Vollrath, *Scientific Amercain*, March 1992, p. 70.

"Biosynthesis and processing of silk proteins," D. L. Kaplan et al., *MRS Bulletin*, October 1992, p. 41.

"A non-periodic lattice model for crystals in *Nephila clavipes* major ampullate silk," B. L. Thiel and C. Viney, *MRS Bulletin*, September 1995, p. 52.

Wood and Cellulose-Based Materials

The Cambridge Guide to the Material World, R. Cotterill, chapter 17 (Cambridge University Press, 1985).

Structural Biomaterials, J.F.V. Vincent (Princeton University Press, 1990).

"Biologically related materials," R. M. Turner, J.F.V. Vincent, and G. Jeronimidis in *EAM*, p. 244.

"Emerging technologies and future prospects for industrialization of microbially derived cellulose," R. M. Brown in *Harnessing Biotechnology for the 21st Century*, ed. M. R. Ladisch and A. Bose (American Chemical Society, 1992).

"Bless their cotton genes," L. Spinney, *New Scientist* 12 October 1996, p. 22.

Microbial Polyhydroxyalkanoates and Polyesters

Plastics from Microbes: Microbial Synthesis of Polymers and Polymer Precursors, ed. D. P. Mobley, chapter 2 (Hanser, Munich, 1994).

"Microbial synthesis and properties of polyhydroxyalkanoates," Y. Doi, *MRS Bulletin*, November 1992, p. 39.

"Biodegradable polymeric materials," E. Chiellini and R. Solaro, *Advanced Materials* **8**, 305 (1996).

"Biomolecular materials," J. G. Tirrell, M. J. Fournier, T. L. Mason, and D. A. Tirrell, *Chemical & Engineering News*, 19 December 1994, p. 40.

Biomineralization and Biomimetics

Biomineralization: Chemical and Biochemical Perspectives, ed. S. Mann, J. Webb, and R.J.P. Williams (VCH, New York, 1989).

"Biogenic inorganic materials," S. Mann in *Inorganic Materials*, ed. D. W. Bruce and D. O'Hare (John Wiley, 1992).

Biomimetic Materials Chemistry, ed. S. Mann (VCH, Weinheim, 1996).

Hierarchical Structures in Biology as a Guide for New Materials Technology, National Materials Advisory Board (NMAB-464, National Academy Press, Washington, DC, 1994).

"Eggshell mineralization: A case study of a bioprocessing strategy," D. J. Fink, A. I. Caplan, and A. H. Heuer; "Biomineralization: Biomimetic potential at the inorganic-organic interface," S. Mann et al.; "Biomimetic ceramics and composites," P. Calvert, *MRS Bulletin*, October 1992.

"Molecular recognition in biomineralization," S. Mann, *Nature* **332**, 119 (1988).

"Innovative materials processing strategies: A biomimetic approach," A. H. Heuer et al., *Science* **255**, 1098 (1992).

"Molecular tectonics in biomineralization and biomimetic materials chemistry," S. Mann, *Nature* **365**, 499 (1993).

"Transfer of structural information from Langmuir monolayers to three-dimensional growing crystals," E. M. Landau, M. Levanon, L. Leiserowitz, M. Lahav, and J. Sagiv, *Nature* **318**, 353 (1985).

"Crystallization of an inorganic phase controlled by a polymer matrix," P. A. Bianconi, J. Lin, and A. R. Strzelecki, *Nature* **349**, 315 (1991).

"Flattery by imitation," S. Mann, *Nature* **349**, 285 (1991).

"Magnetoferritin: In vitro synthesis of a novel magnetic protein," F. C. Meldrum, B. R. Heywood, and S. Mann, *Science* **257**, 522 (1992).

"Lamellar aluminophosphates with surface patterns that mimic diatom and radiolaria skeletons," S. Oliver, A. Kuperman, N. Coombs, A. Lough, and G. Ozin, *Nature* **378**, 47 (1995).

"Synthetic hollow aluminophosphate microspheres," S. Oliver, N. Coombs, and G. A. Ozin, *Advanced Materials* **7**, 931 (1995).

"Skeletons in the beaker: Synthetic hierarchical inorganic materials," G. A. Ozin and S. Oliver, *Advanced Materials* **7**, 943 (1995).

"Crystal tectonics: Construction of reticulated calcium phosphate frameworks in bicontinuous reverse emulsions," D. Walsh, J. D. Hopwood, and S. Mann, *Science* **264**, 1576 (1994).

"Fabrication of hollow porous shells of calcium carbonate from self-organizing media," D. Walsh and S. Mann, *Nature* **377**, 320 (1995).

"Synthesis of inorganic materials with complex form," S. Mann and G. Ozin, *Nature* **382**, 313 (1996).

"Hierarchical inorganic materials," C. T. Kresge, *Advanced Materials* **8**, 181 (1996).

"Spheres of influence," P. Ball, *New Scientist*, 2 December 1995, p. 42.

CHAPTER 5

Biomedical Materials

Biomaterials Science: An Introduction to Materials in Medicine, ed. B. D. Ratner, A. S. Hoffman, F. J. Schoen, and J. E. Lemons (Academic Press, 1996).

"Materials for medicine," R. A. Fuller and J. J. Rosen, *Scientific American*, October 1986, p. 97.

"New challenges in biomaterials," N. A. Peppas and R. Langer, *Science* **263**, 1715 (1994).

"Biomaterials: New polymers and novel applications," R. Langer, *MRS Bulletin*, August 1995, p. 18.

"Report of the committee to survey needs and opportunities for the biomaterials industry,"
S. A. Barenberg, *MRS Bulletin*, September 1991, p. 26.

"Overview of biomedical materials," M. N. Helmus, *MRS Bulletin*, September 1991, p. 33.

"Biomedical polymers," D. F. Williams in *EAM*, p. 251.

Cardiovascular Materials

"Cardiovascular materials," R. C. Eberhart, H.-H. Huo, and K. Nelson, *MRS Bulletin*, Spetember 1991, p. 50.

"Arteries, synthetic," D. Charlesworth in *EAM*, p. 130.

"Implantable pneumatic artificial hearts," K. D. Murray and D. B. Olsen, in *Biomaterials Science: An Introduction to Materials in Medicine*, p. 389, ed. B. D. Ratner, A. S. Hoffman, F. J. Schoen, and J. E. Lemons (Academic Press, 1996).

Artificial Blood

"Materials biotechnology and blood substitutes," N. Kossovsky and D. Millett, *MRS Bulletin*, September 1991, p. 78.

"In vivo measurement of oxygen concentration using sonochemically synthesized microspheres," K. J. Liu et al., *Biophysical Journal* **67**, 896 (1994).

"Small spheres lead to big ideas," R. F. Service, *Science* **267**, 327 (1995).

Bone Substitutes

"Bioceramics," L. L. Hench and J. Wilson, *MRS Bulletin*, September 1991, p. 62.

Handbook of Bioactive Ceramics, ed. T. Yamamuro, L. L. Hench, and J. Wilson (CRC Press, Boca Raton, FL, 1990).

"Ceramics, glasses, and glass-ceramics," L. L. Hench, in *Biomaterials Science: An Introduction to Materials in Medicine*, p. 73, ed. B. D. Ratner, A. S. Hoffman, F. J. Schoen, and J. E. Lemons (Academic Press, 1996).

Tissue Engineering

"Tissue engineering," J. A. Hubbell and R. Langer, *Chemical & Engineering News*, 13 March 1995, p. 42.

"Tissue engineering," R. Langer and J. P. Vacanti, *Science* **260**, 920 (1993).

"Artificial organs," R. Langer and J. P. Vacanti, *Scientific American*, September 1995, p. 100.

"Polymers for tissue engineering," S. P. Baldwin and W. M. Saltzman, *Trends in Polymer Science* **4**, 177 (1996).

"New biomaterials for tissue engineering," K. James and J. Kohn; "The importance of new processing techniques in tissue engineering," L. Lu and A. G. Mikos; "*In situ* material transformations in tissue engineering," J. A. Hubbell; "Tissue engineering of bone by osteoinductive biomaterials," U. Ripamonti and N. Duneas; "Tissue-engineering strategies for ligament reconstruction," M. G. Dunn; "Soft-tissue analogic design and tissue engineering of liver," P. V. Moghe; "Tissue-engineering approaches for central and peripheral nervous-system regeneration," M. Borkenhagen and P. Aebischer; "Growth-factor delivery for tissue engineering," W. M. Saltzman, *MRS Bulletin*, November 1996.

Drug Delivery Systems

"Drug delivery systems," R. Langer, *MRS Bulletin*, September 1991, p. 47.

"New methods of drug delivery," R. Langer, *Science* **249**, 1527 (1990).

CHAPTER 6

Batteries

"Cells and batteries," *Physical Chemistry Source Book*, p. 297 (McGraw-Hill, 1988).
"Batteries: An overview," A. K. Vijh in *EAM*, p. 214.
"Batteries: Materials requirements," C. A. Vincent in *EAM*, p. 218.
"Challenge of portable power," B. Scrosati, *Nature* **373**, 557 (1995).
"A nickel-metal hydride battery for electric vehicles," S. R. Ovshinsky, M. A Fetcenko, and J. Ross, Science **260**, 176 (1993).

Fuel Cells

"Fuel cells," B.C.H. Steele in *EAM*, p. 885.
"Fuel cell," J. Davis, L. Rozeanu, and K. Franzese, in *Physical Chemistry Source Book*, p. 314 (McGraw-Hill, 1988).

Batteries and Fuel Cells for Electric Vehicles

"Check the tyres and charge her up," J. Glanz, *New Scientist* 15 April 1995, p. 32.
"The clean machine," R. H. Williams, *Technology Review*, April 1994, p. 20.
"The different engine," *The Economist*, 5 April 1994.
"Fuel cells hit the road," Daimler-Benz HighTech Report **3**, 13 (1994).

Solar Cells

"The photovoltaic challenge," W. C. Sinke; "Single-crystal silicon: Photovoltaic applications," M. A. Green; "Overview of cast multicrystalline silicon solar cells," H. Watanabe; "Thin polycrystalline silicon solar cells," A. M. Barnett, R. B. Hall, and J. A. Rand; "Recent progress in amorphous silicon solar cells and their technologies," Y. Hamakawa, W. Ma, and H. Okamoto; "CuInSe$_2$ and other chalcopyrite-based solar cells," H.-W. Schock; "CdTe thin-film solar cells," T. Suntola; "III–V materials for photovoltaic applications," L. M. Fraas; "Quantum-well solar cells," K. Barnham et al.; "Iron sulfide for photovoltaics," R. Dasbach, G. Willeke, and O. Blenk; "Nanocrystalline thin-film PV cells," M. Grätzel, *MRS Bulletin*, October 1993.
"Photovoltaic power," Y. Hamakawa, *Scientific American*, April 1987, p. 76.
"Solar cells," K. J. Bachmann and H. J. Lewerenz in *EAM*, p. 2563.
"Solar energy," W. Hoagland, *Scientific American*, September 1995, p. 136.
"A low-cost, high-efficiency solar cell based on dye-sensitized colloidal TiO$_2$ films," B. O'Regan and M. Grätzel, Nature **353**, 737 (1991).

CHAPTER 7

Zeolites and Molecular Sieves

"Solid acid catalysts," J. M. Thomas, *Scientific American*, April 1992, p. 82.
"Synthetic zeolites," G. T. Kerr, *Scientific American*, July 1989, p. 82.
Hydrothermal Chemistry of Zeolites, R. M. Barrer (Academic Press, 1982).
Catalysis at Surfaces, I. M. Campbell (Chapman and Hall, 1988).
"Self-assembling frameworks: Beyond microporous oxides," C. L. Bowes and G. A. Ozin *Advanced Materials* **8**, 13 (1996).
"The smallest chemical plants," R. Pool, *Science* **263**, 1698 (1994).

Clays

"The structural chemistry and reactivity of organic guests in layered aluminosilicate hosts," J. M. Thomas and C. R. Theocharis in *Inclusion Compounds*, vol. 5, ed. J. L. Atwood, J.E.D. Davies, and D. D. MacNicol (Oxford University Press, 1991).

"Clay chemistry," R. W. McCabe in *Inorganic Materials*, ed. D. W. Bruce and D. O'Hare (John Wiley, 1992).

Mesoporous Materials

"Ordered mesoporous molecular sieves synthesized by a liquid-crystal template mechanism," C. T. Kresge, M. E. Leonowicz, W. J. Roth, J. C. Vartuli, and J. S. Beck, *Nature* **359**, 710 (1992).

"Organizing for better synthesis," M. E. Davis, *Nature* **364**, 391 (1993).

"Cooperative formation of inorganic-organic interfaces in the synthesis of silicate mesostructures," A. Monnier et al., *Science* **261**, 1299 (1993).

"Titanium-containing mesoporous molecular sieves for catalytic oxidation of aromatic compounds," P. T. Tanev, M. Chibwe, and T. J. Pinnavaia, *Nature* **368**, 321 (1994).

"Heterogeneous catalysts obtained by grafting metallocene complexes onto mesoporous silica," T. Maschmeyer, F. Rey, G. Sankar, and J. M. Thomas, *Nature* **378**, 159 (1995).

Organic Open-Framework Solids

"Bonding on the molecular building site," P. Ball, *New Scientist*, 5 August 1995, p. 40.

"Self-assembling organic nanotubes based on a cyclic peptide architecture," M. R. Ghadiri, J. R. Granja, R. A. Milligan, D. E. McRee, and N. Khazanovich, *Nature* **366**, 324 (1993).

"Artificial transmembrane ion channels from self-assembling peptide nanotubes," M. R. Ghadiri, J. R. Granja, and L. K. Buehler, *Nature* **369**, 301 (1994).

"An organic solid with wide channels based on hydrogen bonding between macrocycles," D. Venkataraman, S. Lee, J. Zhang, and J. S. Moore, *Nature* **371**, 591 (1994).

"Hollow organic solids," J. S. Moore, *Nature* **374**, 495 (1995).

"Molecular tectonics: Three-dimensional organic networks with zeolitic properties," X. Wang, M. Simard, and J. D. Wuest, *Journal of the American Chemical Society* **116**, 12119 (1994).

"Assembly of porphyrin building blocks into network structures with large channels," B. F. Abrahams, B. F. Hoskins, D. M. Michail, and R. Robson, *Nature* **369**, 727 (1994).

"Scandal of crystal design . . . ," P. Ball, *Nature* **381**, 648 (1996).

"Synthesis from DNA of a molecule with the connectivity of a cube," J. Chen and N. C. Seeman, *Nature* **350**, 631 (1991).

Aerogels

Sol-Gel Science: The Physics and Chemistry of Sol-Gel Processing, C. J. Brinker and G. W. Scherer (Academic Press, 1990).

"Aerogels," J. Fricke, *Scientific American*, May 1988, p. 68.

"Inorganic and organic aerogels," J. D. LeMay, R. W. Hopper, L. W. Hrubesh, and R. W. Pekala, *MRS Bulletin*, December 1990, p. 30.

"Superexpansive gels," J. Fricke, *Nature* **374**, 409 (1995).

"Silica aerogel films prepared at ambient pressure by using surface derivatization to induce reversible drying shrinkage," S. S. Prakash, C. J. Brinker, A. J. Hurd, and S. M. Rao, *Nature* **374**, 439 (1995).

CHAPTER 8

Diamond and Carbon Materials

Diamond, G. Davies (Adam Hilger, Bristol, 1994).

High-Pressure Diamond Synthesis

The New Alchemists, R. M. Hazen (Times Books, New York, 1993).
"Industrial diamonds," M. Gardis in *Synthetic Diamond*, ed. K. E. Spear and J. P. Dismukes (John Wiley, 1994).
"Man-made diamonds," F. P. Bundy, H. T. Hall, H. M. Strong, and R. H. Wentorf, *Nature* **176**, 51 (1955).

Low-Pressure Diamond Synthesis

"Low-pressure, metastable growth of diamond and 'diamondlike' phases," J. C. Angus and C. C. Hayman, *Science* **241**, 913 (1988).
"Metastable growth of diamond-like phases," J. C. Angus, Y. Wang, and M. Sunkara, *Annual Review of Materials Science* **21**, 221 (1991).
Synthetic Diamond, ed. K. E. Spear and J. P. Dismukes (John Wiley, 1994).
Diamond and Diamond-like Films and Coatings, ed. R. E. Clausing, L. L. Horton, J. C. Angus, and P. Koidl (NATO ASI B266, Plenum Press, 1991).
"Diamond chemical vapor deposition," F. G. Celii and J. E. Butler, *Annual Reviews of Physical Chemistry* **42**, 643 (1991).
"Diamond synthesis at low pressure," K. V. Ravi in *EAM*, p. 617.

Diamond Semiconductors and Electronics

"Diamond film semiconductors," M. W. Geis and J. C. Angus, *Scientific American*, October 1992, p. 84.
"Technological applications of CVD diamond," K. V. Ravi; and "Diamond electrical properties and electronic device behaviour," J. L. Davidson, in *Synthetic Diamond*, ed. K. E. Spear and J. P. Dismukes (John Wiley, 1994).
"Diamond-based field-emission displays," J. E. Jaskie, *MRS Bulletin*, March 1996, p. 59.

Hard Ceramics

"A review of cubic BN and related materials," R. C. DeVries in *Diamond and Diamond-like Films and Coatings*, ed. R. E. Clausing, L. L. Horton, J. C. Angus, and P. Koidl (NATO ASI B266, Plenum Press, 1991).
"Boron carbide," F. Thévenot in *EAM*, p. 292.
"Aluminum oxide," F. L. Riley in *EAM*, p. 88.
"Nitrides," D. P. Thompson in *EAM*, p. 1760.
"Silicon nitride," F. L. Riley in *EAM*, p. 2512.
"Silicon carbide," K. A. Schwetz in *EAM*, p. 2455.
"Sialons," T. Ekstrom in *EAM*, p. 2444.
"Prediction of new low compressibility solids," A. Y. Liu and M. L. Cohen, *Science* **245**, 841 (1989).
"Experimental realization of the covalent solid carbon nitride," C. Niu, Y. Z. Lu, and C. M. Lieber, *Science* **261**, 334 (1993).
"Harder than diamond?" R. W. Cahn, *Nature* **380**, 104 (1996).

CHAPTER 9

Polymer Science

Polymer Chemistry, M. P. Stevens (Oxford University Press, 1990).
Polymers: Chemistry and Physics of Modern Materials, 2d ed., J.M.G. Cowie (Chapman & Hall, 1991).
"Polymers," J. Candlin in *The Chemical Industry*, 2d ed., ed. A. Heaton (Chapman & Hall, 1994).

Strong Polymer Fibers

"Organic fibers of high modulus and high strength," M. Jaffe in *EAM*, p. 1843.
"Plastics get oriented—and get new properties," F. Flam, *Science* **251**, 874 (1991).

New Synthetic Strategies

"Functional polymers and dendrimers: Reactivity, molecular architecture, and interfacial energy," J.M.J. Fréchet, *Science* **263**, 1710 (1994).
"Rational design and synthesis of new polymeric materials," H. R. Allcock, *Science* **255**, 1106 (1992).
"Living polymerization methods," O. W. Webster, *Science* **251**, 887 (1991).
"Carboanionic polymerization," M. Fontanille in *EAM*, p. 319.
"Carbocationic polymerization," P. Sigwalt in *EAM*, p. 328.
"Advances in 'living' free-radical polymerization: Architectural and structural control," C. J. Hawker, *Trends in Polymer Science* **4**, 183 (1996).
"Metallocene catalysts initiate new era in polymer synthesis," A. M. Thayer, *Chemical & Engineering News*, 11 September 1995, p. 15.

Dendrimers

"Meet the molecular superstars," D. Tomalia, *New Scientist*, 23 November 1991, p. 30.
"Dendrimer molecules," D. Tomalia, *Scientific American*, May 1995, p. 62.
"Dendritic polymers," R. Spindler and D. A. Tomalia in *EAM*, p. 581.
"Dendrimers, arborols, and cascade molecules: Breakthrough into generations of new materials," H.-B. Mekelburger, W. Jaworek, and F. Vögtle, *Angewandte Chemie* (English ed.) **31**, 1571 (1992).

Microbial Protein Polymers

Plastics from Microbes: Microbial Synthesis of Polymers and Polymer Precursors, ed. D. P. Mobley, chapter 3 (Hanser, Munich, 1994).
"Living factories," P. Ball, *New Scientist*, 3 February 1996, p. 28.
"Biomolecular materials," J. G. Tirrell, M. J. Fournier, T. L. Mason, and D. A. Tirrell, *Chemical & Engineering News*, 19 December 1994, p. 40.
"Polymers made to measure," P. Ball, *Nature* **367**, 323 (1994).
"Elastic biomolecular machines," D. W. Urry, *Scientific American*, January 1995, p. 44.
"Artificial spider silk," R. Lipkin, *Science News*, 9 March 1996, p. 152.
"Genetic engineering of polymeric materials," D. A. Tirrell, M. J. Fournier, and T. L. Mason, *MRS Bulletin*, July 1991, p. 23.
"Genetic production of synthetic protein polymers," J. Cappello, *MRS Bulletin*, October 1992, p. 48.

"Genetically directed syntheses of new polymeric materials: Expression of artificial genes encoding proteins with repeating -(AlaGly)$_3$ProGluGly- elements," K. P. McGrath, M. J. Fournier, T. L. Mason, and D. A. Tirrell, *Journal of the American Chemical Society* **114**, 727 (1992).

"Genetically engineered fluoropolymers: Synthesis of repetitive polypeptides containing p-fluorophenylalanine residues," E. Yoshikawa, M. J. Fournier, T. L. Mason, and D. A. Tirrell, *Macromolecules* **27**, 5471 (1994).

Polymerized Films and Assemblies

"Molecular architecture and function of oriented polymer systems: Models for studying organization, surface recognition and dynamics of biomembranes," H. Ringsdorf, B. Schlarb, and J. Venzmer, *Angewandte Chemie* (English ed.) **27**, 113 (1988).

"Bulk synthesis of two-dimensional polymers: The molecular recognition approach," S. I. Stupp, S. Son, L. S. Li, H. C. Lin, and M. Keser, *Journal of the American Chemical Society* **117**, 5212 (1995).

"Polymerization of nonlamellar lipid assemblies," Y.-S. Lee et al., *Journal of the American Chemical Society* **117**, 5573 (1995).

"Direct colorimetric detection of a receptor-ligand interaction by a polymerized bilayer assembly," D. Charych et al., *Science* **261**, 585 (1993).

"A 'litmus test' for molecular recognition using artificial membranes," D. Charych et al., *Chemistry & Biology* **3**, 113 (1996).

Light-Emitting Polymers

"Plastics get wired," P. Yam, *Scientific American*, July 1995, p. 74.

"After years in the dark, electric plastic finally shines," D. Clery, *Science* **263**, 1700 (1994).

"Light-emitting diodes based on conjugated polymers," J. H. Burroughs et al., *Nature* **347**, 539 (1990).

"Flexible light-emitting diodes made from soluble conducting polymers," G. Gustafsson et al., *Nature* **357**, 477 (1992).

CHAPTER 10

Surfaces and Interfaces

Physical Chemistry of Surfaces, 5th ed., A. W. Adamson (John Wiley, 1990).

Intermolecular and Surface Forces, 2d ed., J. N. Israelachvili (Academic Press, 1992).

Microscopy and Surface Structure

Introduction to Surface Physics, M. Prutton (Oxford University Press, 1994).

Concise Encyclopedia of Materials Characterization, ed. R. W. Cahn and E. Lifshin (Pergamon Press, 1993).

Journeys in Microspace, D. Breger (Columbia University Press, 1995).

Scanning Probe Microscopes (STM and AFM)

Taming the Atom, H. C. von Baeyer (Viking, 1992).

Scanning Probe Microscopy and Spectroscopy, R. Wiesendanger (Cambridge University Press, 1994).

"The scanning tunnelling microscope," G. Binnig and H. Röhrer, *Scientific American*, August 1985, p. 40.

"Scanned-probe microscopies," H. K. Wickramasinghe, *Scientific American*, October 1989, p. 98.

Wetting, Spreading, and Adhesion

"How to make water run uphill," M. J. Chaudhury and G. M. Whitesides, *Science* **256**, 1539 (1992).
"Rapid motion of liquid drops," C. D. Bain, G. D. Burnett-Hall, and R. R. Montgomerie, *Nature* **372**, 414 (1994).
"Adhesion: molecules and mechanics," K. Kendall, *Science* **263**, 1720 (1994).

Nonstick Coatings

"Coming to an unsticky end," R. F. Brady, *Nature* **368**, 16 (1994).
"Water-based non-stick hydrophobic coatings," D. L. Schmidt et al., *Nature* **368**, 39 (1994).

Organic Thin Films

Order in Thin Organic Films, R. H. Tredgold (Cambridge University Press, 1994).
"Self-assembling materials," G. M. Whitesides, *Scientific American*, September 1995, p. 114.
Langmuir-Blodgett Films, ed. G. Roberts (Plenum Press, 1990).
"Surface absorption of monolayers," A. Ulman, *MRS Bulletin*, June 1995, p. 46.
"Self-assembled and Langmuir-Blodgett organic thin films as functional materials," D. H. Charych and M. D. Bednarski, *MRS Bulletin*, November 1992, p. 61.
"Organized monolayers by adsorption. 1. Formation and structure of oleophobic mixed monolayers on solid surfaces," J. Sagiv, *Journal of the American Chemical Society* **102**, 92 (1980).
"A microstamp of approval," C. M. Knobler, *Nature* **369**, 15 (1994).

Nanotribology

"Nanotribology: Friction, wear and lubrication at the atomic scale," B. Bhushan, J. N. Israelachvili, and U. Landman, *Nature* **374**, 607 (1995).
"Friction at the Atomic Scale," J. Krim, *Scientific American* October 1996, p. 74.
"Use of the atomic force microscope to study mechanical properties of lubricant layers," M. B. Salmeron; "Tribological investigations using friction force microscopy," R. Overney and E. Meyer; "Interfacial junctions and cavitation," U. Landman and W. D. Luedtke; "Simulations of nanometer-thick lubricating films," M. O. Robbins, P. A. Thompson, and G. S. Grest, *MRS Bulletin*, May 1993.
Fundamentals of Friction: Macroscopic and Microscopic Processes, ed. J. L. Singer and H. M. Pollock (Kluwer, Dordrecht, 1991).

Atomic-Scale Manipulation of Surfaces

"Atomic-scale surface modifications using a tunnelling microscope," R. S. Becker, J. A. Golovchenko, and B. S. Swartzentruber, *Nature* **325**, 419 (1987).
"Positioning single atoms with a scanning tunnelling microscope," D. M. Eigler and E. K. Schweizer, *Nature* **344**, 524 (1990).
"Field-induced nanometer- to atomic-scale manipulation of silicon surfaces with the STM," I.-W. Lyo and P. Avouris, *Science* **253**, 173 (1991).
"Nanometer-scale recording and erasing with the scanning tunnelling microscope," A. Sato and Y. Tsukamoto, *Nature* **363**, 431 (1993).

"Atomic imaging and positioning," J. Foster in *Nanotechnology: Research and Perspectives*, ed. B. C. Crandall and J. Lewis (MIT Press, 1992).

"Imaging standing waves in a two-dimensional electron gas," M. F. Crommie, C. P. Lutz, and D. M. Eigler, *Nature* **363**, 524 (1993).

"Quantum constructions," M. Reed, *Science* **262**, 195 (1993).

"Confinement of electrons to quantum corrals on a metal surface," M. F. Crommie, C. P. Lutz, and D. M. Eigler, *Science* **262**, 218 (1993).

"There's plenty of room at the bottom," R. Feynman, *Engineering and Science*, February 1960; reprinted in *Nanotechnology: Research and Perspectives*, ed. B. C. Crandall and J. Lewis (MIT Press, 1992).

F I G U R E   C R E D I T S

Fig. I.1 adapted from *The Encyclopedia of Advanced Materials* (ed. D. Bloor, R. J. Brook, M. C. Flemings, and S. Mahajan) (Pergamon, 1994); fig. 2.1 adapted from *Physics Today*, April 1995; fig. 3.20 from *Trends in Polymer Science*, April 1993, p. 98; fig. 5.5 from *The Encyclopedia of Advanced Materials* (ed. D. Bloor, R. J. Brook, M. C. Flemings, and S. Mahajan) (Pergamon, 1994), p. 94; fig. 5.10 adapted from *Chemical & Engineering News*, 13 March 1995, p. 47; fig. 5.11 from *Diabetes* **40**, 1511 (1991); fig. 10.9*b* from the cover of *Langmuir*, 1 May 1996.

INDEX

Aastrom Biosciences company, 235
abrasives, 314, 321, 323, 338–340
absorption, 285
Acetobacter xylinium, 186, 187
Acheson, A. G., 338, 339
acids, 145, 146; solid, 285, 292, 296
acoustic insulation, 282
actin, 165, 166
actuators, 105, 108, 115
adaptive optics, 118
adenine, 152, 153
adhesion, 404–407, 416; cells, 369, 371
adhesives, 356, 367, 407
adrenalin, 216
adsorption, 397, 399, 400, 416
advanced ceramics, 315, 338–343
advanced materials, 4–6, 143
Advanced Tissue Sciences company, 228
Aebisher, Patrick, 233
aerogel, 306–312
aerospace engineering, 104, 118, 142, 267, 275, 347
air pollution, 244, 245
Aisenberg, S., 323
alanine, 146, 180–182, 372
albumin, 213, 214, 308
Alcaligenes eutrophus, 190
Alexander, Harold, 229
alkaline fuel cell, 261
alkoxides, 297, 309–311
alkylthiols, 410–413
Allara, David, 410
all-optical networks, 57, 58
α-helix, 147, 148, 175–177
alumina. *See* aluminum oxide
aluminophosphates, 206, 288, 303
aluminosilicates, 283–292, 302
aluminum oxide, 221, 222, 224, 251, 338
Amborite, 340
amino acids, 145–148
amorphous metal alloys, 121, 122
amorphous silicon, 272–275
amphiphiles, 162, 163, 172–175, 363, 375–379, 410, 414, 415, 419–421
amplifiers, optical, 28
amylopectin, 188
amylose, 188
Angus, John, 326–328
animal experiments, 211, 212

anionic living polymerization, 353, 354
ankyrin, 165, 166
anode, 247
Anthony, Thomas, 329
antibodies, 149, 230, 231, 235, 242
antiferromagnetism, 69, 77
apatite, 192, 200, 223
Applied Solar Energy Corporation, 280
aragonite, 192–194, 197, 206–208
aramids, 347, 348
Archibald, Douglas, 204
armor, 347
arsenic, 21
arteries, 211, 212, 372
artificial organs, 209–217, 230–234, 242, 243
ASEA, 319, 325
Asea Brown Bovery Ltd., 251
Ashby, Michael, 10
asperities, 404, 406, 416
associative memories, 98, 102
aster, 169
atactic polymer, 351, 352
atherosclerosis, 212
atomic force microscopy, 393–396, 416–418
atomic-scale imaging, 386–396, 422–427
atomic structure, 68
austenite, 126–130
automobiles, 244–266
Avouris, Phaedon, 423, 424

bacterial fermentation, 102, 150, 186, 187, 189–191, 234, 370
bacterial polypeptide synthesis, 150, 151, 367–374
bacteriorhodopsin, 95–102
Baekeland, Leo, 344
Bakelite, 344, 345
Ballard Power Systems, 263, 264
band gap, 19–23, 42, 60, 269, 317, 333, 381
band-gap engineering, 33, 38, 43
Bardeen, John, 17
barium ferrite, 79
barium titanate, 91, 110–112
Barrer, Richard, 286, 292
batteries, 245–259, 266, 383
Beck, Jeffrey, 293
Becker, Russell, 422
Bednorz, Georg, 12
Bein, Thomas, 299

Bell, Alexander Graham, 17
benzene, 393, 394
Berzelius, Jons Jacob, 345
bicontinuous phases, 358, 359
bilayer, lipid, 162, 163, 170–175
binary data, 66, 67, 71, 86
Binnig, Gerd, 391, 393–395
bioactive materials, 211, 224–226, 229
Bioglass, 224, 225
biomaterials, 143
biomedical ethics, 210–212, 243
biomedical materials, 135, 138–142, 172, 173,
    209pp, 348, 369–372, 374
Biomer, 216, 217
biomimetics, 199–208, 295
biomineralization, 192–202
Biopol, 190
bioprosthetic devices, 216
bioreactors, 234, 373, 374
biosensors, 175, 373, 374, 378, 379
biotin, 136, 137, 173
Birge, Robert, 98–102
Bisceglie, Vincenzo, 232
block copolymers, 356–359
Blodgett, Katharine, 414
blood, 212, 234, 235
blood, artificial, 217–221
blood vessels, 177, 178, 212
blood vessels, artificial, 212--214
Boeing Corporation, 280
Bolton, W. von, 325, 331
bone, 144, 177, 192, 198, 200, 208, 221–226,
    229, 282
bone marrow, 234, 235
bone replacement, 210, 211, 221–226, 234
bone sutures, 223
Borazon, 340
boron, 21, 333, 334
boron carbide, 339
boron nitride, 335, 339, 340
boundary lubrication, 419–422
Bovenkerk, Harold, 319
Bowden, F. P., 419
Bragg, William, 388
Bragg, W. Lawrence, 388
brain, 230
brain cell transplants, 233
Brattain, Walter, 17
Broglie, Louis de, 388, 393
Brown, Malcolm, 186, 187
Bridgman, Percy, 319–321, 326, 339
Brinker, C. Jeffrey, 309–311
Brintzinger, Hans, 352

β-sheet, 180–182, 369, 372
Bundy, Francis, 319

C_{60}, 94
cadmium, 252, 276, 382
cadmium selenide, 203, 204, 275
cadmium sulfide, 275, 276, 299
cadmium telluride, 276
Cahn, Robert, 342
calcite, 192, 193, 196, 197, 202
calcium carbonate, 188, 192–197, 202, 224
calcium phosphate, 223–226
calcium sulfate, 223
cancer treatment, 233–237, 239, 241
Canham, Leigh, 59–62
Capasso, Federico, 43, 44
capillary waves, 397
Cappello, Joseph, 368, 369
carbide ceramics, 328, 338, 339
carboanions, 354
carbocations, 354
carbohydrates, 165, 177
Carboloy, 321, 339
carbon, 316–318
carbon fibers, 223, 229, 230, 318
carbon nitride, 341–343
Carothers, Wallace Hume, 345
Carrel, Alexis, 212
cartilage, 227, 229, 230
catalysis, 148, 149, 172, 261, 263, 285–292,
    296–298, 301, 305, 318, 322, 349, 351, 352,
    364, 365, 377, 399
cathode, 247
cationic living polymerization, 354, 355
cell, 145, 161pp
cell membrane, 95, 96, 161–166, 170, 171, 175
cellophane, 186, 200
celluloid, 186
cellulose, 184–187, 345, 346
cellulose acetate, 71, 186
cellulose nitrate, 186
cermets, 339
CFCs, 8
Chabot, R., 323
Chang, Thomas, 219
chaos, 427
chaperonin, 154, 155
Chardonnet, Louis Hilaire de, 186
Charych, Deborah, 379
Chaudhury, Manoj, 407, 409, 410
chemical vapor deposition, 37, 272, 340, 342,
    400, 401; of diamond 324–336
chemisorption, 399, 410

Cheney, James, 319
chirality, 146, 147, 365
chitin, 187, 188, 193
chlorophyll, 278, 305
chloroplast, 165, 201, 267, 278
cholesterol, 162, 219
chromatin, 156, 157
chromium, 69, 77, 79, 256
chromium dioxide, 75, 76, 79
chromophores, 93, 94
chromosomes, 156, 157, 168, 169
cilia, 170
Cima, Linda, 234
Clarke, Arthur C., 332
claw, 175–177
clays, 290–292, 302
cloverite, 289
close packing, 303
cobalt, 77, 79, 83, 84, 221, 339
cobalt oxide, 253, 254
coccolithophores, 204, 207
codon, 154, 155
coercivity, 72, 74–76, 78, 80, 83
Cohen, Marvin, 341, 342
coherent light, 29
cold-cathode diode, 337, 338
collagen, 177, 192, 198, 213, 214, 216, 225,
    227–229, 231, 372
colloids, 218, 219
comb polymer, 356
compact-disk technology, 34, 66, 74, 79, 80,
    85
compass, 66
composite materials, 14, 125, 129, 177, 184,
    192pp, 200, 201, 223, 225, 226, 264, 339,
    347, 359
computers, 17, 57, 58, 64, 65
condensation polymerization, 349
conducting polymers, 93, 94, 299, 374, 380–
    383
conduction band, 19–23, 269, 317, 333, 337,
    380
Constable, Ed, 364
contact angle, 401, 402, 409
contraception, 236, 238, 241
controlled release, drugs, 237–242
convergent synthesis of dendrimers, 360–362
coordination compounds, 303–306
copolymers, 92, 140, 152, 190, 223, 239, 240,
    241, 346, 356–359
copper, 17, 77, 304, 328, 335
copper indium diselenide, 275
coral, 224

cornea, 177
corundum, 338
cotton, 186
covalent bond, 19, 396
covalent ceramics, 338–343
Crick, Francis, 152
Crommie, Michael, 426
Cronstedt, Axel, 283
Crookes, William, 318
crystallography, 386–388
cubic phase, 174, 377
Curie, Jacques, 108
Curie, Pierre, 71, 108
Curie temperature: ferromagnetism, 71; ferro-
    electricity, 110, 111
cutting tools, 314–316, 324, 332, 338–340
cysteine, 176, 177
cytochrome c, 151
cytoplasm, 149
cytoskeleton, 165, 166, 168, 170, 200
cytosine, 152, 153
cytosol, 164
Czochralski growth, 271

Dacron, 212, 213, 215–217, 229
Daimler Benz, 264
Darwin, Charles 152
data storage. *See* information storage
Davisson, Clinton, 388
Dawkins, Richard, 161
De Beers, 322, 323
deCarli, Paul, 324
DeGrado, William, 150, 151
DeLuchi, Mark, 266
demagnetizing field, 72, 74, 75, 78, 83, 84
Democritus, 386
dendrimers, 359–366
de novo protein design. *See* protein design
dentine, 177
deoxyribose, 152, 153
Dermagraft, 228
dermis, 227, 242
Deryaguin, Boris, 326–329, 335
detergent, 173, 174
deVries, Robert, 329
diabetes, 140, 175, 230–233, 236, 240, 243
diamond, 302, 313, 398; atomic structure, 316;
    CVD, 324–336; films, 314, 324–338; micro-
    electronics, 333–338; natural, 313–316; semi-
    conducting, 314, 333–336; single-crystal
    films, 334, 335; synthetic, 315, 318–338
diamond anvil cell, 313, 314
diatoms, 204

diffraction, 87, 386–388
digital data, 65–67, 71
Digital Equipment Corporation, 18, 57
digital video technology, 15, 89
diode, 23, 24, 51, 336, 337, 381, 382
direct band gap, 272
directional solidification, 7
dislocations, 128
dispersion, 27
disulfide bond, 176, 177
divergent synthesis of dendrimers, 360, 361
DNA, 152–160, 368–370
DNA polymerase, 158, 159
Dodecasil 3C, 286
Donne, John, 314
dopamine, 233, 235
drug delivery, 135, 138, 139, 172, 188, 211,
    235–242, 364, 376
Duracell, 255
dysprosium, 119, 121

eggshell, 192, 196, 197
Eigler, Donald, 422–426
elastin, 177, 178, 191, 372
elastomers, 178, 191, 345, 346, 356, 357, 375,
    402, 412
electric vehicles, 245–266
electrochemical cell, 247, 248, 289
electrochemistry, 246
electroluminescence, 58–60, 380, 381
electrolyte, 247, 248
electromagnet, 71–73
electromagnetic induction, 67, 68, 71–73, 76,
    120
electromagnetic radiation, 24, 67
electron diffraction, 38, 388
electron microscopy, 86, 388–391
electro-optic effect, 53, 54, 92
electrorheological fluids, 130–134
Electrosource, 249, 250
electrostriction, 116, 117
enamelin, 192
enantiomers, 146, 147
endoplasmic reticulum, 155, 163, 164
endothelial cells, 213, 214
Energy Conversion Devices Inc., 256, 258, 274
energy levels, 29, 30
energy technology, 244
entropy, 136
enzymes, 139, 140, 147–149, 154, 177, 182,
    189, 190, 218, 237, 285, 364, 369
epidermis, 227, 228, 242
epitaxial growth, 36, 194, 201, 335
epoxides, 296, 297

erythrocyte. *See* red blood cell
Escher, M. C., 407, 408
Escherichia coli, 151, 190, 369, 370, 372
eukaryotes, 154–156, 163–165, 170
Eversole, William, 325, 326
exons, 155
exoskeletons, 204, 205

Faraday, Michael, 66, 67
Fauchet, Philippe, 62
Fedoseev, Dmitri, 327
Feldman, Bernard, 342
ferritin, 204
ferroelectric materials, 91, 110–119
ferromagnetism, 69–71, 77, 111, 119, 120
Feynman, Richard, 422, 425, 427
Fiat, 252
fiber optics. *See* optical fibers
fiber-reinforced composites, 195, 339
fibers, synthetic, 179, 180, 186, 191, 318, 346–
    348
fibroblasts, 216, 227
fibronectin, 229, 369, 371
fibrous proteins, 175–183, 229
Fischer, Emil, 149
flagellum, 164, 170
flat-screen displays, 379–383
Fluosol, 220
Ford Motor Co., 251
Franklin, Rosalind, 152
Fréchet, Jean, 361, 363, 367
free-radical polymerization, 349–350, 352
frequency doubling, 52
friction, 332, 416–422
friction force microscopy, 417, 418
Friend, Richard, 381, 382
fuel cells, 259–266
functional proteins, 145, 147–150

Gabor, Dennis, 86
gallium aluminum arsenide, 32–35, 45, 48
gallium arsenide, 32–35, 41, 45, 48–50, 58,
    280, 338, 380
gallium nitride, 34, 35, 52
Gardner, Nelson, 326, 327
garnet, 31, 84
Garnier, Francis, 383
gasoline, synthetic, 285, 288
Geis, Michael, 335, 338
gelatin, 177
General Electric, 319–325, 329, 339
General Motors, 245, 249, 258, 259
gene therapy, 235, 237, 242
genetic code, 152–155

genetic engineering, 102, 136, 137, 150, 190, 368–374
genetics, 152–155
Genzyme Tissue Repair, 228
Gerber, Christoph, 393, 394
Germer, Lester, 388
Ghadiri, M. Reza, 300, 301
giant magnetoresistance, 77
Gilbert, William, 66
glass. *See* silica
glass-ceramic composites, 225
global warming, 245, 246, 267
globular proteins, 147–150, 360
glucose, 140, 175, 184, 185, 188–190, 231–233, 240
glucose oxidase, 140, 240
glutamic acid, 146, 372
glycerophospholipids, 162
glycine, 146, 177, 180, 182202, 203, 372
glycoprotein, 177, 197, 200
gold, 315, 328, 410, 411, 413, 414
Golgi apparatus, 164, 165
Goodyear, Charles, 191
Goretex, 213
graded-index optical fibers, 27
Graftskin, 228
gramicidin A, 171
graphite, 196, 215–217, 254, 255, 264, 271, 316–318, 324, 326–330, 339; structure, 317
Gratzel, Michael, 278, 279
Grove, William, 259, 261
guanine, 152, 153
gun cotton, 186

hair, 175, 176
Haldane, J.B.S., 145
half-cell, 247, 249, 250, 254, 257, 259, 260
Hall, Tracy, 319, 321
Halobacterium halobium, 95
Hampp, Norbert, 98
Han, He-Xiang, 342
Hannay, James Ballantyre, 318, 319
hard ceramics, 315,
Hardy, William, 419
heart, 134
heart, artificial, 211, 214–217, 230
heart transplants, 210, 216
heart valves, 214–217
heat engines, 125, 267
Heeger, Alan, 382, 383
Heerden, Pieter van, 86
hemoglobin, 218–220
hemophilia, 235
Hench, Larry, 224

heparin, 213, 214
hepatocytes, 231
heredity, 152
Herron, Norman, 299
Hershey, J. Willard, 318
Hesselink, Lambertus, 88
heteroepitaxial growth, 335, 336
heterojunctions, 272, 276, 277
heterostructures, 34–39, 41–45, 48, 358, 389
hierarchical materials, 147, 156, 157, 176, 177, 180, 181, 184, 185, 198, 200, 206, 258
Higashimura, T., 354
high-temperature superconductors, 11–13, 290
hip replacement, 210, 221, 222
Hirose, Yoichi, 331
histones, 156, 157
Hoffman, Allan, 136, 137
holography, 86–92, 98
homoepitaxial growth, 335, 336
Honda, H., 328
Hooke, Robert, 117, 179
Horbett, Tom, 240
horn, 175, 226
hot-filament-assisted chemical vapor deposition, 329, 330
hot isostatic pressing, 7
holes, charge carriers, 21–23
Hubbell, Jeffrey, 232
Hyatt, John and Isaiah, 186
hybrid circuits, 47, 48
hydrodynamic lubrication, 419
hydrogels, 135, 138, 139
hydrogen, 326–330; solid, 313
hydrogen bonds, 110, 111, 136, 147, 148, 152, 153, 181, 182, 300–302, 396
hydroxyapatite 198, 199, 226
hydroxylcarbonate apatite, 225
hyperbranched polymers, 366, 367
hysteresis, 126, 127

Iga, Kenichi, 40
immune system, 149, 162, 209, 210, 219, 227, 230
immunoglobulins, 149
indirect band gap, 59, 271
indium tin oxide, 272, 382, 383
industrial ecology, 8, 9
industrial waste, 7–9
information technology, 15, 63
information storage, 63, 422–426
Inganas, Olle, 382
injection molding, 375
insulators, 19, 20, 333
insulin, 140, 230–233, 235, 236, 240

integrated circuits, 18, 40, 338
integrated optoelectronic circuits, 48, 49, 53, 54, 62
integrated photonic circuits, 49, 50–58
intelligent materials. *See* smart materials
interatomic forces, 394, 397, 405, 406
intercalation compounds, 252–254, 256
interference, 387, 395, 396
interpenetrating networks, 302–306
introns, 154, 155
ion channels, 95, 134, 170, 171
ion exchange, 284
ionic bond, 19, 396
ion implantation, 334
iron, 68–71, 75–77, 83, 119, 318, 321
iron disulfide, 276, 277
iron oxides, 66, 71, 75, 76, 79, 204
islets of Langerhans, 231–233
isotactic polymer, 351, 352
Israelachvili, Jacob, 405, 406, 420
Iwasaki, S., 79

Jamieson, John, 324
Jarvik-7 artificial heart, 216, 217
Jewell, Jack, 41
Joule, James, 119

Kajiwara, Kenji, 138
Kamerlingh Onnes, Heike, 12
Kaminsky, Walter, 352
kaolinite, 291
Kaplan, David, 371
KDP, 110, 111
Keggin ions, 292
Kendall, Kevin, 195, 406
keratin, 175–177
Kerr effect, 80, 82
Kerr, George, 286, 287, 293
Kevlar, 179, 229, 347, 348
kidney, artificial, 230, 234
kidney dialysis, 210, 230
kidney transplants, 210
Kim, Young, 366
Kimberley, 315
kimberlite, 315
Kirchoff, Constantine, 188
Kistler, S. S., 306–309
Kost, Joseph, 240
Koten, Gerard van, 365
Kresge, Charles, 293

Lahav, Meir, 202, 203
lamellar phase, 174, 295

Lander, James, 327
Langer, Robert, 228, 230, 233, 240, 241
Langmuir-Blodgett films, 379, 414, 415, 419
Langmuir films, 202, 203, 414
Langmuir, Irving, 293, 414, 415, 419
lasers, 28–35, 40–44, 80–88
laser diodes, 34, 35, 40, 58, 80–82, 91
Laue, Max von, 386
Lavoisier, Antoine, 318
layered chalcogenides, 252, 253
lead, 248, 249
lead-acid battery, 248–250, 258
lead magnesium niobium oxide. *See* PMN
lead lanthanum zirconate titanate. *See* PLZT
lead zirconate titanate. *See* PZT
lecithin, 162, 163
leucine, 151
Levi, Primo, 386
Lieber, Charles 342, 343
ligament, 177, 211, 225, 229, 372
light-emitting diodes, 45, 58, 62, 204, 338; polymeric, 379–383
lignin, 184, 185
liposomes, 172, 219, 237, 376, 379
liquid crystals, 87, 88, 347, 410; displays, 338, 379, 382
lithium, 252–256
lithium battery, 252–256
lithium niobate, 50, 52–54, 86–88, 91
Liu, A.-Y., 341
liver, 210, 220, 230
liver, artificial, 230, 231, 233, 234
liver transplants, 210
living polymerization, 352–356
low-energy electron diffraction, 388
lower critical solution temperature, 136, 137
lubrication, 187, 216, 217, 318, 339, 415–422
Lundblad, Erik, 319
Lutz, Charles, 426
Lux, Benno, 325
Lyo, In-Whan, 423, 424
lysine, 151, 177, 178, 228
lysosome, 164

Mach-Zehnder interferometer, 54
macromolecules, 345
magnetic data storage, 64
magnetic disk, 64, 72, 73
magnetic domains, 69, 70
magnetic moment, 67–71, 78, 119, 122
magnetic multilayers, 77, 84
magnetic recording, 66, 71–79
magnetic tape, 65, 71–75

magnetic thin films, 75, 76
magnetite, 66, 71
magneto-optic data storage, 34, 80–85
Magnetophon, 65
magnetoresistance, 76, 77
magnetorheological fluids, 134
magnetostriction, 119–122
magnets, 66
Maiman, H., 29, 31
Mamin, John, 423, 426
manganese, 69, 256
manganese oxide, 253, 254
Mann, Stephen, 202, 204, 206, 207
Marconi, Guglielmo, 17
martensitic transformation, 126–130
materials processing, 6, 7, 12
materials selection charts, 10
Mathiowitz, Edith, 239
Matsuda, Atsushi, 140
Maxwell, James Clerk, 67
McDonnell Douglas company, 118
MCM-41, 293–298
Meijer, E. W., 364
Meincke, Hans, 325
membrane proteins, 151, 162, 165, 166, 170,
    171, 175, 200, 229, 237
Mendel, Gregor, 152
mercury, 250
mercury battery, 250
mesoporous materials, 293–299
messenger RNA, 154, 155
metal, 19
metallocene catalysts, 352
metallurgy, 7
metalorganic chemical vapor deposition, 38
meteorites, 324, 332
mica, 405, 406
micelles, 174, 175, 293, 295, 296, 298, 363,
    364
microchips, 17, 18, 47–49, 271
microcontact printing, 412–414
microemulsion, 207
microfibril, 176, 177
microlasers, 40–42
micromotors, 390
microporous materials, 283–292, 300–306
microscopy, 386, 388–396, 416–418
microtubule, 168–170
microwave-plasma chemical vapor deposition,
    329, 330
Miller, S. E., 48
mitochondrion, 163, 164, 171
mitosis, 169

mitotic spindle, 169
Mitsubishi, 256
Moerner, W. E., 92
Moissan, Henri, 318, 319, 321, 338, 339
Mok, Fai, 86
molecular-beam epitaxy, 35–39, 43, 44
molecular chaperones, 155
molecular sieves, 284, 285, 293, 295, 297, 377
Moli Energy, 253
mollusk shell, 192–196
molten carbonate fuel cell, 264
molybdenum, 256, 275
monomer, 144, 345
montmorillonite, 290, 291
Moore, Jeffrey, 304
Morrison, James, 327
MOSAIC process, 335
mother-of-pearl. *See* nacre
Müller, Alex, 12
Mullis, Kary, 158
multicrystalline silicon, 271
multiplexing, 56, 57
muscle, 135, 165–167, 175, 229
myoglobin, 218
myosin, 165–168

nacre, 193–196, 201
Nafion, 262, 263
Nakamura, Shuji, 35
nanocrystals, 203, 204, 278, 279
nanotechnology, 283, 298, 299, 422–427
nanotribology, 416–422
Natta, Guilio, 351, 352
NEC, 115
negative electron affinity, 337
neurotransmitters, 233
Newkome, George, 363
Newnham, Robert, 113, 114, 116
Newton, Isaac, 117
nickel, 69, 256, 261, 335
nickel-cadmium battery, 252, 255, 256, 258
nickel-metal hydride battery, 256–259
N-isopropylacrylamide, 135–138
Nissan, 252
Nitinol, 128
nitride ceramics, 338–343
nitrosoureas, 239
nonlinear optical loop mirror, 55, 56
nonlinear optics, 51–55, 85–95, 415
Norplant system, 238
n-type doping, 21–23, 333, 334
nuclear power, 267, 280
nucleic acids, 145, 152–160

nucleosome, 156, 157
nucleotide, 152, 153
nucleus, of cells, 155, 163, 164
Nuzzo, Ralph, 410
nylon, 144, 180, 218, 219, 288, 346

O'Brien, David, 377
Organogenesis, 228
organ transplants, 209, 210
open-framework materials, 300–306
optical data storage, 66, 79, 85
optical fibers, 16, 24–28, 55–57, 142
optical microscopy, 386
optoelectronics, 16, 47–49, 53, 54, 58, 62, 204,
    268, 286, 298, 299, 375
organelle, 163–165
organ replacement, 209, 210
organ transplants, 209, 210
Osada, Yoshihito, 140
osteoblasts, 198, 221, 226
osteoclasts, 198, 223
Ovchinnikov, Yuri, 96
Ovonic Battery Co., 256–259
Ovshinsky, Stanford, 256
Ozin, Geoffrey, 204–206, 208, 286, 299

packing density, 303
pancreas, 140, 230, 231
pancreas, artificial, 231
paper, 3, 186
parallel processing, 102
Parkinson's disease, 233
Parsons, Charles, 319
Parsons, J. Russell, 229
Pauling, Linus, 147
PCR. *See* polymerase chain reaction
peptide bond, 146, 300, 349
peptide drugs, 240, 241
peptides, synthetic, 150, 151, 300, 301, 367–
    374
perfluorocarbons, 219–221
perovskite structure, 110–112
perpendicular magnetic recording, 78–80
petrochemical industry, 284–288, 295–298
Peyghambarian, Nasser, 94
PHAS. *See* polyhydroxyalkanoates
PHB. *See* poly(3-hydroxybutyrate)
P(HB-HV), 190
phase separation, 95, 241, 356
phase transition, 182, 183
phloem, 184, 185
phosphatidylcholine, 162, 163
phospholipids, 162, 163, 172–174, 200, 219,
    237, 376, 377

phosphor displays, 338
phosphoric acid fuel cell, 261–263
phosphorus, 333, 334
photoconductive materials, 46
photocycle, 97–99
photodetectors, 46, 47, 58, 80, 82, 87, 88, 105,
    268, 272
photodiodes, 46–49
photoelectrochemical cell, 277–279
photolithography, 39, 40, 50, 375, 413
photoluminescence, 58–60, 380, 381
photon, 24
photonic materials, 15
photorefractive effect, 89–91
photorefractive materials, 89–95
photostriction, 122–124
photosynthesis, 95, 151, 165, 184, 278, 279
photovoltaic cell, 267–281
photovoltaic effect, 122, 268, 269
photovoltaic power plants, 281
physisorption, 397, 399
piezoelectric materials, 108–119, 122, 312, 392,
    395
pillared clays, 291, 292
Pinnavaia, Thomas, 296–298
Pinneo, Michael, 330
Plante, Gaston, 248
plasmid, 369, 370, 373
plastic deformation, 128
plastics, synthetic, 144, 186, 189–191, 345–347
platinum, 247, 261, 262, 263, 277
Pliny the Elder, 314
Plunkett, R. J., 402
PLZT, 122, 123
PMN, 117
p–n junction, 22–24, 46, 269, 270, 280
Pockels, Agnes, 414
poling, 94, 113
Pollack, Martin, 48
polyacetylene, 376, 380, 381
polyacrylates, 54, 55
polyamides, 346
polyamines, 240
polyaniline, 299, 383
polyanhydrides, 239
polydiacetylene, 376–379
polyesters, 189, 212, 213, 346, 349
polyethylene, 144, 180, 191, 222, 226, 345,
    348, 349, 352, 376, 380, 402
polyglycolic acid, 223, 228, 233, 238, 241
polyhydroxyalkanoates, 189–191, 234
poly(3-hydroxybutyrate), 189
polyimides, 49, 54, 55
polyisoprene, 346, 359

polylactic acid, 223, 228–230, 233, 238, 241
polymerase chain reaction, 157–159, 369
polymer gels, 135, 138–142, 177, 214
polymerization processes, 349–356, 360–362
polymer-membrane fuel cell, 262, 263
polymers, 144, 344; biomedical, 212; blends,
    359, 382; conducting, 93, 94, 299, 374,
    380–383; drug delivery, 237–242; fibers,
    179, 180, 186; industrial, 344–347; ion-
    conducting, 262, 263; lubrication, 419, 420;
    luminescent, 379–383; membranes, 232, 233;
    microelectronic packaging, 49; natural, 143;
    nonlinear optics, 54, 55, 91–95; optical
    fibers, 28; optical waveguides, 50; photo-
    refractive, 91–95; photoresists, 39, 40; piezo-
    electric, 119; smart, 135–142
poly(methyl methacrylate), 28, 224, 375
polypeptides, 147–151, 367–374
poly(phenylene vinylene), 381, 383
polypropylene, 180, 189, 190, 191, 345, 352
polysaccharide, 184, 185, 187, 188, 193, 227,
    232, 233, 241
polysilanes, 59
polystyrene, 144, 345, 350, 352, 359, 369
poly(styrene-butadiene), 346, 357
polytetrafluoroethylene, 212, 213, 215, 229,
    253, 261, 262, 373, 400, 402, 403, 416
polythiophene, 374, 381, 382
polyurethanes, 54, 191, 213, 216, 217, 224,
    346, 367, 402
polyvinyl chloride, 345, 349, 350
population inversion, 31
porous materials, 59–62, 224, 248, 250, 261,
    282pp, 377
porous silicon, 59–62, 203, 298, 299, 303
porphyrins, 305
potassium dihydrogen phosphate. *See* KDP
potassium niobate, 52
Poulsen, Valdemar, 65
Prasad, Paras, 94
primary battery, 250
primary structure, proteins, 147, 148
prokaryotes, 163, 164
proline, 177
ProNectin F, 369, 371
prostacyclin, 213, 214
prosthetic limbs, 209,
protein design, 150, 151
protein folding, 147, 148, 150, 154, 300
proteins, 95, 145–155, 367, 368
protofibril, 176
proton-exchange-membrane fuel cell, 263, 264
Psaltis, Demetri, 88
pseudointima, 213, 216

PTFE. *See* polytetrafluoroethylene
p-type doping, 21–23, 333
pyrite. *See* iron disulfide
pyrolitic carbon, 215, 217
PZT, 112, 122

quantum-cascade laser, 42–44
quantum confinement effects, 60, 298, 427
quantum well, 33–35, 40–44, 426
quartz, 108, 109, 271, 341
Quate, Calvin, 393, 394
quaternary structure, proteins, 147, 148

radiolarians, 204–206
random coil, 178
Ratner, Buddy, 240
Ravi, K. V., 333
Rayleigh, Lord, 414, 419
rayon, 186, 200, 345, 346
recombination, 21–23, 33, 58–60, 271, 272,
    278–280, 380–382
red blood cell, 161, 165, 166, 204, 218–221
red giant stars, 332
Regen, Steven, 376
Regenerex company, 231
refractive index, 25, 51, 53, 54, 56, 87, 90–92,
    98, 314
Renault, 252
replication, 156, 158, 160, 168, 169
reptation, 346
retrovirus, 160
reverse micelle, 174, 175, 204
ribosomal RNA, 155
ribosome, 154, 155, 164, 368, 373
Ringsdorf, Helmut, 376
RNA, 153–155, 160, 368
RNA editing, 154, 155
robotics, 128, 138, 142
Robson, Richard, 304, 305
Rochelle salt, 109, 110
rocking-chair battery, 254–256
Röhrer, Heinrich, 391, 393
Ron, Eyal, 234
roughening temperature, 397
Roy, Rustum, 309, 329
rubber, 191, 308, 345, 357
ruby, 338

Sagiv, Jacob, 410
Sanyo Electric Co., 275
sapphire, 221, 338
scanning electron microscopy, 390, 391
scanning tunneling microscopy, 391–394, 416,
    422–426

Schmellenmeier, H. Z., 325
Schmidt, Paul, 323
Schonbein, Christian, 186
Schrödinger, Erwin, 145
Schweizer, Erhard, 422, 423
secondary battery, 248
secondary structure, proteins, 147, 148, 175–182
second-harmonic generation, 52
self-assembled monolayers, 410–415
self-assembly, 150, 160, 161, 172–175, 200, 201, 204, 295, 296, 300–306, 358, 375–379, 410, 411
semiconductors, 17–24, 268, 314, 333, 380
sensors, 108, 113, 118, 119, 121, 122, 142, 172, 175, 243, 289, 324, 378, 379
serine, 146, 180, 182
Setaka, Nobuo, 324, 328, 330
shape-memory alloys, 124–130, 140
shape-memory effect, 126–130, 140–142
shell, 192–197, 201, 226
sialons, 340
sickle-cell anemia, 161, 221
Siemens, 276
Silent Power Ltd, 251
silica (silicon dioxide): aerogels, 306–311; Bioglass, 224; biominerals, 192; gels and suspensions, 132; mesoporous, 293–298; optical fibers, 25–28, 50; silicon passivation, 398. *See also* quartz
silicon, 17–22, 39, 58–62, 203, 268–275, 299, 303, 328, 333, 334, 338, 387, 394, 398, 399, 407–410, 417
silicon carbide, 196, 261, 272, 273, 318, 332, 335, 338, 339
silicone polymers, 192, 215, 224, 227, 238, 346, 402, 410
silicon nitride, 340, 341, 395
silicon solar cells, 268–275, 281
silk, 178–183, 346, 347, 369, 371
silk gland, 182, 183
siloxenes, 60
Sinke, Wim, 280
sintering, 226
skin, 175–178, 226–230, 241, 372
skin, artificial, 211, 226–230
smart materials, 103, 240
smart structures, 105–108
smectic clay, 291
smoke, 306
soap, 174
sodium nitrite, ferroelectricity, 111
sodium, 251
sodium-sulfur battery, 251, 252, 258

solar cell. *See* photovoltaic cell
solar energy, 172, 246, 266–281, 311
solar thermal power, 267, 311
sol-gel processing, 307–309
solid-oxide fuel cells, 264, 265
solid-phase polymer synthesis, 150, 371
solid-state ion conductors, 265
sonar sensors, 110, 113, 121, 312
Sony Corporation, 255
spacecraft, 245, 261, 271, 272, 280, 324
spandex, 191
spectrin, 165, 166
spectroscopy, 396
sphingolipid, 162
spiders, 179–183
Spitsyn, Boris, 326, 327
sputtering, 75, 79, 83
starch, 188
star polymer, 355, 356, 359
Staudinger, Hermann, 345
Stayton, Patrick, 136
steel, 4, 5, 126, 179, 210, 221, 226, 340, 347
stepped-index optical fibers, 25–27
stick-slip behavior, 420, 421
storage battery, 248
streptavidin, 136, 137, 173
Strong, Herbert, 319, 321, 324
structural proteins, 147, 175–183, 368–372
Stucky, Galen, 295
sugars, 165, 177, 184, 185, 188, 189
sulfides, microporous, 289, 290
sulfur, 251
Sun, Anthony, 232, 233
superacids, 285
superconductivity, 11, 12
supercritical drying, 61, 308–311
supercritical fluids, 9, 308
superlattice, 43, 44, 77
supernova, 332
surface coatings, 402–404, 407–415, 419–422
surface electronic states, 426, 427
surface excess energy, 396, 397, 401–404, 407, 409, 412, 414, 415. *See also* surface tension
surface force apparatus, 405, 406, 419, 420
surface reconstruction, 394, 398–400
surface science, 384
surface tension, 307, 308, 310, 397
surfactants, 174, 175, 202–207, 219, 293–296, 359, 364
Suslick, Kenneth, 219
Symbion Co., 216
syndiotactic polymer, 351, 352
synthetic diamond. *See* diamond, synthetic
Szwarc, Michael, 354

Tabor, David, 405, 406
tacticity, 351, 352
Tanaka, Toyoichi, 138
tectons, 302, 303, 306
Teflon, 373, 400, 402. *See also* polytetrafluoroethylene
Telegraphone, 65
television, 338, 379, 383
Tencel, 187
tendon, 175, 177, 229, 230
Tennant, Smithson, 318
terbium, 119, 120
Terfenol-D, 121
tertiary structure, proteins, 147, 148
Terylene, 189
textiles, 186, 187, 191
thermal conductivity, 311, 314, 324, 332
thermal insulation, 282, 311
thermistor, 105, 106
thermoplastic rubber, 357
thermostats, 105, 125
thixotropy, 187
Thomas, John, 290, 291, 297
Thompson, D'Arcy Wentworth, 206
III-IV semiconductors, 32, 38, 39, 43, 50, 52,
  280
thrombin, 213
thrombosis, 213–216, 227
thrombus, 213–216
thymine, 152, 153
Tirrell, David, 368, 372–374
tissue engineering, 226–235, 242, 243, 369
titania. *See* titanium oxide
titanium, 210, 215, 217, 221, 226, 256, 296–
  298, 352
titanium oxide, 224, 278, 279
titanium silicalite (TS-1), 296–298
titanocene, 297
tobacco mosaic virus, 160, 161
Tomalia, Donald, 360
tooth, 177, 192, 199
tooth implants, 221
Townes, Charles, 29
tracheid cell, 184, 185
transcription, 154, 155, 163, 368
transdermal drug delivery, 241, 242
transfer RNA, 154, 155, 368
transistors, 17, 48, 51, 53, 336, 383
translation, 153–155, 368
transmission electron microscopy, 388, 389
transpiration, 184
tribology, 332, 415–422
Tributsch, Helmut, 278
tubulin, 168–170

tungsten carbide, 321, 339
tunneling, quantum-mechanical, 392, 393
Twaron, 348
Twieg, Robert, 92
II-VI semiconductors, 34, 52
two-way shape-memory effect, 130
tyrosine, 177

Uchino, Kenji, 116, 118, 122–124
Ulhir, A. and I., 59, 60
ultrasound, 219, 220, 240, 242
United Solar Systems Corporation, 274, 275
uracil, 154
urban pollution, 244, 245
Urry, Dan, 372
U.S. Advanced Battery Consortium, 252, 258,
  259
U.S. Food and Drug Administration, 216,
  241

vaccine release, 242
vacuum tubes, 17, 336
Valasek, J., 110
valence band, 19–23, 268, 269, 317, 333, 337,
  380
vanadium, 256, 277
van der Waals force, 318, 394, 396, 399, 402,
  425
vapor-phase epitaxy, 35–39
Varnin, Valentin, 327, 328
Varta, 255
vaterite, 202
vesicles, 172, 173, 201, 205, 206, 237, 364,
  375, 377
vibration damping, 106, 107, 113, 114, 118,
  121, 126, 129, 132, 142
Viney, Christopher, 180
vinyl polymerization, 349–356
viruses, 149, 159–161, 378, 379
volume transition, 138, 139
VPI-5, 288
vulcanization, 191

Wacker process, 271
Walsh, Dominic, 206, 207
Watson, James, 152
waveguides, 50, 53, 57, 58
wave-particle duality, 388
wear, 332, 340, 415–418
web, spider's, 178–183
Webster, Owen, 366
Wentorf, Robert, 319, 321, 324, 339, 340
Westinghouse, 265
wetting, 400–402, 412, 415, 416

Whitesides, George, 375, 407, 409, 410, 412–414
Wilkins, Maurice, 152
wind power, 267
Winslow, Willis, 132
Wilson, W. B., 329
Wong, Mike, 219
wood, 143, 183–185
wool, 177
Wuest, James, 301–303, 306

X-500 fibers, 348
xanthan gum, 187
Xanthomonas campestris, 187
xerogel, 307, 309
X-ray crystallography, 386–388

X-ray diffraction, 300, 372
xylem, 184, 185

Yannas, Ioannis, 227
Young, Thomas, 401
Yu, Luping, 94, 95

Zasadzinski, Joseph, 172, 173
zeolites, 283–291, 293, 298, 299, 302, 303, 306
Ziegler, Karl, 349, 351, 352
Ziegler-Natta polymerization, 349, 351, 352
zinc, 250
zinc selenide, 34, 35, 52
zirconium, 256, 352
zirconium oxide, 221, 265
zsm-5, 287, 288, 293, 303

Philip Ball is an associate editor for physical sciences at *Nature*.
He regularly contributes articles on all fields of science to the popular
and academic presses, and is the author of *Designing the Molecular
World* (Princeton).